Why 'Where' Matters:

Understanding and Profiting From GPS, GIS, and Remote Sensing

Practical Advice for Individuals, Communities, Companies and Countries

Dr. Bob Ryerson, CMS, FASPRS and Dr. Stan Aronoff

Printed in Canada

Published by:
Kim Geomatics Corporation,
Box 1125, Manotick, ON.
Canada K4M 1A9

Library and Archives Canada Cataloguing in Publication

Ryerson, Robert A.
Why 'where' matters : understanding and profiting from GPS, GIS, and remote sensing / Bob Ryerson and Stan Aronoff -- 1st ed.

Includes bibliographical references and index.
ISBN: 978-0-9866376-0-5

1. Geospatial data. 2. Geospatial data--Economic aspects. 3. Geospatial data--Social aspects. 4. Global Positioning System. 5. Geographic information systems. 6. Remote sensing. I. Aronoff, Stanley. II. Title.

G70.2.R93 2010 910.285

Cover design, book design, and production by Yannis Souris.
This book is typeset using Adobe Garamond Pro, Conduit ITC and Stone Sans typefaces.

TABLE OF CONTENTS

ACKNOWLEDGEMENTS

This book owes so much, to so many, that it is impossible to catalogue them all without fear of missing someone.

For Ryerson, it all started with hurricane Hazel in June of 1954 when I spent several days in the safe haven of the library in the interior of my Grandfather's solid brick house surrounded by National Geographic magazines. He later gave me a globe that I studied – and that I still have more than 50 years later. Then there were two high school teachers who really opened my eyes to the wonders of geography: the late Bill Bell and Al Kingston. High school was followed by McMaster University where my classmates (Class of '69) inspired me as did Professors Lloyd Reeds, Harold Wood, Bill Found (later President at York University) and Phil Howarth (later at U of Waterloo). That excellence continued at the University of Waterloo with Dieter Steiner, Ralph Krueger, and Chris Bryant who provided guidance and continued to fan the geographic fire that burned within me. External advice on my doctoral work (and geography in general) came from Jim Anderson, Chief Geographer of the USGS and Ernie Hardy of Cornell University.

Since then I have been blessed with working with and/or being inspired by some of the finest minds in the field. They are listed here in alphabetical order: Frank Ahern, Nick Chrisman, Josef Cihlar, Bob Colwell, Jon Fairall, Steve Franklin, Lee Godby, Dave Goodenough, Barry Haack, Changchui He, Tom Lillesand, Larry Morley, John MacDonald, John McLaughlin, Salim Mehmud, Shunji Murai, Bob O'Neil, Warren Philipson, Keith Raney, Dianne Richardson, Leo Sayn-Wittgenstein, Vern Singhroy, Al Stevens, Suvit Vbulsresth – and of course, my co-author Stan Aronoff. The list could go on and on to include more in both my generation and younger members of the field – people like Jeff Labonte, Kevin Lim, and others. I have also been blessed with wonderful support staff too numerous to name. Last, but not least my wife, Ann Blair, who has provided stability and a keen editorial pen helping to make this more understandable to the reader than it would otherwise have been.

The path to a career in a developing field is often dependent on chance encounters as much as determined effort. For Aronoff it was a perceptive senior scientist, Delphine Maclellan who as my undergraduate advisor at McGill University suggested I consider doing a Masters degree in Environmental Design at the University of Calgary and introduced me to one of the program's professors. There I found air photo interpretation and remote sensing valuable in applying ecological principles to environmental planning and impact assessment. This led me in 1976, with the encouragement and guidance of Professors Val Geist, Bill Ross, and Grant Ross to secure funding and embark on a study of the application of remote sensing for environmental monitoring of the Alberta oil sands. The study was funded by an agency of the Alberta government with substantial support from the Canada Centre for Remote Sensing (CCRS) in Ottawa in the form of subsidized imagery and multi-sensor over-flights of my study area, use of equipment and facilities, and a good deal of mentoring from the applications scientists at CCRS.

All of the satellite image analysis for that project was done in Ottawa at CCRS under the guidance of my co-author for this book, Bob Ryerson, who was then a Senior Environmental Scientist at CCRS. He also offered to serve as a member of my thesis committee and became a valued friend and colleague ever since. I was privileged to complete my Ph.D. at Berkeley under Robert (Bob) N. Colwell where I was exposed to the operations of large remote sensing programs such as AgriStars at the Space Science Lab and his unique and very effective style of positive motivational and powerfully-understated leadership. After graduating it was Paul Pearl, the President of a fledgling image processing company developing satellite image analysis systems who took a chance and hired me to head up remote sensing applications development which I later morphed into projects integrating remote sensing and GIS analysis tools. Some of these projects drew on the assistance of Jack Dangermond and evolved into a career which has come to include diverse consulting and book projects in the GIS and remote sensing fields. There are many others who have helped me along the way and though I cannot list them all here I am no less appreciative of their efforts.

This book was Bob's idea and I was pleased to contribute as a co-author. I would like to thank the many people and organizations who generously allowed us to use their imagery and illustration materials. In many cases they produced enhanced versions of their material specifically to suit our needs and are acknowledged in the figure captions. It was particularly valuable to have Ann Blair willing and able to critically read and comment on the manuscript. I appreciate the many long hours she spent reading multiple drafts and suggesting ways to combine the contributions of the two authors. I feel it has made the work accessible to a wider readership.

Finally, I would like to acknowledge the support of my family and friends whose encouragement has always been so important to sustaining the effort to complete my book projects. This book owes so much, to so many, that it is impossible to catalogue them all without fear of missing someone.

Bob Ryerson and Stan Aronoff
Ottawa
June 2010

INTRODUCTION TO THE GEOECONOMY

1. Welcome to the GeoEconomy.

This book is written for the individual, the business person, communities and for governments (politicians and bureaucrats) everywhere who want to become or remain successful in the new economic era in which we have just entered – the GeoEconomy.

The GeoEconomy is how we define the economy that is increasingly being driven by, and dependent on, geospatial or geographic information – information that is tied to or linked to a geographic location. This location based information is termed geospatial or geographic information or geo-information. (These terms are used interchangeably in this book.)

Most, if not all, of the so-called developed world finds itself rapidly entering this new paradigm – and it is a new paradigm in the true sense of that overused word.[1] The developing world is also beginning to wrestle with the implications of the GeoEconomy with varying levels of success ranging from active use of geographic information to failure – almost no use at all.

Many international agencies have clearly embraced the importance of geospatial. We will show some of the ways they have promoted its use – helping some developing countries leap into the GeoEconomy and the associated economic and quality of life improvements. However, some bi-lateral aid agencies seem to have missed the point in recent years – knowingly or unwittingly impeding the development of the geospatial capabilities that poor nations desperately need – with dramatic consequences for the poverty stricken the world over. Some developing countries seem to have made the leap into the GeoEconomy, while others have not. We explore the reasons for the successes – and the failures.

In many ways we are returning to our pre-historic roots. In some significant ways the GeoEconomy is a throw-back to prehistoric times when individual and societal survival depended on them having a thorough understanding of the geography of the places in which they lived. However, in the past much of the geographic advantage was based on what we call geo-luck – those in ancient times that happened to have the advantage of being located close to water and good soil, or who had an easily defended home were lucky – those that did not, died or lived so close to the subsistence level that they could not develop more advanced civilizations. The difference today is that we can all have access to this geo-information on a scale not dreamed of even a decade ago.

Simply put: today those who "get it," those who understand the GeoEconomy and how to use it to their advantage, will do well. Those who do not will, as our ancestors did, depend on geo-luck. Luck, geo or otherwise, like hope, is not a strategy,[2] whether for the success of us as individuals, as societies, or ultimately for our success as a species. It has been that way for all of history.

As societies did develop, specialists in geographic understanding came to exist. At the outset these "specialists" were simply those who had an innate understanding of the importance of geography. Eventually this was to lead to special technologies and approaches to mapping, planning and related geographic disciplines. We now have a far better understanding of geography and its impact. Jared Diamond has certainly illustrated this in his land mark books *Guns Germs and Steel* and *Collapse.*[3] as has the *World Development Report 2009: Reshaping Economic Geography.*[3] The latter clearly shows the links between geography and development: "the best predictor of income in the world today is not *what* or *whom* you know, but where you work." As we say in the title of the book, 'Where' matters.

However, today individuals, businesses, and governments at all levels can now make geo-decisions based on better information that is more easily accessible than at any time in history. Not only do we all have access to the data, we also have ready access to low or no cost tools that enable the non-specialist to use that data. Anyone capable of accessing and using the Internet can access and use geospatial information relevant to themselves, their community, business, country, or the global environment. Whether producing driving directions and a map to your destination across town, or viewing satellite images of atmospheric ozone over Antarctica, or a global map of current and past sea surface temperature – we all have access to a vast repository of geospatial information freely Internet accessible to anyone at anytime.

Ready access to at least some current geospatial data for any location on Earth in the form of satellite images and maps has profoundly changed the way we know our world – what we can see and know – and what we can hide. This unprecedented access to global geospatial information provides unprecedented transparency – helping to detect potentially damaging activities such as pollution, illegal harvesting, military build up and the like, before they overtake us.

However, the explosion of geospatial data collection and dissemination also raises important issues of what others might be doing with all of the information that is being collected about us and the places in which we live. The information about us could be tied to our postal code or street address as before. But now information about us can also be easily tied to an image obtained by a satellite surreptitiously circling the earth, or camera equipped aircraft, or by a van hired by today's version of "Big Brother" that is driving down your street taking pictures of your home and front yard.

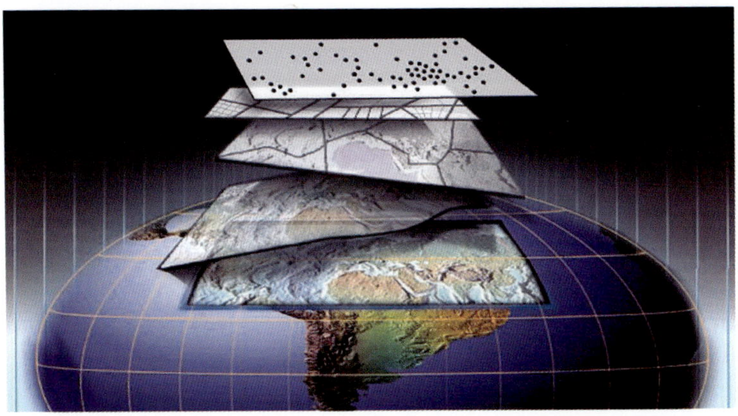

A World of Geospatial Data. Satellite imagery and maps are now readily available online for virtually any location on Earth. Source: C. Stokes

While geospatial is a new word, it is an old concept. Geospatial information is (and has been) dramatically increasing in importance for individuals, businesses, communities (however they are defined) and governments – be they local, state or national. However, now, we no longer have to rely on luck, nor must we always call on specialists to help us use and benefit from this information. To those with this data and the tools to use it, geo-luck is replaced by having a geo-advantage. With such information we are no longer subject to the tyranny of our geography.

Our experience suggests that many of those presented with geospatial information for the first time often find or develop new applications of the technology that were not previously considered. Examples of such novel applications – including those of the mushroom pickers and soil erosion specialists – are given in Chapter 6. We expect that the readers of this book will develop new applications and ideas for the use of geo-information – and we invite them to send them to our web site www.GeoEconomy.com where we will post those that we believe will be of the broadest interest.

How to Read This Book

This book is written for those who want to have a better understanding of this Geo-Economy paradigm and the technologies that support it. But it is also written for those with an interest in geography, as well as for those simply wanting to better understand what others might be doing with all of the information that is being collected about us. We have presented material in this book in several distinct and what might be regarded as stand-alone sections. For the reader who may want to follow up on an idea or who may want to verify a source, we have provided detailed notes (over 275) and more than 85 references at the end of the book. By so doing we hope to have retained the flow of the text, while providing sufficient detail for those who may want it.

Chapter 2 introduces the concept of geographic or geospatial knowledge. It covers the history of the use of geographic information, using some well known examples and some that may be somewhat more obscure – and thus perhaps somewhat more interesting.

Chapter 3 – "Now!" provides information on the technologies in use, the sort of data we now have access to, and how (and why) the need for geographic information is increasing. This technology section may be heavy going for the non-specialist – and be seen as too cursory by the specialist! While parts of this section may be of less interest to some readers, we suggest that all readers should at least skim this section and look for highlighted text or text boxes that provide some of the key points – and look at the illustrations. As is the case for those providing geographic information – you cannot please everyone all of the time!

Chapter 4 is another more general section that draws together information on a number of converging technologies, policies, and disciplines that influence geo-spatial technologies (and data), leading to an assessment in general terms of the current impacts of geospatial data on our society.

Chapter 5 introduces what we see as the primary future issues faced by our species, our environment and our society – however that society and environment may be defined and wherever it may be located. The importance of geographic information in formulating policies to deal with these key issues leaps off the pages where we link these issues and geographic information.

Chapter 6 identifies how the individual, communities, companies and countries can and should benefit from the intelligent use of geospatial or geographic information. While the assessment is not exhaustive, we have provided a number of examples that should help these different audiences begin the quest to develop their geo-awareness to do better – however better may be defined.

For those who may regard this as a "self-help" book, we caution the reader to seek competent advice before investing or following what may be seen as advice. We mention a number of companies and products by name. Mentioning a company or not mentioning a company should not be seen as a criticism, or necessarily as an endorsement.

We have tried to only mention companies with which we are familiar or which have been newsworthy in a positive way. Some have been clients, some we have worked with, some we have studied, and some have been worthy competitors. But all of them have played a role in the GeoEconomy.

Highlighted Text and Text Boxes

We have used highlighted text and text boxes in this book to draw the reader's attention to what we believe are important or interesting points. Readers can scan the book quite quickly using these text boxes – stopping to read more detail on whatever catches their eye.

We have used highlighted text or text boxes like this to draw the reader's attention to what we believe are important or interesting points.

GEOSPATIAL KNOWLEDGE: KEY TO SURVIVAL

2.1. Introduction

The development of geospatial knowledge has had major impacts on the speed and course of human history. It is a topic that could fill many volumes. This chapter will highlight important developments that illustrate the way geospatial capabilities grew, in fits and starts, from important centres at different times – from prehistoric mapping to the advanced geographic information systems (GIS) of today.

The value of geospatial skills and knowledge was often not fully recognized or developed in the locations where they first emerged. In the case of map making, the skills and knowledge were first developed by the Greeks and Romans, but were lost and then subsequently re-discovered. Yet, geospatial capabilities afforded significant economic and military advantages to those leaders who recognized their value. As a result, significant developments in geospatial knowledge tended to be associated with those groups who found ways to use those skills most effectively to gain competitive advantage.

2.2. The Nature of Geospatial Knowledge

Location, geographic location, has always been critical to human survival. From earliest times human societies have depended for their survival on their ability to identify, locate, ensure access to, and manage critical resources.

> ***Geographic location has always been critical to human survival. From earliest times human societies have depended for their survival on their ability to identify, locate, ensure access to, and manage critical resources.***

At their most basic, these resources included water, food, shelter, and the means to ensure safety of both the individual and the group. The more developed the skills and technologies to identify, access, and manage resources, as well as to organize the efforts of the group, the greater the benefits.

Fundamental to a society's ability to grow and prosper was its ability to identify the geographic location of the resources on which the people depended, and when they would be present. This type of information is what we today refer to as geospatial or geographic information.

For the northern hunter, knowing where to hunt caribou during their migration and when during the year the caribou will be there can be a matter of life or death. To the farmer, knowing which areas are better for different crops directly affects the quantity and quality of produce returned. The military planner seeks to capture and hold locations that provide optimal cover and maximum control of the surrounding terrain. The roadway engineer seeks to optimize road designs by finding route locations where soils are well-drained – whether in ancient Rome or modern New York State. All are examples of where more accurate and/or timelier knowledge of the geographic location of valuable resources offers an increased advantage.

In the course of human history, it has been this advantage that determined which human enterprises survived, be they competing societies or competing business enterprises. The ability to recognize, obtain and use geospatial knowledge has been a major factor in determining which competitors succeeded and which did not. We are just as dependent today on geospatial information for our survival as in the past, but the information we need has become much more complex and that information changes much more rapidly.

To early man, geospatial knowledge was a matter of life or death.

To early man, geospatial knowledge was a matter of life or death. With the development of larger organized societies the individual's spatial knowledge abilities were less critical, so long as the society had the specialized experts to provide those skills. Geospatial knowledge skills in effect became institutionalized. The world has now come full circle. With today's widespread availability of, and easy access to, geospatial data the individual person (or enterprise) can now make decisions using geospatial data to better control their fate – be it the fate of a person, a company, a community, or a country.

2.3. The Impact of Geography

The development of human societies has been fundamentally influenced by the geography in which they live. Jared Diamond in his landmark book, *Guns, Germs,* and *Steel*[4], provides a cogent and well researched argument for the geographic imperative. Diamond's book examined the question of why human development proceeded at such different rates on different continents. He concluded that the reason people in some regions of the world developed intensive agriculture, large nation states, powerful armies, and technology, while others remained stone-age hunter gatherers had nothing to do with their innate intelligence and everything to do with the opportunities afforded by their local climate and natural resources – what we call their geo-luck.

Those societies that developed in regions with plants and animals suitable for domestication, and a climate and landscape suited to agriculture, developed a food package (a combination of domesticated plant and animal species) that could produce a food surplus. Agricultural societies could feed more people on an area of good agricultural land than could hunter gatherer societies. For this reason, in good agricultural areas, societies developed the food surpluses to support segments of the society not directly involved in food production. These individuals became the craftsmen, rulers, bureaucrats, and other specialists who developed and administered the technologies and systems needed to support and expand societies with high population densities. Some of these also became the earliest geographers and mapmakers – the first geospatial specialists.

Within a region, societies competed for resources. Those better skilled, better armed, and/or with larger forces tended to be more successful at defending their territories and capturing other territories. Those societies successful in defending their territory remained, while those unsuccessful ultimately disappeared or were restricted to areas that competing societies did not want.

As societies became larger and developed larger non-food producing specialist groups, the need to manage the production and distribution of food and the organization of other

The Impact of Geography. Societies that inhabited areas more favourable for agriculture could produce more food with less effort, which gave them a competitive advantage over those less fortunate. Source: S. Aronoff

important tasks such as the construction of irrigation systems, the raising of armies, and the collection of taxes, required the development of systems to keep track of important resources. Whether by coercion or for mutual self-interest, to sustain large populations a society had to find a way to organize food production based on cooperative effort, and the sharing of food production among producers and those who performed other roles deemed useful in that society.

It is not surprising that the earliest examples of writing have been associated with these earliest, large, organized, agricultural societies: in Mesopotamia (the Fertile Crescent) before 3000 BC, Egypt around 3000 BC, and China by 1300 BC. Sumerian cuneiform arising in Mesopotamia is the oldest writing system for which there is detailed historical evidence. Yet before it was formalized, people in farming villages of the Fertile Crescent had been using clay tokens for accounting, recording numbers of sheep and quantities of grain. This led to the use of clay tablets and the use of pointed instruments to neatly press marks into the tablet. Evidence suggests the earliest writing was used to record quantities of goods, essentially a method of accounting – a recording of ownership of assets.

2.4. Prehistoric Maps: Early Depiction of Geospatial Information

The use of a graphic (be it a rudimentary map or some advanced technology) to represent geospatial information is wide-spread among human societies. Why this is so is explained by Robinson and Petchenik (1976) in their book *The Nature of Maps.* They concluded that "the concept of spatial relatedness … is a quality without which it is difficult or impossible for the human mind to comprehend anything."[5] From clay tablets, papyrus, or paper of the Fertile Crescent and China, to the driftwood, pebble and sealskin maps of Greenland Inuit and the stick charts of South Sea Islanders, human societies, even those without a formal written language, have developed ways to record and convey geospatial information.[6, 7]

This idea, that spatial concepts are fundamental to the way humans think and understand their world, suggests that evidence of spatial understanding should be found among early human artefacts. The drawing of symbols to represent features distributed to convey their relative positions, what we call a map, appears very early in human history. It has been argued that the precursors of maps as we know them can be found among the cave paintings of prehistoric man.

We have always had maps because we need them – to both understand our surroundings and, some would argue, to understand our place in the world in which we live.

Maps are essentially pictures that show the relative geographic position of features the mapmaker wishes to communicate.[8] A map is a particularly useful means of recording and presenting geospatial information because the spatial relationships among all the features are shown and can be understood together. A written list of features and their locations may contain the same information, but without the graphic presentation it is difficult to comprehend the spatial relationships.

The earliest direct evidence of maps as defined by cartographers has been found in the area of the Fertile Crescent. A wall painting dated to about 6200 BC from Çatal Höyük is commonly credited as the oldest existing picture map.[9] Discovered in the late 1950s, and first excavated between 1961 and 1965, this large Neolithic site near the present-day city of Konya, Turkey, under active archaeological research, is being considered for UNESCO World Heritage status.

We have always had maps because we need them – to both understand our surroundings and, some would argue, to understand our place in the world in which we live.

Maps are essentially pictures that show the relative geographic position of features the mapmaker wishes to communicate.

Based on the similarity of the layout of houses excavated by archaeologists, the 3 metre long wall painting is considered to depict, in plan-view, the layout of the ancient settlement. Also depicted behind the settlement is a representation of the twin peaks of a mountain located about 160 km from the site with its volcano erupting. This map significantly predates the first appearance of writing (around 3000 BC).[10]

The Çatal Höyük settlement is exceptional on many counts. It is believed to be among the first settlements to have made the transition from hunting-gathering to become an agrarian society, around 7400 BC. The city that evolved dates back to 6800-5000 BC and is considered to have been the most developed urban center of the Near East and Aegean regions. Excavations indicate that the city was built according to a designed plan, with rectangular-plan houses next to one another around courtyards.

The population of the settlement was large for that time, believed to have numbered in the thousands. It is considered to have then been the most populated city in the world and the world's first urban centre. Archaeological evidence suggests that this site saw the development of the first fabric, mirror, wooden bowl, and methodical system of agriculture and cattle farming, as well as the emergence of an organized religion.[11, 12]

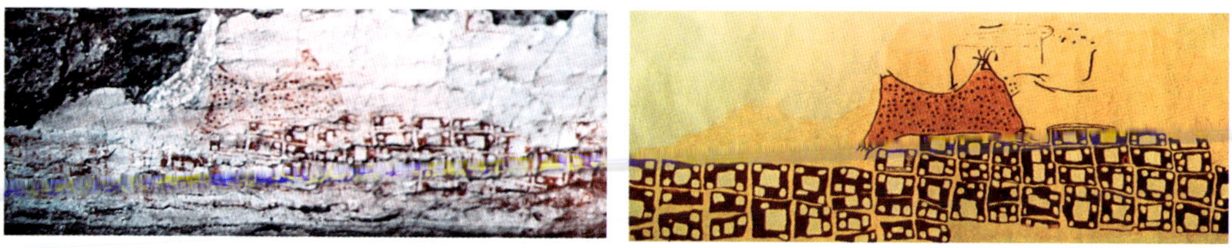

Map from 6300 BC. Enhanced photograph (left) and artist's rendition (right) of a drawing on the wall of a shrine unearthed at the ancient city of Çatal Höyük. Source: Photos of artefact and drawing courtesy of J. Siebert.

The large population size and high density of this city would require a substantial food supply, provided by the agricultural base. It would also have required the society to develop the organization to distribute resources to non-food producers such as craftsmen and officials, and to organize cooperative effort for large projects such as defence or city construction and maintenance. To do so would require the means of planning and communicating the plans to those supervising the activities. The ability to make and use maps, as suggested by the archaeological findings, provided a means of recording and communicating geospatial or geographic information. We believe this may have been an important factor in giving this society a significant advantage over its neighbours, and enabling it to become much larger, better organized, develop new technologies, and gain access to more resources. The central importance of map information, whether in historical societies or in modern development, has been recognized by the United Nations, as is discussed in Section 3.2.1.

Prehistoric maps often seem to have had some function in recording assets of value – such as the location of dwellings in villages. In this sense they were much the same as today's

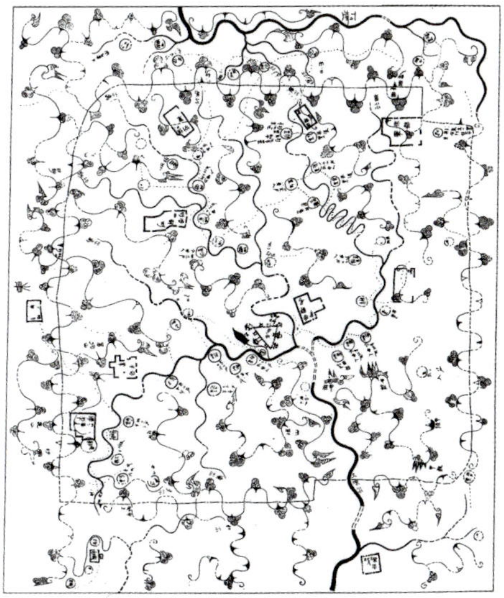

Reconstruction of an Ancient Chinese Military Map. This reconstruction from fragments of a Han Dynasty map (206 BC – 220 AD) excavated at Mawangdui, China shows army camp locations and prominent geographic features. Source: Cultural Relics Press, Government of China, Beijing

cadastral or land ownership maps. Maps inscribed on clay tablets found at Nuzi in Northern Iraq have been dated to 2300 BC. They show settlements, streams, and mountains along with written information indicating the dimensions of cultivated plots of land. Other clay tablets from the same period contain surveying notes for the purpose of taxing properties. Maps of real estate (cadastral maps) seem to have been common in ancient Mesopotamia and Egypt. It is thought that areas of the Nile valley were mapped in ancient times to recover property lines after the annual Nile flood. Several examples of clay maps survive showing property plans in ancient Babylonia dating to about 2000 BC. A clay map dated to 1300 BC describes several privately owned fields with records of the owners' names.[13] The first military topographic maps appear to have been developed in China some time before 168 BC.[14]

2.5. Greek and Roman Mapping: Geospatial Knowledge and Frame of Reference

Geospatial information, such as maps, is not an entirely objective record of information since such maps are also influenced by the compiler's point of view. For example, in selecting what information to record and what to leave out in creating a map, map makers choose what is important and what is not. Their decisions are influenced by their understanding of their society's world view or their frame of reference. That frame of reference, or world view, influences what will be observed and how it will be interpreted. There is truth to the phrase "If I didn't believe it I wouldn't have seen it." It is a complement rather than a contradiction to the common phrase "Seeing is believing."

Our frame of reference can be defined as the filter through which we see and interpret the meaning of the world around us. Our frame of reference not only provides a context to deduce meaning from what we observe, it also affects what we will observe in the first place. Depending on that frame of reference, a particular event could be seen as significant or given no notice at all. For example, to a novice driver, driving a car requires considerable concentration. But with experience the mental processing load is much reduced, as one's mind learns which features are important to watch and which can be ignored. We refine our frame of reference to include only the important information and ignore most other detail.

Thus, in affecting what we observe and how we understand its significance, our frame of reference affects the range of opportunities available for us to discover.

A relevant historical geospatial example was the choice to depict the Earth as flat or round – which in turn was a reflection of the map maker's geographic understanding or belief. To early man the sight of a boat becoming smaller as it sailed away was of no particular import. But to the Greeks, noticing the way the lower part of the boat disappeared from view first and the top of the mast last, it was confirmation of their belief that the Earth was a sphere. Their frame of reference of a spherical Earth led them to observe and interpret details to which others had paid little regard.

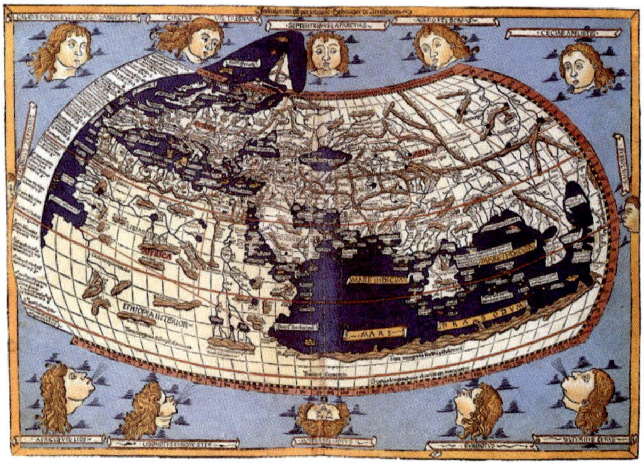

Ptolemy's World Map. This map, produced in 1482, is one of many versions produced from the Roman cartographer's world map description in his book "Geographia" published around 150 A.D. The original manuscript was lost but transcribed copies of the book were rediscovered in the 1300s. Source: Jonathan Potter Ltd., London

A frame of reference can fundamentally affect a society's development of new possibilities. Columbus sought to reach the Orient (which was known at that time to be "in the east") by sailing west because he believed that the Earth was round, at a time when many Europeans still believed the Earth to be flat – and his ideas absurd. For those who believed the world to be round, not only was the idea of sailing west to reach the east possible, this idea of the Earth as a globe encouraged a flood of new ideas, new avenues of exploration and development, and new solutions. It led to the revolutionizing of navigation and cartography, the flow of colonizing people, the spread of disease, the conquering of lands, and the exploitation of natural resources.

Acceptance of the Earth as round was an idea that had ramifications on human society far beyond changing its depiction on a map. It changed people's frame of reference, the context within which new ideas are generated and new solutions are found: the richer that context the greater the opportunity for innovation and, potentially, for breakthrough discoveries and inventions. The increasing integration of geospatial knowledge into the societal knowledge infrastructure is a catalyst that not only offers the possibility of individual gain but will also, we believe, be a determinant of the quality of life enjoyed by a society's population. In Chapters 4, 5, and 6 we will deal with the impact of geospatial information and technology on things as diverse as security of the food supply, management of disease, security of a country, and prosperity.

Returning to the earth as a sphere, in the sixth century BC the Greek mathematician Pythagoras and his followers concluded that since the sun and moon were spherical, the Earth's perfection should therefore be expressed in the form of a perfect sphere. With this frame of reference, and the technical skills of geometry, mathematics, and an inquiring mind, a Greek scholar named Eratosthenes (276 – 196 BC) in Alexandria sought to measure the circumference of the Earth. He had been told that in the city of Syene (today Aswan) to the south on the Nile River, at noon on June 21, the longest day of the year, the sun shone straight down to the bottom of a well. He knew that in Alexandria the sun was not directly overhead on that day and reasoned the difference must be due to the curvature of the Earth. By measuring the angle of the sun in Alexandria at noon on that day and knowing the distance to Syene he was able to calculate the circumference of the Earth. His estimate was about 16% too large but it was to be more than 1500 years before a more accurate estimate was obtained.[15]

The Greeks developed a scientific approach to map making using mathematics with the aid of astronomy. Greek map making practice is considered to have culminated in Roman times with the work of cartographic thinker and practitioner Claudius Ptolemy (AD 90 – 168) working in Alexandria. His 8 volume work *Geographia,* written as a manual for map makers, summarized the practice of map making with discussions on map scale, descriptions of map projections, latitude and longitude, and the use of an index of place names with their coordinate positions.

The legacy of the Greek and Roman principles of cartography was to have a profound influence on the development of societies throughout the world – and in some respects continues to do so.

The works of Ptolemy survived in the hands of the Islamic and Byzantine civilizations. For some time, western scholars rejected scientific inquiry and Ptolemy's work was only rediscovered in 1406 from Byzantine Latin translations from the original Greek. The first printed translations appeared in 1475 and later editions contained reconstructions of Ptolemy's maps based on his writings.

Ptolemy's work contained some notable errors. He rejected Eratosthenes' estimate of the Earth's circumference in favour of that of the Greek astronomer Poseidonius, who underestimated the size to be about three quarters of its actual size. It was this underestimate that in 1492 led Columbus to believe that his voyage would take him to India, not realizing the Earth was large enough to have a major land mass in between. By the turn of the sixteenth century contemporary knowledge identified many of the shortcomings of Ptolemy's work and European mapmakers developed more accurate methods of cartography or map making.[16]

The legacy of the Greek and Roman principles of cartography was to have a profound influence on the development of societies throughout the world – and in some respects continues to do so. The map making skills, and the geospatial frame of reference it embodied, influenced world exploration. However for centuries there was not the combination of resources and vision to fully capitalize on this knowledge. When these conditions reappeared, the ability to record, communicate, and make use of geospatial information in the form of maps and written accounts offered significant advantage.

2.6. The European Age of Geography

The era, often described as the "European Age of Discovery" or the "European Age of Exploration," might be more accurately referred to as the "European Age of Geography" – when Europeans used mapping to both enhance their power and justify their colonization.

It describes the period from the mid 15th century, when European countries began making voyages to distant parts of the globe both for colonization and in search of resources to exploit. After 1280 AD, the presence of a powerful Ottoman Empire to the east restricted European expansion eastward. Travellers such as Marco Polo had taken lengthy journeys as far as China and prompted interest in other routes to reach the Far East. Portuguese expansionist and colonial policies prompted exploration of the African coast.

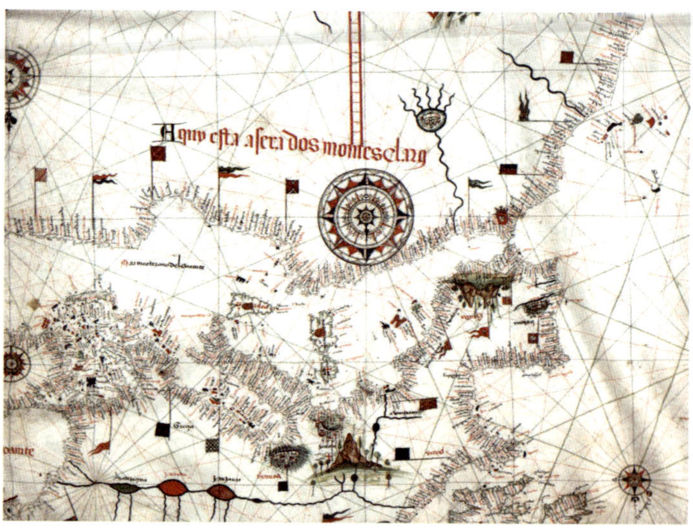

Portuguese Nautical Chart 1492. This detailed and for its time highly accurate chart of the Mediterranean and West African Coasts by Jorge de Aguira was published in 1492. North is down. The detailed geography of the West African coast was considered sensitive military information and was kept secret by Portuguese authorities at the time. Source: Beinecke Rare Book and Manuscript Library, Yale University

The European Age of Discovery (or Age of Geography as we prefer to call it), as dependent as it was on ship building and sailing skills, could not have happened without the development of geospatial information skills in the form of map making and use. If you are going to venture far afield in search of riches in distant lands you need to have some basis on which to decide the correct direction to go. You also need a means of knowing where you are and a way to record where you've been so that you can find the places again – and your way home! You also would benefit by knowing the safest and fastest routes to travel from one place to another.

In short, the European's "Age of Discovery" was dependent on their ability to collect geographic information and use it effectively. They had to become expert at making and using maps, particularly maps of ocean and coastal areas – often termed charts. The collection of charts a ship carried was, in effect, its geospatial data set, without which travel of any significant distance was too risky and uncertain to consider.

The era, often described as the European Age of Discovery or the European Age of Exploration, might be more accurately referred to as the European Age of Geography.

Fifteenth century Portuguese charts showed remarkable detail of the African coast and by 1497 Vasco da Gama had rounded the Cape of Good Hope opening a sea route to Asia. As we look back from the vantage point of today, there is a remarkable similarity between how these Portuguese maps were treated and how data from military spy satellites are treated now. The Portuguese charts, better than anything else available at the time, were treated as closely guarded and valuable military secrets – as they were at the time. According to Seed (2008) the Portuguese tried to prevent their charts from being distributed. She describes how a spy was able to smuggle a Portuguese chart of the world out of Portugal – the Cantino Planisphere. Columbus had charts when he sailed in 1492 – probably with information from the Portuguese maps.

Europe's "Age of Discovery" was dependent on their ability to collect geographic information and use if effectively: they had to become expert at making and using maps.

Spain sponsored expeditions by Columbus in 1492 and others followed finding the landmass of the Americas. As early as 1500, detailed charts became available showing the coastlines of North and South America. These charts were designed for navigation and showed little detail of the land.

In 1569, the Flemish cartographer Gerhardus Kramer, Latinised as Mercator, produced a map projection that kept angles and bearings accurate in every part of a map. This allowed mariners to chart a straight-line course of constant bearing on the map. To do so, the scale on the map varied, distorting the size of landmasses, greatly magnifying temperate landmasses, and diminishing those in tropical zones. This was not a problem for European rulers and merchants as it exaggerated their domains. What's more, Mercator placed Europe at the top center of his map, giving the northern hemisphere primacy over the south and giving the southern hemisphere less than half the map. Familiarity with Mercator projections and its utility for maritime navigation led to its continued use into the 20th century. The exaggerated size of European countries was a symbol of their growing power in the world.[17]

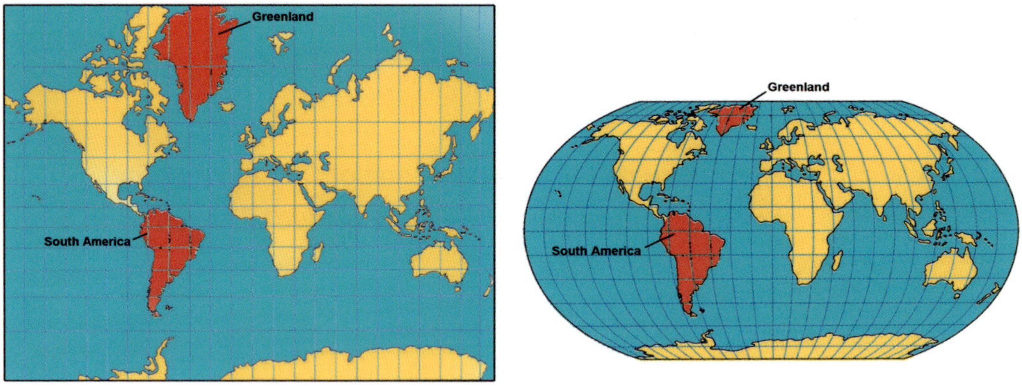

Mercator's Projection. The projection (a mathematical conversion to draw maps of the spherical earth on a flat piece of paper) developed by Mercator, a 16th century Flemish cartographer, made ocean navigation much easier and revolutionized nautical charting. The distortion of area it introduced exaggerated areas closer to the poles making Northern European countries look larger in proportion to areas near the equator. South America is more than 8 times the area of Greenland yet appears smaller when mapped with a Mercator projection (left). The Robinson projection (right), now commonly used for maps of continental areas, more faithfully represents areas across all latitudes. Source: S. Aronoff

The following two hundred years saw extensive voyages of exploration and expansion by European powers. The development of ever more accurate mapping methods was instrumental to this expansion. Not only did expeditions conduct mapping, they gathered information on the plant, animal, and mineral resources of the lands they visited. Their mapping methods gave them the means to project power around the globe. Their sense of purpose and belief in their right to claim lands and resources occupied by non-Europeans was reflected in the way these new lands were depicted on maps as unclaimed land. Such land was available to be exploited in competition with other mostly European powers. Decisions on the division of these foreign lands were made in the halls of power in Europe – far from the people whose future was being decided.

In this competition for new lands and new resources, geospatial information was critical in providing the means to assert power and maintain control over colonies located vast distances from the home country. By 1765 the British had developed a timepiece accurate to 1/14th second per day, sufficient to accurately determine longitude on ocean voyages. The timepiece, termed a chronometer, was immediately adopted by Captain Cook for his return voyage to the Pacific to explore and chart new territories.

Early on, controlling powers mapped their colonies for both economic and military purposes. When Martinique and Guadeloupe were mapped by the French after the Seven Years War (1756-1763) their maps recorded details of the plantation system of their colony, the sugar, coffee, and cotton crops produced, and the names of plantation owners. The mapping, carried out by engineers, was also designed to consider the placement of fortifications in the event of future hostilities with Britain.[18]

> ***The difference between how the Ordnance Survey and the United States Geological Survey do business goes a long way towards explaining how the UK, a nation that was once the world's leader in mapping, has declined to a position of importing almost all of their mapping technology and even much of their base data.***

It is not surprising that Western nations became dominant in map production. Not only in the production of maps by the state, but also in commercial or quasi-commercial map production. Beginning in 1823 the British Admiralty offered an extensive range of charts for public sale. This trend has continued with Western nations dominant in the quasi-commercial world of map production, as well as state sponsored mapping, subsidies, and distribution. Mapping projects and institutions receive far greater public subsidies and greater commercial opportunities among industrialized nations than in the less developed ones.[19]

What is interesting is that in many instances (including the Ordnance Survey in the UK) public or quasi-public entities have often competed with the private sector using what is in effect a tax-supported base and, in some cases, what amounts to a government mandated monopoly. Other agencies like the United States Geological Survey have provided data at low (or no) cost and have studiously avoided competition with industry. They see their role to support the use of geospatial data and development of industry through the provision of low cost and high quality geospatial data. Our studies have indicated that competition with industry and charging high prices for data severely limits the development of the private sector, while restricting access to geospatial information and increasing costs.

Ship Chronometer – Late 18th Century. This innovative timepiece, designed by John Harrison, could remain sufficiently accurate over a long and rough ocean voyage for ocean navigators to calculate their geographic position. Source: Bjoertvedt

In our opinion, the quasi-public Ordnance Survey's competition with the private sector explains why the UK, a nation that was once the world's leader in mapping, has declined to a position of importing almost all of their mapping technology and even much of their base data. By contrast the United States Geological Survey's policy of not competing with industry and providing low cost high quality geospatial data has made it a trusted data source widely used by Americans and the world.

Another interesting observation on the use of maps in developing and developed countries was made by Brandenberger. He suggested that "Developed countries do not have good maps because they are developed; they are developed because they have good maps." [20]

European Imperial countries not only used the power of their geospatial capabilities to discover and control foreign lands, their drawing of boundaries on the maps they produced asserted political control and legitimacy derived from that power. By a cartographic sleight of hand they superseded indigenous spatial boundaries.

European Imperial countries not only used the power of their geospatial capabilities to discover and control foreign lands, their drawing of boundaries on the maps they produced asserted political control and legitimacy derived from that power. By a cartographic sleight of hand they superseded indigenous spatial boundaries. A number of countries have recently tried to redress this early geospatially based larceny through the settlement of land claims with the remaining indigenous populations.

Military surveyors mapping areas where European forces were active also recorded information on the vegetation, relief, settlements, mineral deposits, and possible trade routes. Their maps presented Africa, Oceania, and the Americas as empty or "uncivilized" thereby "legitimizing," to European sensibilities, European control of those foreign lands.[21] Indeed, ancient maps often placed the words "here be demons" or "here be dragons" on lands (or oceans) unknown to the mapmaker.

The incentives for rapidly developing and perfecting the use of geospatial information, at that time principally in the form of maps, were several. It enhanced a nation's ability to project power and thereby gain control of new territories, access to resources, and capabilities for commerce over great distances.

Geospatial information was also powerful as propaganda – influencing perceptions and political outcomes. The choice of map projection and orientation to exaggerate the size and importance of European countries is an image of the world still widely used in maps today and is still powerfully influencing public perceptions. Maps were also an expression of societal objectives and purpose, whether it was to colonize and introduce religious beliefs to new lands, exploit resources, or to capture or defend territory.

2.7. China's Focus on Geography

China provides an interesting contrast to the European experience of a society's development and use of geospatial information. In the final analysis, the world view (or lack of it) of China's leadership several centuries ago influenced its choices in the use and development of geospatial data.

China was one of the few areas where the domestication of plants and animals arose independently and food surpluses were produced. Large organized societies developed as early, or perhaps earlier, than those of the west in the Fertile Crescent. The culture and traditions of the Fertile Crescent later gave rise to the European societies of Western Europe.[22] Why then was it that the Europeans and not the Chinese came to dominate maritime exploration, colonization, and trade in the 17th, 18th, and 19th centuries? It was not for lack of technology or skill. It was a social and political choice.

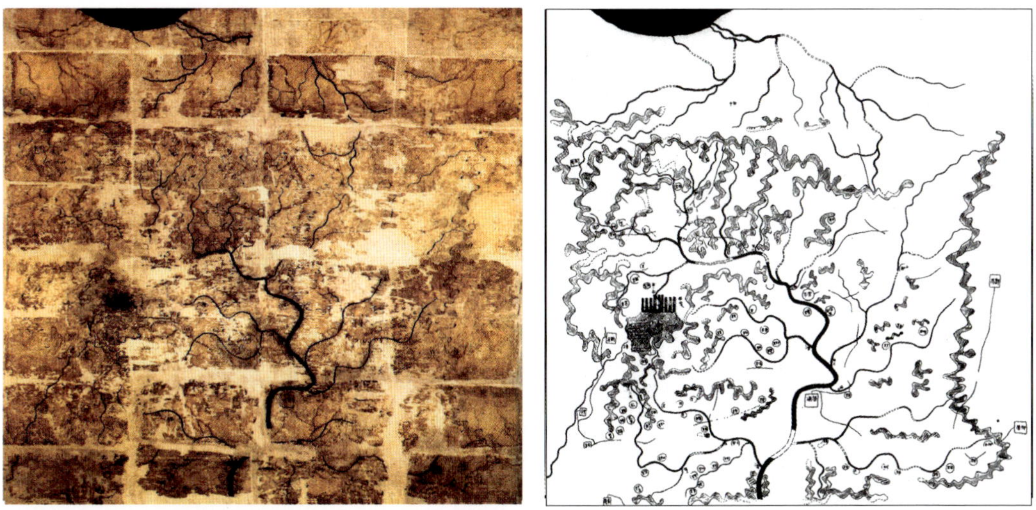

Early Chinese Topographic Map. The artefact (left) and reconstruction (right) of a topographic map drawn on silk that was recovered from a tomb at Mawangdui has been dated to the early Western Han period (183–168 BC). Source: Cultural Relics Press, Government of China, Beijing

Chinese writing had developed by about 1300 BC and the first known map from China dating to 2100 BC is on the outside of a cooking pot. Maps of a graveyard dated to 323 – 315 BC were discovered in a tomb in 1977, and topographic maps were apparently in routine military use shortly after – by the time of the Western Han Dynasty (206 BC – 9 AD). Maps in China were common and were frequently used in various publications for administrative purposes. By the first century AD the Chinese had developed the scaling of distances and a rectangular grid system for maps. They subsequently developed the compass and the printing of maps using wood block techniques before these existed in Europe. These advances were adopted by Islamic traders and subsequently passed to Europeans at about the same time that the cartography of Ptolemy was being rediscovered in the 15th century.[23]

The development of European methods and approaches (particularly those related to cartography) to geospatial information handling that contributed to the spread of European power drew to some degree on Chinese roots. However the focus was different. In the case of China the focus was on increasing access to resources by consolidating control within China and her immediate surroundings. Among European countries, powerful neighbours made local expansion costly and dangerous. Instead they believed opportunity and their destiny were to be found in the discovery and colonization of distant new lands distributed across the globe, from which they could obtain foreign resources to increase their own wealth, power, and competitive advantage.

China had an environment less dry and less ecologically fragile than the Fertile Crescent. While its environmental problems are increasing today, it is a region that has supported intensive agriculture for nearly 10,000 years. These advantages and an early adoption of agriculture enabled medieval China to lead the world in technology. It was where cast iron, the compass, gunpowder, paper, and printing were first developed. China also led the world in political power, navigation, and control of the seas. In the early 15th century China sent merchant fleets, numbering hundreds of ships up to 400 feet long, across the Indian Ocean reaching the east coast of Africa. These exploits were the topic of the major exhibit at the National Museum on Tiananmen Square in Beijing in 2005. This ocean-going venture occurred some decades before Columbus' much smaller vessels made the shorter crossing of the Atlantic to the east coast of the Americas. With their navigational skill and superior ships, why was it not the Chinese that rounded Africa's southern cape, instead of the Portuguese who launched Europe's colonization of East Asia? Or why was it not China that ventured across the Pacific to colonize the America's west coast?

The answer appears to lie in internal politics. Simply put, the faction identified with sending and captaining the fleets lost the political struggle to those whose vision did not include maritime expansion. Those opposing expansion stopped the fleets, eventually dismantled the shipyards, and forbade ocean going shipping. That temporary decision became irreversible as no shipyards remained to produce new vessels that could prove the folly of losing that technology.

By contrast, in Europe power was fragmented among many competing states. When explorers were denied support by one state they switched their allegiance to others until a suitable sponsor could be found. Columbus, an Italian by birth, switched allegiance to the Duke of Anjou in France, then to Portugal, and finally to Spain in order to find sponsorship for his voyages that took him to the east coast of the Americas.[24] Competition among multiple players ensured that a race to develop better ships and navigational skills would persist.

Thus despite its lead in geospatial technology and skills to navigate to distant shores and the organizational skills to manage the operation of expeditions over long distances, China's choice was to project power locally rather than to expand across the globe. Ultimately, global expansion proved to offer a competitive advantage to Europe, providing access to a greater range and quantity of resources. The competition for resources among European nations spurred the development of the technologies used to project power, and secure and manage foreign lands. Geospatial information in the form of navigational charts, maps of foreign lands, and information about their resources were instrumental in both military operations and commercial success. And they still are.

In 1984 China was the first nation to completely map the land use of such a large country using satellite imagery. Since then China has aggressively pursued the modernization of mapping and related education.

The perception of the value of maps and geospatial information in China in a domestic sense has continued to the present. Interestingly enough, in 1984 it was China, under the leadership of Li Boheng and He Changchui[25] of the National Remote Sensing Center under the State Science and Technology Commission of China, that was the first nation to completely map the land use of a large country using satellite imagery. Since then China has aggressively pursued the modernization of mapping and related education.

Wuhan, China was home to the Wuhan Technical University of Surveying and Mapping which was once the largest educational institution in the world dedicated to geospatial disciplines including mapping, surveying, remote sensing, geographic information science and others that we include in the term geomatics. It has since been merged with several other universities to form Wuhan University – which has 5000 faculty members and 45,000 students. A number of the best scientists and entrepreneurs in geomatics now in Canada and the USA have at least one degree from Wuhan. It has been identified as a State Key Laboratory of Information Engineering in Surveying, Mapping and Remote Sensing. An interesting approach in Wuhan has been the naming of many international scholars to its staff as guests or part-time faculty. A few other leaders in the field in North America have one or more degrees in remote sensing from Beijing University, which at one time had a dozen or more faculty members in the field.

2.8. Application of the Geospatial Perspective

Maps were an important form of geospatial information used to project power and administer large complex societies from the 17th century onward. While applications that offered military, political, and commercial gains garnered the greatest investment of money and talent, as familiarity with maps grew other applications were also developed.

> ***Maps were an important form of geospatial information used to project power and administer large complex societies from the 17th century onward.***

In the late 1700s, cities in the eastern United States experienced fearsome epidemics that killed thousands of people. In 1793, about 10% of Philadelphia's population died in an epidemic. Valentine Seaman, a surgeon at New York Hospital in the city of that name, investigated the yellow fever outbreak in his city. When he published his results in 1796 the cause of yellow fever was not known. However from his experience, Seaman suspected that there was a geographic relationship – the nearer people lived to the shipping docks and urban waste sites (sewage, garbage and cesspools) the greater their risk of contracting the disease.

> ***In the late 1700's geospatial concepts (in the form of a map) were used to analyze and eventually combat the outbreak of yellow fever in New York City. A map proved far more effective in promoting general understanding, leading to effective action, than simple statistical information could ever have done.***

He included maps in his published paper which gave his work immediacy and an accessibility that the written article alone did not have. However, the map was not merely an illustration of disease occurrence; it was a carefully considered argument about the incidence of yellow fever and the environments that promoted it. Presenting his argument in a geospatial context not only confirmed his analysis, it also presented the evidence in a manner anyone could understand, which prompted the local authorities to take action.

Yellow fever is a viral infection that was being spread by mosquitoes imported by the sailing ships, something Seaman could not have known with the knowledge of that time. He thought the cause was the putrid smell of the waste areas rather than them being a favourable habitat for mosquitoes. Nonetheless, he correctly deduced that cleaning up these sites would reduce the disease occurrence and his work led local authorities to clean them up.[26]

The presentation of geo-spatial information as a map or in some other form is more than simply recording what is "there" at some location. Rather it is a way of thinking that is inherently multi-dimensional. For example, in constructing a map, one or more sets of features, such as roads, rivers, or buildings are shown in positions representing their locations relative to each other. Presenting them together suggests that they may be related in some way. Thinking of information in this way fosters a perspective that is both relational and spatial at the same time. In so doing, information often can be deduced because the spatial distribution suggests relationships among the data. Making a map is a highly creative enterprise that can lead to conclusions based on the empirical evidence. This explains how Seaman deduced the means to control yellow fever without understanding the true mechanism of the disease and its transmission.

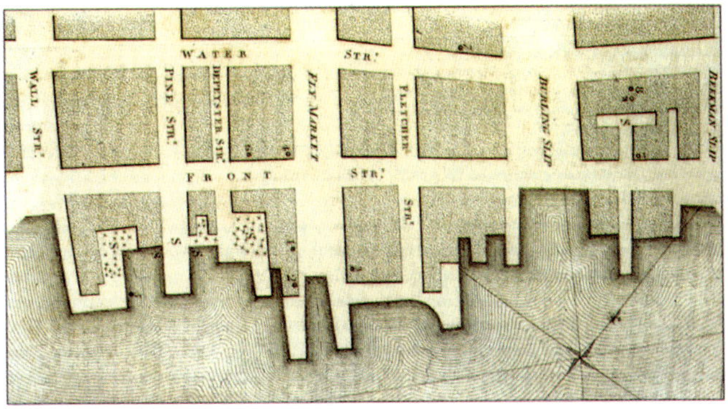

Early Mapping of Disease. This map of yellow fever incidence by Seaman was included in his 1798 publication."Inquiry into the cause of the prevalence of yellow fever in New York". Source: Osler Library, McGill University

The effectiveness of a map depends as much on what is left out as what is included. If there is too much information the map will be illegible; if there is too little there may be insufficient information for its intended use. For that reason a map is not simply a representation of an objective reality as much as it is selected information that advances a point of view – a statement about a particular area as described by information selected for inclusion by the map maker. Indeed, the fact that maps can be used to advance ideas that are not correct has led a respected US-based author, professor, and researcher, Mark Monmonier, to write a book titled *How to Lie with Maps.*[27]

A map is not simply a representation of an objective reality. It is selected information that can advance a point of view and provide a statement about a particular area as described by the information selected for inclusion by the map maker.
A map can be biased and can even be constructed to lie.

One of the interesting changes in map-making in recent years has been the use of an image acquired from either satellite or aircraft as background – allowing the map user to also extract information from the interpretation of the image as is needed. Not only does this save the cost of encoding or developing information few users require, it also allows a skilled user to deduce information that might have been too costly to interpret and map from an image. In effect, an image can be seen as the first abstraction from the "real world," while a map is a second abstraction and statistics that have been derived from measurements from the image, map or other source can be considered to be a third level of abstraction.

The broader understanding of mapping and the spatial reasoning it entails has resulted in its application to an ever broader range of planning and analytical activities – those relevant to the administration and management of large populations living in dense groupings such as cities and states. But the principle drivers in developing the more advanced technologies for geospatial applications tended to be those that provided competitive advantage for significant material benefits, be they greater military, political, or commercial success.

2.9. Technological Escalation

2.9.1. Introduction

Improvements in geospatial technologies typically produced incremental changes in capability – evolutionary development. For example, improvements in surveying instruments enabled greater accuracies in mapping. Less frequently, new technology provides revolutionary capabilities that eclipse previous methods. The marine chronometer enabled longitude to be accurately determined at sea; previously that was not possible. This type of technological change is revolutionary in that a significant barrier is overcome. Suddenly something can be done that could not be done before, or something can be done so much more quickly or so much less expensively that not only does the new technology replace previous methods but it changes the way the work can be done. Over the past few years the rapid growth in the development of new geospatial technologies has led to revolutionary changes in how map information is made, provided, and used.

> ***Three important geospatial technologies have developed in a revolutionary way in recent decades: aerial photography and remote sensing, geographic information systems, and satellite positioning technologies such as the GPS (Global Positioning System).***

Three important geospatial technologies have developed in what we argue is a revolutionary way in recent decades. These are aerial photography and remote sensing, geographic information systems (GIS), and satellite positioning technologies, such as GPS (Global Positioning System). Aerial photography, remote sensing, and geographic information systems are introduced in this chapter. Satellite positioning (GPS), satellite and airborne remote sensing, and GIS are more thoroughly addressed in the next chapter.

2.9.2. Aerial Photography and Remote Sensing [28]

The collection, analysis, and presentation of geospatial information was revolutionized by the development of aerial photography. Later, the development of non-photographic sensor systems such as electronic scanners and radar, led Evelyn Pruitt, in 1958, to coin the term *remote sensing*[29] to include both photographic and non-photographic imaging sensors. These technologies enabled, for the first time, information about an area to be captured in the form of an image, at an instant in time, representing features together with their spatial relationships.

Early Aerial Photography. This aerial view of San Francisco was taken from a kite just weeks after the 1906 earthquake devastated the city. It not only documented the city's condition, the image sold briskly to the public as both a historical document and the novelty of aerial photography, something most people of the time had never seen. Source: Photo by George R. Lawrence, Courtesy Library of Congress

Stereoscope. Analyst using a pocket stereoscope to view aerial photographs in stereo – or three dimensions. Source: S. Aronoff

Aerial photography began soon after the invention of the camera. French photographer and balloonist, Gaspar-Felix Tournachon, known as "Nadar" is credited with taking the first aerial photograph from a tethered hot air balloon 80 metres above the ground in 1858. Perhaps the most famous early aerial photographs were the panoramic views of San Francisco taken by George Lawrence just after the 1906 earthquake and subsequent fires devastated the city. He used a 22 Kg camera of his own design suspended from a kite flown 600 metres above the ground to capture the entire city of San Francisco in a single 43 cm by 122 cm contact print.

Not only was the subject matter dramatic, the viewpoint was novel. The average person had never seen an aerial photograph. What's more the enterprise was lucrative. Selling for $125 each he earned some $15,000 which today would be worth over $350,000.[30]

As innovative as aerial photography from hot-air balloons and kites may have been, it was the invention of the airplane and powered flight that enabled aerial photography to revolutionize mapping. Only with powered flight could the camera be reliably positioned where it was needed and flown in a relatively straight line to capture a series of overlapping images of an area. Not only did this provide complete coverage, it enabled the collection of a series of images with a controlled amount of overlap. When two images with the correct amount of overlap are viewed with an instrument called a stereoscope, the images appear as a single perspective or 3-D view. Features above the terrain such as mountains or buildings appear to extend up towards the viewer and low areas (such as valleys or quarries) appear to extend down into the image.

Aviator using early aerial camera. Source: American Society for Photogrammetry and Remote Sensing

This 3-D effect is perceived because the relative positions of features on successive photos are shifted differently depending on their elevation. The field of photogrammetry developed techniques to accurately measure these displacements and calculate horizontal position and elevation. From these measurements, reasonably accurate[31] topographic maps could be produced from aerial photographs with much less field work. As photogrammetry developed and aerial camera technology advanced, topographic mapping of very high accuracy could be produced from aerial photography. Aerial photography revolutionized map making to the point that virtually all geographic mapping today makes use of aerial photography or other remotely sensed imagery.

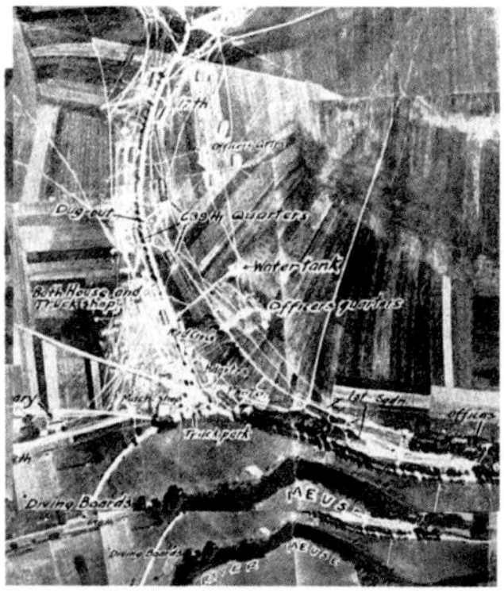

Aerial photograph of military facilities in France during World War 1. Source: American Society for Photogrammetry and Remote Sensing

As is often the case with developments in mapping, it was a war that accelerated the development of aerial photography for use in mapping. At the beginning of World War I, aircraft carried observers over the battle front to draw sketch maps of military positions. It was only during the war that the notion of using aerial photographs to do this job was tried. By the end of the war air-photo coverage of the entire battle front was produced on a daily basis, with additional photography obtained as necessary.

Allied military photo-interpreters at work during WWII. Source: R.C. Nesbit

Between the wars aerial photography was applied for peaceful applications. For example, the production of city maps could be done much faster and less expensively from aerial photographs than by traditional field work. Large areas of previously unmapped terrain, for example in northern Canada, were flown and mapped for the first time. Methods of air-photo interpretation began to be developed for forestry, geology, soils mapping, and other applications. But it was wartime that once again provided the incentive and the resources to rapidly develop aerial photography for mapping, the analysis of terrain conditions, and the collection of military and industrial intelligence.

Germany had successfully developed advanced techniques of mapping from aerial photographs and the interpretation of terrain conditions from aerial photography before the outbreak of World War II. At the opening of the war, Germany had the best photo-reconnaissance mapping and photographic interpretation capabilities in the world. Germany flew extensive photography missions and made detailed maps of the target countries and their military installations before they attacked. It was the German Chief of Staff, General Werner, who, in 1938, remarked that the coming war would be won by the side with the best photo-reconnaissance. He was proven correct. The General was killed during the war and without his leadership German photo-reconnaissance operations were hampered. At the same time, the Allies developed effective analysis techniques and superior capability to collect images and rapidly generate maps and useful information, such as the location of enemy positions and strength, assessments of combat effectiveness, and identification of targets. The more extensive and more current geospatial knowledge derived from recent aerial photography, together with ground observations and other intelligence information, gave the Allied forces an important advantage in the conduct of the war.

Airphoto Acquisition for Mapping in 1924. The Vickers flying boat was widely used over the vast sparsely populated regions of northern Canada to acquire aerial photography that was used to produce some of the first topographic maps of the region. Source: Library and Archives Canada

During the war, research produced improvements in photographic films, including color photography and the development of black and white infrared film. "False color" infrared films were also developed for military purposes. The name "false" was attached because the colours are not what one would naturally see with the naked eye, inasmuch as the color assignments include red, green and near infrared portions of the electromagnetic spectrum – not the blue, green and red of natural color film. False color infrared film, also known as camouflage detection film and colour infrared (or "CIR"), was developed to show green painted vehicles as blue and vegetation as red. After its commercial release by Kodak in 1960, this film proved to be of value for vegetation assessment and was applied in forestry, agriculture, and environmental monitoring.

As is often the case with developments in mapping, it was a war that accelerated the development of aerial photography for mapping.

With the Cold War, the major military powers had an urgent need for intelligence gathering methods. This urgency provided the impetus and the funding to develop new types of sensors. These new sensors included radar (which enabled imaging at night, through cloud cover, and because of its side-looking geometry, from the other side of borders) and thermal infrared (which could produce images at night, detecting the temperature of objects such as the heat given off by military vehicles). The imagery from these new sensors complemented that collected with conventional photography.

Film for Specialized Applications. The deciduous and coniferous trees are different shades of green in the normal color photograph (left). In the false color infrared image (right) the deciduous trees appear red and the coniferous trees are green-blue. The truck is yellow because it reflects both red and near infrared light Source: S. Aronoff

Specialized high-altitude aircraft were developed by the United States such as the U-2 and SR-71, capable of flying higher or faster than pursuit aircraft or missiles of that era could reach. The development of space photography was a revolutionary advance in aerial reconnaissance. Although the satellite imagery could not capture details as fine as those collected by aircraft, they provided broad coverage over hostile terrain – without the potential for a pilot to be shot down over hostile territory. On May 1, 1960, a covert American reconnaissance U2 aircraft was shot down over the Soviet Union and the pilot, Gary Powers, was captured. The international incident that ensued led President Eisenhower to accelerate the top secret Corona satellite mission with its high resolution camera system. The first successful Corona mission lasted only a few days but provided the United States with image coverage over more of the Soviet Union than all of its previous aerial reconnaissance missions combined.

The ballistic missile installations of concern were large and easily seen from space. Analysis of the photographs from this and subsequent missions showed the Soviet Union had far fewer ballistic missiles than it claimed, and the much feared Missile Gap (difference in number of US and Soviet missiles) was actually in America's favour.

The 1962 Cuban Missile Crisis was sparked by the identification, by American intelligence analysts on U-2 aerial photography, of nuclear missile sites on Cuban soil. While the more detailed aerial photography discovered the threat, knowledge of overall Soviet missile capability, aided by information from space photography, strengthened the US bargaining position.

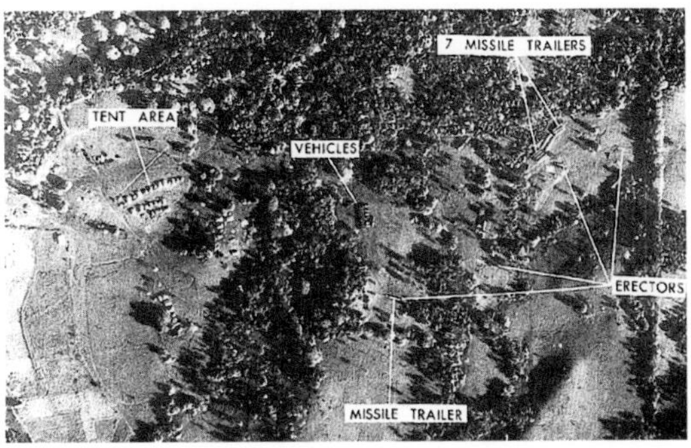

Aerial Reconnaissance During the Cuban Missile Crisis. Air photos collected by U2 reconnaissance aircraft (right), produced images from which expert military interpreters could extract a wealth of information (left). Source: Central Intelligence Agency

The resolution, i.e. the smallest feature visible on the images of the early top secret Corona space photography, varied during a mission according to atmospheric conditions. (Resolution is discussed in detail in Section 3.2.3.1 in the next chapter). The best resolution, at first about 8 metres in 1960, had improved to about 2 metres in 1963.[32] Today earth observation satellites constantly provide publicly available imagery of the earth, some with resolutions better than 1 metre. What military systems can now acquire is, of course, considered secret – although considerable conjecture does exist as

Satellite Imagery Then and Now. The satellite photograph of the Pentagon, a vast building complex in the United States, is shown on an early space photograph (left) acquired by a highly classified military satellite during the early 1960s. It has a spatial resolution of about 8 m. The digital satellite image (right) acquired by the commercial QuickBird satellite in 2001 has a spatial resolution of about 0.6m. Source: Central Intelligence Agency Record Group #263 (left), DigitalGlobe (right)

to current capabilities.[33] The imagery is used in applications ranging from military reconnaissance to environmental conservation. Satellite and aerial imagery are now ubiquitous. The public has become used to seeing television weather and news displayed over remotely sensed images. There is a plethora of imagery available online for purchase, and much of it viewable at no charge.

A few years ago Google Inc. introduced a free online service allowing users to search street locations and view them with an image background from satellite or air-photo imagery. Subsequent refinements and the introduction of a more powerful online service, Google Earth, and a competitive offering called Virtual Earth from Microsoft Inc., rapidly introduced the public to a very intuitive means of presenting geospatial information – from driving directions to Aunt Martha's to the location of the nearest Pizza restaurant. More recently (May 2007), Google introduced Street View – which provides a view of the street from a moving van. The information captured on camera ranges from the mundane – the view of parking lots and streetscapes – to the more exciting – in one instance in San Francisco the camera seems to have caught a street fight! One of the interesting debates sparked by such imagery relates to concerns of the right to privacy.

Earth observation imagery has developed from a specialized information source to become a part of our geospatial infrastructure, accessible to an increasingly well informed general public.

2.9.3. Geographic Information Systems

Geographic information systems comprise one of the critical technologies to make effective use of remotely sensed imagery and other geospatial information.[34] This technology provides a means to rapidly analyze geospatial data from diverse sources, and generate detailed information products to communicate results tailored to the needs of the recipient. A geographic information system (GIS) is an integrated set of computer software tools and hardware designed to collect, store, analyze, and output geospatial information – i.e. information related to geographic locations.

The development of geographic information systems began in the 1960s, before there was the computer technology to support the type of data handling its original developers envisioned. A major driver in the development of GIS was the idea of "overlay." In an overlay analysis one simultaneously overlays maps of different characteristics of an area. The purpose of this is to identify locations that have combinations of characteristics making them uniquely suited to certain types of use – be it agriculture, forestry, or urban land use planning. (Tomlinson, 1987)

Planners and landscape architects use the overlay technique to find optimal locations for housing developments or the route of a new highway. For example, one might want to locate a new road through a forested area that contains a unique animal. The road must not pass within 100 metres of the particular type of forest known to be the habitat of that animal – areas that are within 20 metres of a stream. At the same time, the road must avoid slopes greater than a certain amount, and also be as close as possible to sources of gravel. Each of these characteristics comes from a specific map. Overlaying all of these maps will lead to a best location for the road – that is a location with the fewest limiting factors and the most beneficial ones.

Geographic information systems comprise one of the critical technologies to make effective use of remotely sensed imagery and other geospatial information.

In the 1960s, this method of overlay analysis was popularized by Ian McHarg in his book *Design with Nature* published in 1969.[35] He used multiple copies of a base map drawn on clear plastic sheet material. Each overlay represented the suitability of a different characteristic such as

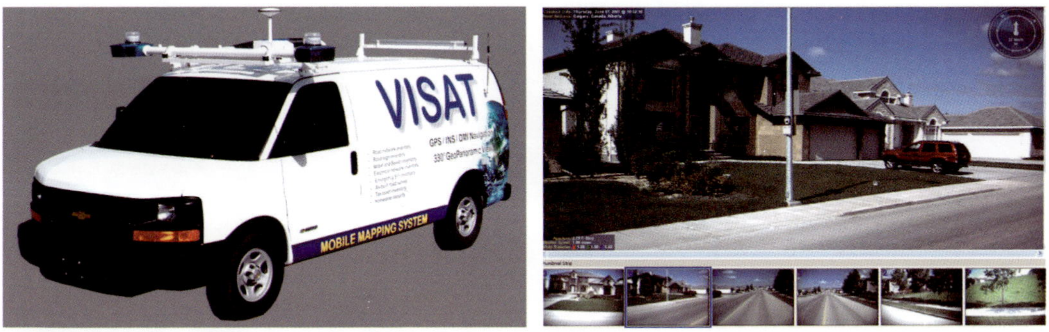

Mobile Ground-Based Image Capture. Vans equipped with cameras capturing multiple views and real-time positioning technology like GPS are used to continuously acquire geo-referenced imagery along a travel path. Source: Dr. Nasar El Shiemy

soil drainage, geology, vegetation, and so on as it pertained to the specific project at hand. The characteristics on a particular overlay were represented as shades of grey, the lighter the shade the more suitable the area for the particular project. For example, in siting buildings and roadways well-drained soils are more suitable so they would be represented as light shades, while boggy poorly drained soils would be shown as dark shades. Once this type of overlay was constructed for each factor, the overlays could be placed one over the other on a light table and the lighter areas would indicate those most suitable for the project. A planner still had to use judgement to take into account factors not represented on the overlays, but the process dramatically improved what had been a difficult map comparison task.

The development of GIS tools and the geospatial data that they use has proceeded at an accelerating pace since the adoption of GIS by governments and industry began in earnest during the 1980s.

It was to perform this type of overlay analysis that geographic information systems developed. The idea of using separate layers for each class of feature was retained. Instead of shades of grey, feature characteristics were represented by numerical values allowing the same data layer to be color-coded differently to suit any project.

In the course of developing this overlay analysis capability, a broad range of other spatial operations were introduced as well. Today's GIS handles virtually any form of data that is referenced to a geographic location – be they data conventionally shown on a map, such as vegetation, geology, soils, political boundaries, or street networks, or geospatial data not usually thought of as "mapable" such as an audio file telling a legend about a particular place, a video clip of a ceremony performed at a particular location, or a set of coordinates and attributes representing all the street lamps in a city. What has made all of this geospatial information, now in digital form, so useful are the processing tools available, particularly those developed for GIS. The use of GIS has broadened even further with the development of what are called geospatial data infrastructures – formed to allow the seamless sharing of geospatial data across different jurisdictions.

The development of GIS tools and the geospatial data that they use has proceeded at an accelerating pace since the adoption of GIS by governments and industry began in earnest during the 1980s. The development of GIS technology is fuelled in part by both the ever growing range of uses and the very practical benefits that its application brings. The better the GIS tools the more useful the geospatial data, the more useful the data the more readily money is spent increasing the quality, quantity, and affordability of the data.

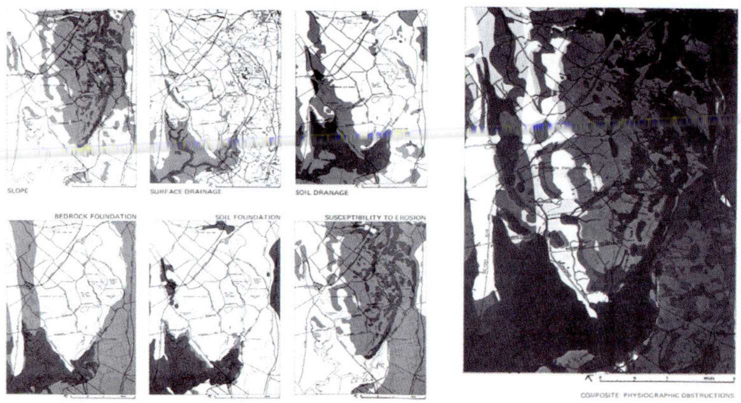

Data Integration Using Maps. Ian McHarg in his 1969 book "Design with Nature" promoted the use of maps to simultaneously assess multiple environmental factors. In his method, each environmental factor was mapped onto clear plastic in three shades of grey representing the suitability for the specific development being planned. The individual maps (small maps) were overlayed and viewed on a light table. The light areas in the combined map (large map) indicated areas most suitable for the proposed development. "Design with Nature". 25th Anniversary Edition by I.L. McHarg. © 1992 John Wiley and Sons, Inc. Reprinted with permission of John Wiley and Sons, Inc.

The better the quality and quantity of data and the more affordable it is, the more reason to invest in the GIS tools, and skilled personnel to use that data. GIS seemed to be on an upward spiral of increased use – and that attracted more investment, more applications, and more interest from those outside the so-called "traditional" geospatial field.

Over the past two decades there has been a massive investment in the geospatial data infrastructure in most industrialized nations. Geospatial information, previously stored on paper maps, was digitized to convert it to digital form suited for use in GISs and other computer-based technologies. At the same time, GIS technology and related geospatial technologies, including remote sensing and satellite positioning or GPS, have developed rapidly. Today a GIS is used not only to optimize planning, it may be an integral part of the information system used to run an enterprise such as a city, province or country. In that capacity, a GIS is incorporated into such services as keeping track of tax roles, issuing permits, scheduling maintenance, planning of road closures, and a wide range of other services.

Increasingly, what has come to be considered GIS functionality is becoming part of the information technology infrastructure of organizations. This evolution of geospatial information processing, from specialized separate systems to functional inclusion in the information infrastructure of an organization, is reflected, for example, in the offerings of major database software providers such as Oracle and its competitors. They now promote geospatially-enabled database products capable of handling all of the data needs of an organization – both spatial and non-spatial.

It is not that geospatial data has suddenly become important to government and industry. The geospatial component was always important. Shipping distances, the location of service vehicles, and locations where road maintenance will be done all add important spatial dimensions to the decisions they influence. What has changed is that by making geospatial data collection and storage digital, more information is available and it can be analyzed more quickly, in more diverse ways, and at lower cost than ever before. Applications are now possible that were previously impractical or, in some cases, not even dreamed of in the past.

For example, a GIS can easily produce a map containing any combination of data to which it has access. Since maps can be produced on-demand (See Section 3.2.4) within minutes, analyses can be progressively refined in an iterative manner. For each iteration, a new map can be displayed on a computer screen or printed as a paper-based map. The process allows a planning session where multiple stakeholders can each formulate and test their competing scenarios over a few hours.

Another critical question often overlooked is whether the data have been changed, intentionally or otherwise, to render them inaccurate for our purposes.

Digital collection and storage of geospatial data has enabled the data to be maintained more efficiently. Data can be continuously updated so that instead of the traditional map that presented information compiled months or years before, a GIS enables maps and other outputs to be generated within minutes of an analysis, representing the most current information available at that time. In essence, the GIS provides a window to view a constantly changing database of geospatial information and to extract, analyze and output those data in a wide range of formats to suit different needs.

Of course this capability introduces another issue – the age of the data. How old are they? Are the data still useful for our purposes? Another critical question to assess is have the data in some way been changed to render them inaccurate for our purposes? It is a question commonly overlooked, but inaccurate data can render an analysis completely wrong. Worse than no data at all, incorrect data thought to be accurate can prompt wrong actions to be taken with conviction. Consider the implications of false data when in March 2010 after a major earthquake in Chile, coastal residents were reassured that a Tsunami would not occur. Frightened residents taking refuge on high ground returned to their ocean side homes only to be struck within hours by a Tsunami, destroying homes with loss of life. In this case, with no information residents may have remained wary, stayed on high ground, and been safe.

Geographic information system technology has become a part of our geospatial infrastructure – and that in turn has become part and parcel of our other infrastructures – be they corporate, municipal, state, or national. Whether we are performing an analysis using a GIS or doing an online search of travel routes that are displayed as a list of routing instructions and a map, GIS technologies have been driving the applications.

Geospatial technologies have changed the way we think about our world and make decisions that affect our health, safety, and prosperity.

They have not only improved the efficiency, quality, and timeliness of the information we depend upon in our work and our leisure, geospatial technologies have changed the way we think about our world and make decisions that affect our health, safety, and prosperity.

2.10. Conclusion

Geospatial information has been important for success, if not survival, throughout all of human history. The relative importance of the role played by geospatial information in securing access to resources, ensuring safety and security, and projecting power depended on the cost of use (in time, money, or other resources), the accuracy and timeliness of the information produced, and the importance of the decisions that information would affect.

With the rapid development of geospatial capabilities over the last two decades the number of applications and range of conditions where they are cost-effective has expanded rapidly. Often what was once deemed to be impossible or impractical becomes routine within a few years or less. Those who recognize and seize these opportunities first, be they individuals, private enterprise, or government organizations, have the opportunity to benefit first – and most.

Geospatial information has been important for success, if not survival, throughout all of human history.

The benefits range from the mundane to the critical. To the individual it might mean not getting lost on a road trip – or finding an alternate route to a hospital when the way is blocked by a bridge that has just collapsed in an earthquake. To an emergency response crew it could mean more lives saved. To a resource company it might be a matter of obtaining resource rights to a rich property ahead of the competition, and to the military it might mean the difference between a successful defence and a rout. Knowing the location of all the resources critical to an enterprise, wherever they may be, changes the speed and effectiveness with which opportunities and threats can be recognized and acted upon.

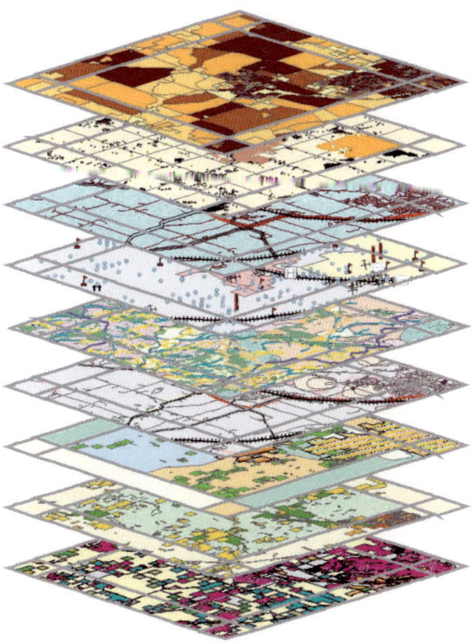

Data Integration. Computer-based geographic information systems enable a virtually unlimited number of data sets or data layers pertaining to the same geographic area to be registered and analyzed together, as illustrated here. Source: Land Information and Computer Graphics Facility, University of Wisconsin, Madison, Wisconsin

As the use of geospatial information becomes more widespread and sophisticated it changes not only the way individual enterprises operate, it changes the nature of a society's economy and way of life.

The pace of such change today, combined with the wide disparity in access to and skill in using geospatial information and technology, create significant disparities – with those having greater access reaping a significant advantage over their less knowledgeable competitors – what we term a geo-advantage. The following chapter reviews important geospatial technologies that now impact our lives in today's GeoEconomy.

Another factor adding to the complexity in geospatial data is the concept that was introduced in the mid-late 1980s: cost recovery. The concept is simple. Those requiring and using the data should pay for the costs of creating and supplying the data. A KPMG study (Sears, 2001) in which we were involved found that cost recovery policies in geospatial data severely limited the use of location data for some players and some applications – especially those related to governments and the "public good." Another interesting finding was that the cost to administer cost recovery is often greater than the funds received. Furthermore, there are significant costs associated with the economic impacts resulting from less use of the data. On average, the cost recovery seemed to return just 3% to 30% of the actual data production and distribution costs, depending upon the country. Furthermore, it was also found that cost recovery can inhibit the development of innovative technologies, applications, and industry development in geospatial: the higher the level of cost recovery, the lower the level of development of the industry. On publication of these results that compared the generally lower data prices in the USA to Canada and Australia, both Canada and Australia began to change their cost recovery policies toward lower data prices.

As the use of geospatial information becomes more widespread and sophisticated it changes not only the way individual enterprises operate, it changes the nature of a society's economy and way of life.

The conclusion on cost recovery is, in our opinion, clear. Given that most information pertaining to human activities has a location component, and given that economic activities can be made more efficient when location data is available, it only makes sense to provide such data at low or no cost. To do otherwise is to deny citizens' access to the GeoEconomy.

NOW!

3.1. Introduction

We have explained how geospatial knowledge has always been important – even if those participating in past economies were not always fully (or even partially) aware of that fact.

Today the situation is far different than it has been at any other time in the recorded history of man. Today there is more geo-awareness – and those who are both aware of and using geospatial technologies enjoy a significant advantage – the geo-advantage.

Today the situation is far different than it has been at any other time in the recorded history of man. Today there is more geo-awareness – and those who are both aware of and using geospatial technologies enjoy a significant advantage – the geo-advantage.

Geo-advantage comes from not only being aware of the technologies, but also of the data or information available, having access to that information, and knowing how the information can be used to provide a competitive advantage in the GeoEconomy. Together these can collectively be referred to as the geospatial infrastructure.

It is the geospatial infrastructure that ultimately sets the limit to how well the GeoEconomy will perform and the degree to which individuals, industry, communities, and governments reap the benefits of the geo-advantage – or fail to reap the benefits. The more developed the infrastructure, the more the possibilities. This is where it becomes complex for governments and industry alike.

We must all be smart and forward thinking to build the right infrastructure at the right time – whether it is a road or a set of base maps. We must use the right technologies, acquire the right data, and do so in a coordinated fashion.

We must all be smart and forward thinking to build the right infrastructure at the right time – whether it is a road or a set of base maps. We must use the right technologies, acquire the right data, and do so in a coordinated fashion.

This chapter introduces the reader to the most important of the geo-technologies that make up that infrastructure, discusses the data available and how they are used along with the technologies in the GeoEconomy today, the policies of governments that make it all possible (or impossible in some countries), and the expanding need for information. The rapid convergence of technologies, new data needs, and new data policies are covered in Chapter 4. For those who may wish to skip over some of the technical bits on technology and data types introduced in the next two sub-sections, we suggest going on to the discussion of government policy (Section 3.4) – including geospatial data policy – an area too often ignored by politicians, bureaucrats and even some geospatial experts.

Geospatial technologies and the data they produce release us from the tyranny of our location and our geography.

Together these technologies, and how they are being or can be applied, release us from the tyranny of our location and our geography. It is this release that has pushed us into the GeoEconomy, and it is the integration of all of the geo-technologies that brings us the full impact of the GeoEconomy. What all this means with regard to its impact on geo-luck and the future will be discussed in Chapters 4 and 5 respectively.

3.2. Technology Factors

3.2.1. Introduction to the New – and Not-so-New Geo-Technologies

When did the "new" GeoEconomy actually begin? We believe it began when the United Nations, in a motion adopted in 1999, included geospatial information as an integral part of a national infrastructure.

The motion came from a meeting involving some of the leading policy-oriented minds in the remote sensing world – with an emphasis on developing nations.[36]

The motion states:

> "whereas maps and geospatial data derived from a combination of Earth observation information and other data are as much a part of a nation's infrastructure as the transportation network, the health-care system, telecommunications and education, the creation of a national geospatial infrastructure should be accorded the same level of support as the other elements of national infrastructure. It was noted that the need for geospatial infrastructure suggested that Governments should become knowledgeable customers and users.[37]"

Others might link the beginning of the GeoEconomy to the 2000 removal of the intentional degradation from the USA's GPS system, or President Clinton's support for geospatial data.[38]

We believe that the GeoEconomy began in 1999 with the adoption of a motion on geospatial data by the United Nations.

The rise in importance of the geo-technologies since 1999 has followed the converging development of other elements of the technology infrastructure available to a modern society. We include in this infrastructure the rapid advancement in accessibility to computer power, and the power of the Internet to provide rapid access to a broad range and quantity of information. To these we might also add the ability to make precise measurements in time, distance, and in the physical interactions that occur in both the natural and built or man-made environments.

While the importance of computer power and accessibility, the Internet, and increased precision in measurement are considered undeniable and important to the growth of the GeoEconomy, detailed analysis of each of these factors is beyond the scope of this book. While assumed to be "given," together they do provide the foundation that allows the geospatial technologies to now drive the GeoEconomy.

The full power of the Internet with regard to geo-information is one that is both promising and yet somewhat scary in its implications.

The full power of the Internet with regard to geo-information is one that is both promising and yet somewhat scary in its implications. We can more easily monitor our individual and collective environments – and are more easily monitored by others. One very positive thing that the Internet has made possible is the use of geospatial data infrastructures – a mechanism for sharing and distributing geospatial information. These will be discussed further in Section 3.2.6.

Most of us are familiar with at least certain aspects of geo-technology. Most of us have heard about GPS or the Global Positioning System – using signals from satellites for locating yourself – or your stolen car. Most of us have also seen satellite imagery – in the late-night news showing war zones or the results of natural disasters, or in more benign daily weather reports. We have seen satellite imagery in Google Earth and Microsoft's Virtual Earth, or perhaps in a movie or in the *National Geographic.* Such imagery has become common in most literate societies – and especially so in geo-literate societies. We have all used maps of one sort or another – be they roadmaps, topographic maps, or a map on the back of an envelope on how to get to Aunt Sue's in Sheboygan.

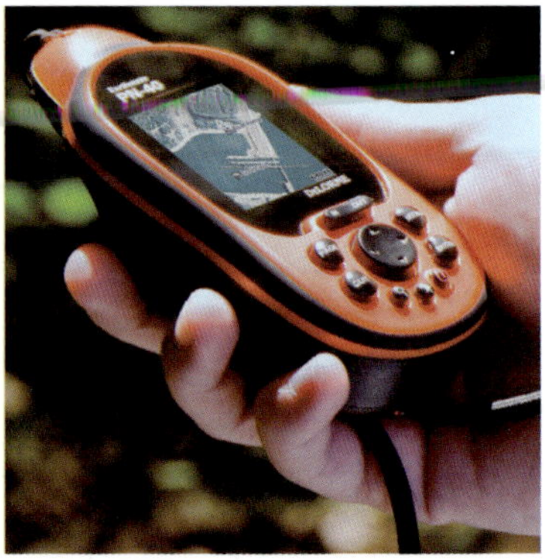

Handheld GPS. Source: DeLorme

Fewer of us have seen or been exposed to the new ways of making maps, or how various data "streams" from geo-technologies can be integrated in a geographic information system, or a decision support system embedded in an enterprise management system.

The balance of this section introduces five of today's most important and relevant geo-technologies – satellite navigation (GPS), remote sensing, mapping, geographic information systems, and geospatial data infrastructures. For each we provide:

- An explanation of what it is;
- What it is used for; and
- Some of the broader implications of using the technology.

3.2.2. GPS

3.2.2.1. What is it?

The Technology

The Global Positioning System (GPS) is a satellite-based navigation system used to determine where on or near the surface of the earth a GPS receiver (and anyone or anything close to it) is located.

The GPS consists of a network of some 24 Earth orbiting satellites. Circling twice a day about 12,000 miles above the Earth, these satellites constantly transmit information which is used by GPS receivers to determine the location of the receiver. That receiver can be a mobile device in your hand, on a truck bringing oranges from Florida, or on a car picking up the kids at school. The GPS receiver compares the time a signal was transmitted by a satellite with the time it was received. From this time difference the distance to the satellite is calculated. With the signals from at least four satellites the device continuously calculates the receiver's position and changes in its position over time. GPS is both weather and location independent – it works everywhere in almost any kind of weather.

The Global Positioning System (GPS) is a satellite-based navigation system. It helps determine where on the surface of the earth the receiver (and anyone or anything close to it) is located.

Since 2000, GPS receivers have become more compact, far less expensive and used in far more ways than was ever considered when the GPS system was first conceived. GPS receivers can be small hand-held devices or even combined with other technologies. Most new mobile telephones (cell phones) now contain a GPS chip – in effect a mini-GPS receiver. GPS receivers vary in their accuracy. The most accurate models are somewhat larger – more backpack transportable than the familiar hand-held devices. However, the technology continues to advance rapidly providing ever more capable devices at lower cost.

There are several different approaches to using and improving GPS with high technology sounding names such as Differential GPS, Wide Area Augmentation System, etc. All we need to know here is that each has a different accuracy in terms of identifying the location of the receiver – ranging from a few metres to fifteen metres. Close enough to find your stolen car – but not close enough to locate a survey marker in downtown New York City.

Competing Systems and Policies

The first and best known satellite navigation system, which has become synonymous with the technology, is the Global Positioning System described previously. A project of the U.S. Department of Defense, it was started in 1978 and became fully functional with 24 satellites in 1994.

Since the first GPS program there have been others developed, under development, or proposed. The most significant in terms of the GeoEconomy is the European Galileo program which has already begun to launch satellites. The Russian GLONASS system was developed and then fell into disrepair – by the end of 2008 it was once again operational over Russian territory. Eventually it will have 30 satellites. Not to be outdone, China began to launch its COMPASS navigation system in April 2007, while India approved the development of the Indian Regional Navigation Satellite System in September of 2006, although no satellites have been launched as this is being written. Literally billions of dollars are being invested in GPS technology to make these systems work – and many more billions in the technology to use them.

> ***Literally billions of dollars are being invested in GPS technology to make these systems work – and many more billions in the technology to use them.***

GPS was originally designed for military use, but in the 1980s, the US government made the system available in a degraded form (called Selective Availability) for civilians to use. Selective availability reduced the accuracy of the location measurements. In a Presidential Directive on May 1, 2000, that degradation was removed,[38] allowing civilians full access to the power and accuracy of the system. Originally the GLONASS system had 100 metre accuracy with its deliberately degraded signals and 10-20 metre accuracy with its military signals. The difference between how the Russians and Americans handled data access is both interesting and instructive. The Russians degraded the accuracy and did not release accurate information, while the Americans did. There was no benefit to using the less accurate Russian system. As a result the Americans came to dominate the commercial GPS market and the Russians never really entered the commercial GPS world.

With the American system present and working well why have there been these other systems? Part of the reason is concern for reliable military capability – independent nations do not necessarily want to rely on the US for a technology of such importance to their military. But there are other considerations as well. One analysis [39] suggested that Europe was "showing an almost American sense of sovereign economic and technological determinism" in launching Galileo....a closer look at the plan reveals solid economic and technological goals that do not threaten the U.S. system nor American military concern, but will certainly challenge its place as the premier global satellite navigation technology.

The Galileo system will be run by a private consortium. Price Waterhouse Coopers estimates the annual cost over the planned 20-year operation of Galileo will reach €220 million ($266.6 million) each year, money slated to come from the companies running the program. The funds are expected to come from revenue streams related to commercial operations. Position or location is not only militarily important – it is also seen as commercially important.

The American GPS system is government funded consistent with other data and services paid for by the US tax payer as a "public good." A similar example is weather satellite images and data provided free of charge by the US National Oceanographic and Atmospheric Administration (NOAA). US government funding and policy has meant that there are no costs to use GPS.

3.2.2.2. What is it used for?

Introduction

As with most high technology, GPS receivers (often just called "GPS") have seen a dramatic drop in price – from thousands of dollars a decade ago, to a few hundred dollars a few short years ago, to under $100 today for a basic hand held system one can buy at almost any hardware, automotive, electronics, computer or department store.

> ***The drop in price of GPS receivers has seen a tremendous increase in the range and number of applications – it is no longer the purview of large corporations or specialized land surveyors.***

The GPS "Imperative"

The drop in price of GPS receivers has seen a tremendous increase in the range and number of applications – it is no longer the purview of large corporations or specialized land surveyors – although many of these have embraced the technology in ways that even they did not envision a few years ago. To some extent, adoption of GPS technology has become a requirement to stay in business. Cost, the level of specialized knowledge required, and the availability of data have always shaped the level of use and range of applications for geospatial data. It is instructive to examine the way a now common vehicle tracking application developed. What follows is a simple example of a theme that stretches throughout human history.

In the early 1990s, the United States had not yet launched its satellite-based navigation system, the Global Positioning System, now used for applications ranging from specialized surveying and field data collection to consumer in-vehicle navigation systems. The cell phone networks we use today had not yet been implemented and mobile communications were done by relatively expensive mobile phones or short range transceivers, as those used in taxis or CB radios. At that time, a company operating trucks nationwide tracked each vehicle by estimating its location based on its departure time, route, and periodic telephone reports from the driver. As valuable as more accurate and timely location information for each vehicle would have been, there was no practical cost effective means of providing that data. All trucking companies faced the same limitation.

GPS is used for vehicle/fleet management, agricultural and forest management and harvesting, ship navigation, and field work in a variety of fields including environmental management as well as land surveying for property ownership.

With the development of GPS and cell phone networks and the availability of inexpensive technology and service, it became possible to track the current location of each vehicle within about 100 metres, display that location on a map, and also maintain current vehicle operating information such as current speed of travel, maximum speed at which the vehicle was operated, time since last stop, duration of last stop, and other data useful in ensuring the efficient and safe operation of the vehicle. Having this information is not essential to operate a fleet of trucks, the old ways still work. However, companies that found an effective way to bring geospatial technology and information into their operations could lower their costs, improve their competitiveness, and increase their income, thereby giving them a competitive advantage. It increased the likelihood that the business would survive and prosper where some of their competitors would fail.

It is not that geospatial information was not important to the trucking industry until GPS and cell phone technology was available. Geospatial data were always important. It is just that the cost, accuracy, and timeliness were such that the more limited tracking method was the best option at the time. The new technology changed that situation by enabling more accurate and timely information to be provided at an affordable price.

Once an industry begins adopting a geospatial technology, if the technology provides a significant advantage, then competitors are forced to adopt it or risk being less efficient, less profitable, and possibly out of business. For the transportation industry, as for many others, it has become imperative to use GPS.

Applications

Military: We have briefly mentioned the military heritage behind GPS. GPS has long been used on the battlefield – whether to call in airborne strikes, cruise missiles, or artillery – or to locate friendly (and unfriendly) troops on the ground. Much has been written on military applications of GPS in recent military history, as well as in more popularly aimed books on the subject.[40] Suffice to say, GPS has been and continues to be used in military applications.

Vehicle Navigation and Tracking: Perhaps the civilian GPS application best known over the past few years in North America is the General Motors OnStar System. The OnStar service offered by General Motors equips a vehicle with GPS tracking and two way voice communication. A live operator can respond to an emergency request or an automatic crash response notification (distress) signal, generated if the vehicle detects that a qualified crash (emergency condition) has occurred. The OnStar system provides the vehicle's location and details about the crash to the dispatcher who can communicate with the occupant, call for emergency services and direct them to the vehicle's precise location. We can attest that it works – I accidentally hit the OnStar button on my new car and was instantaneously greeted by a friendly voice ready to help! What this tells us is that GPS has gone main stream. GM has advertised that some of its new vehicles will have a theft deterrent capability that will literally stop a reported stolen car from moving, on a signal from GM.

GPS Tracking for Emergency Assistance. Services such as OnStar use GPS and wireless communications to give emergency dispatchers the vehicle's location.
Source: General Motors

Car rental agencies offer GPS systems so that you will never be lost – whether traveling in the USA – or southern Africa.[41] A range of services are now available that make use of in-vehicle GPS units. Some companies provide real-time information on traffic jams, others on weather and related road conditions. The company GPS North America offers a service that can track and even immobilize a stolen vehicle,[42] although the more common use is to track company vehicle movements to ensure that vehicle use is authorized – as well as to help in billing clients who have received service calls.

At the commercial level, fleet operators and even small companies attest to savings of up to 25% in fuel, and greatly reduced operating costs through the monitoring of vehicle movement and through better planning of trips.

Trucks are not the only elements of the transportation system using GPS. Ships now routinely use GPS for both navigation and security: countries can obtain both the location and identification of ships approaching their shores using GPS and what is referred to as the "Automated Identification System."

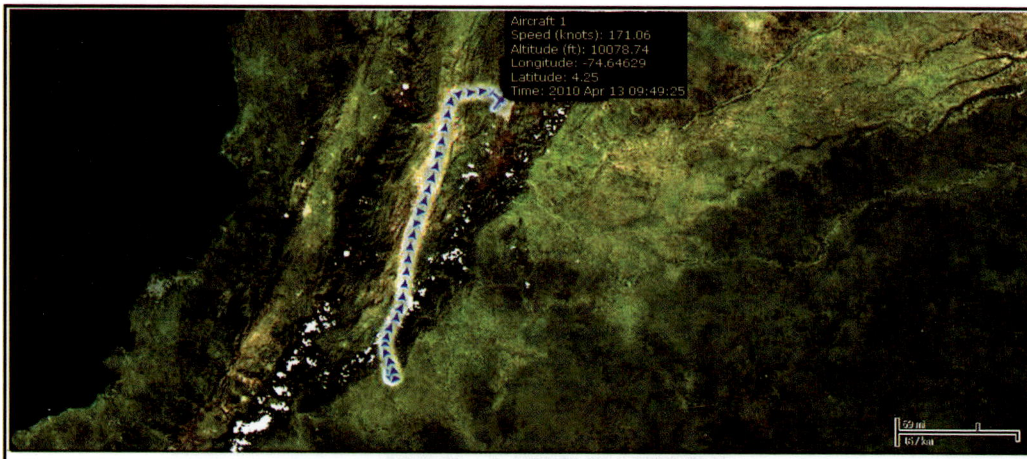

Activity Report - 2010-04-10 To 2010-05-31

	Trip	Activity	Start		End		Duration (hh:mm:ss)	Distance (nm)
			Date	Location	Date	Location		
☐	Trip 2010-04-13 21:46:46	Mobile	2010-04-13 17:46:46	-75.48555, 1.47216	2010-04-13 19:02:45	-74.20602, 4.75386	01:15:59	254.088
☐	Trip 2010-04-13 18:15:46	Mobile	2010-04-13 14:15:46	-74.6431, 4.25502	2010-04-13 15:09:35	-75.48967, 1.49276	00:53:49	186.487
■	Trip 2010-04-13 13:02:17	Mobile	2010-04-13 09:02:17	-74.15238, 4.70986	2010-04-13 09:49:25	-74.64629, 4.25	00:47:08	81.052

Real-Time Vehicle Tracking Application. This display shows the current location and trip path of an aircraft equipped with a GPS-enabled tracking device and operator system. Past trips can be selected for display and analysis from the list. Source: Guardian Mobility, Corp., Ottawa

Emergency Response: Perhaps the best (and one of the first) applications of GPS is in emergency vehicles such as ambulances and fire trucks. They use GPS both to locate the emergency and to manage the fleet. In responding to natural disasters, identifying where emergency resources are located, where supplies are needed, and directing the deployment and delivery of personnel and equipment requires accurate location information. Observers on the ground or in the air need to be able to identify the location of conditions they assess, to communicate useful information.

Tracking to Reduce Loss and Theft: A Canadian company, Guardian Mobility Corporation,[43] provides a monitoring service to let you know about all movements of your aircraft, car, boat, or RV for both asset management and security. In a discussion with US officials at NASA Stennis in southern Mississippi in 2007, we were told that one of the post-Katrina issues they had was that the US Government had sent some 28,000 mobile homes to the region as temporary homes – and now they don't know where they are. After some time, many of them have been moved – in some cases back to home sites that had been destroyed by the storm. A simple GPS system and transponder embedded in each of the homes would have solved the problem.

Precision Agriculture: Another important if not yet high volume user is the farmer practicing what is called precision agriculture, "an integrated information – and production – based farming system that is designed to increase long term, site-specific and whole farm production efficiency, productivity and profitability, while minimizing unintended impacts on wildlife and the environment."[44] Conditions within an agricultural field vary. Applying the same level of herbicide, pesticide or fertilizer treatment over an entire field usually means that some areas are over treated and some are under treated. For precision agriculture the conditions within a field are first assessed and mapped. This mapping is done using GPS to locate coordinates of areas requiring different treatments – whether the field is producing corn, carrots or chardonnay. By using specialized control devices and a GPS unit on the tractor, these treatments can be automatically adjusted as they are applied. As a result the treatment application is optimized and yield is increased.

In recent research in the Niagara Region of Canada one vineyard was mapped using a GPS to precisely locate each vine along with its basic attributes – including its health. This in turn was used in the development of a management plan, including the placement of wind machines that reduce winter damage to delicate rootstock.[45]

Industrial Operations: GPS is also used in forestry to position and identify areas to be harvested, for construction of logging roads, as well as to guide harvesting equipment and

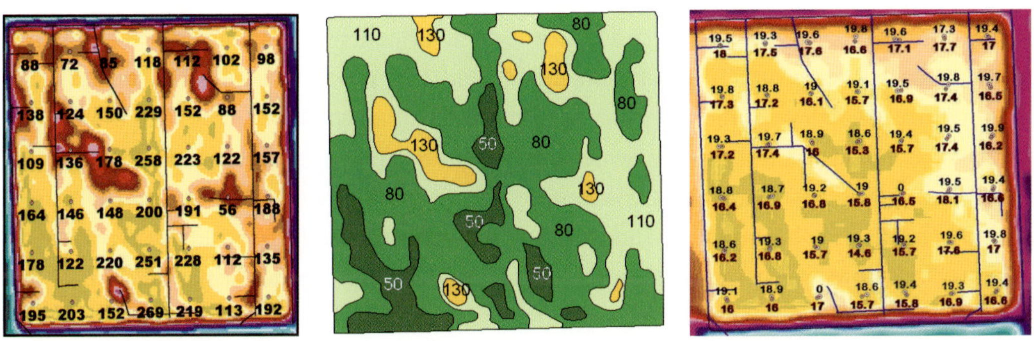

Precision Farming - Sugar Beets. Variations in plant vigour (NDVI index) are interpreted from a Landsat 7 satellite image (left) shown with field measurements of measured leaf nitrogen in pounds per acre before using precision farming. Using this plant vigour image, a spread map (center) is generated that prescribes varying the application of fertilizer to even out the soil nitrogen levels across the field. This plant vigour image of the same field when precision farming was used shows a more even and higher average vigour (right). The values compare the sugar content of field samples from the same field with (upper value) and without (lower value) precision farming methods. Higher values are better. Source: A.W.G. Farms, Inc.

planting. The operation of large heavy equipment such as used in open pit mines can be monitored and even controlled using GPS. GPS technology is also used in various types of construction activities including buildings, roads, dams, and electrical transmission lines.

Field Work: GPS has revolutionized the conduct of many types of fieldwork. Geologists and others who do field work use GPS to locate and record field sample sites on the ground and on the water – again typically used with maps and databases. The authors have first hand experience of finding and correctly recording the location of sample sites before the advent of GPS. Hours of a field day could be spent checking maps, airphotos, and taking measurements to ensure that the location of the field sites were indeed correct and correctly annotated on the map and field sheets. It's a task now done instantly with a GPS. In fact, with GPS equipped hand held or laptop computers, your field location can be viewed on a map of your choice, the data entered and be immediately included in an analysis and then the results displayed.

Tracking People and Animals: GPS can also be used to track the movement of animals and people – whether cattle moving from farm to abattoir, or an endangered form of wildlife to which a GPS receiver has been attached. They are also used in law enforcement. GPS tracking devices that can be locked onto a person and transmit their location can be used to monitor prisoners on day parole, or to enforce court orders restricting someone from being in certain areas. For example, a paroled pedophile might be restricted from approaching specific neighbourhoods or school yards.

Land Surveying: So far most of the applications we have discussed here assume that there is some form of movement over time and through space. One of the most important but static applications of GPS to the economy is in land surveying related both to cadastral survey (property boundaries) and construction. Here of course one cannot use systems that are accurate to a few metres: measurements of land and property boundaries must be far more precise. Higher cost, higher quality, and more accurate receivers can be used to position survey markers related to property ownership. Of course, property ownership has long been recognized as the basis of capitalistic economies. Hernando de Soto asked: "Why has the genesis of capital become such a mystery? And why have the rich nations of the world not explained to other nations how indispensable a formal property system is to capital formation?"[46] De Soto is, of course, recognizing one of the underpinnings of the GeoEconomy. And GPS offers the potential to more quickly develop property ownership maps than any other methods – as the advertising of the major players in the industry will quickly tell you.

The GPS Industry

One of the leading companies providing high accuracy GPS technology and applications is Trimble Inc. With 3400 employees and 2007 revenues approaching $1 billion, Trimble "provides positioning solutions enabling professionals in engineering and construction, surveying, agriculture, fleet management and field service to be more productive by revolutionizing their work processes."[47] Trimble claims (and is quite correct in saying) that it "is transforming the way work is done through the application of innovative positioning." While well known in the field of surveying and mapping, Trimble is far from alone in the market, as we will see in the discussion below.

As we consider the market, it is interesting to note that a government military report on the technology in the 1980s predicted worldwide demand for no more than 25,000 GPS units.[48] This projection, like many others concerning technology and involving a major paradigm shift, has proven to be far, far off base. The shipment of recreational GPS devices doubled from 2002 to 2004 to over 6.5 million, and sales were projected to grow annually at over 30% to the end of 2009.[49] This of course pales in comparison with cell phones, of which some 3 billion are in use. As more and more cell phones become GPS enabled, the market for GPS services will once again explode. Studies suggest that "GPS production value globally is expected to grow to more than US$30 billion in 2008, up from US$13 billion in 2003."[50]

A more personal indication that also tells us that GPS has gone main-stream is the fact that within a few minutes' drive from my home, in my OnStar equipped vehicle, I can buy GPS receivers ranging from under $100 to over $400, made by manufacturers including such well known trade names as Garmin, Magellan, SiRF, and Nextar. While many of these are sold for use in vehicles and boats, there are many other personal level applications. To most non-professional users, personal use applications are among the most interesting as well as diverse. As this is being written, the technology is daily seeping deeper and deeper into the middle class through stand-alone receivers, such as those sold in the electronics store in my nearby village. We have seen the use of GPS to precisely locate that favourite hunting spot, where you took that wonderful photograph, or where you are now in relation to that restaurant where you are meeting your friend for lunch – be it in Maputo, Mumbai, or Milwaukee. These applications gain even more value when linked to the maps discussed in Section 3.2.4 below.

3.2.2.3. What are the Implications?

Introduction

"Always knowing where you are is a fundamentally exciting idea," said D. Noah Eckhouse, an entrepreneur who has been working with GPS since 1992, when he was working in an MIT research lab.[51] But the implications of GPS use go far beyond a fundamentally exciting idea. In the simplest analysis, GPS provides the ability to locate a person, animal or object in relation to a fixed coordinate system (or map). From this geographic positioning capability, sophisticated applications are developed for navigation, tracking, emergency response, and others discussed previously. The implications are many and profound in the

GeoEconomy. Literally hundreds of uses have been identified – and many more are being developed as this is being written. At the same time, privacy advocates have expressed concern that too much information is being made available about where we, and our belongings, are located – especially in real time.

> ***GPS provides the ability to instantly know the geographic location of a person, animal or object. The implications are many and profound in the GeoEconomy.***

The economics of GPS are clear. Aside from the billions of dollars being spent on GPS receivers, the use of GPS can lead to tremendous efficiencies in the management of both human and other resources.

> ***The economics of GPS are clear. Aside from the billions of dollars being spent on GPS receivers, the use of GPS can lead to tremendous efficiencies in the management of both human and other resources.***

The Implications

> ***GPS technology:***
>
> 1. *Allows us to always know where we are;*
> 2. *Helps us manage our resources (including people, equipment and natural resources) better and more efficiently;*
> 3. *Helps ensure the safety and security of ourselves and our property;*
> 4. *Can lead to the invasion of our privacy by governments, industry, and/or other citizens; and*
> 5. *Releases us from the tyranny of our geography.*

For each of these implications we can provide some real-world examples.

Always Know Where We Are

GPS applications can be as simple as preventing one from getting lost while hiking in the woods or providing, through your cell phone, Blackberry or other personal communications device, information on where the nearest pizza restaurant is located relative to your location, where your best friend is as you stroll through the Ginza area of Tokyo, or where that favourite fishing hole is at the lake. Applications can be as complex as calling for support in an emergency situation or knowing the location of the minefield borders close to where you are working.

Helps Manage Resources Better and More Efficiently

GPS not only helps us by letting us know where we are, but it also increases the efficiency with which governments perform tasks on our behalf. GPS can also help the environment in many ways. It allows field workers to locate observations and sample plots so that they can return to exactly the same location for further study or remediation. It is GPS that has allowed us to precisely monitor the decrease in size of glaciers with all that implies – as well as the damage associated with events such as hurricanes and tsunamis. GPS allows farmers to more precisely target the application of chemicals to the benefit of the environment – and lower the cost of production.

In the City of Edmonton we can cite two examples from a study we performed a few years ago. We were told that with hand-held devices, including GPS, building inspectors were no longer required to come into the office to get their assignments. Their work was given out based on the location of the home-base of each inspector and they could immediately start work without having to come into the central office. Once the inspection report was done, it was returned to the central office by e-mail – with GPS location attached. Efficiency was increased 25%: inspectors could do 25% more work with the same resources.

In another Edmonton example, those responsible for maintaining trees and parks were given their assignments tied to an enterprise management system, with a suggested optimal route to take them from place to place to avoid areas where there may be traffic issues at certain times of the day. They were also given the expected time to accomplish specific tasks. In another city we are told that when GPS receivers were put on city trucks and the precise location of each truck could be mapped, they no longer seemed to congregate for an early morning coffee as they once did. Again, efficiencies increased.

It is clear that such management of resources does lead to efficiencies provided, of course, that the input data are accurate. This question of accuracy of input data is a major problem in terms of geo-information – one we consider further in several places in this book.

Ensuring the Safety and Security of Ourselves and Our Property

With a GPS you can always know where you are, where your car or boat is, or where that shipment of boats is on the truck going from the boat manufacturing facility in Pipestone, Minnesota to Pirate's Cove Marina in the Village of Kemptville, Ontario. As noted above, Guardian Mobility offers technology that will let you know whenever your boat (or RV – or any other large possession) is moved. If the movement is expected, the alert can be ignored. If you fear unauthorized use, you can call the authorities – and the company can track the movement.

Avis Car rental has long advertised that it uses GPS in all of its rental cars to deter car theft. Police have used cell phone usage histories in tracking the location of criminal suspects. The result is clear – reduced losses, lower insurance premiums and a greater sense of security.

Invasion of Our Privacy

While there are many obvious benefits to the precise monitoring of location, some may consider that such monitoring has an element of "big brother" associated with it. If you carry a GPS equipped cell phone your location is continuously updated and can be recorded – thereby documenting everywhere you have been. That may be very comforting to a concerned parent worried about their child. But do we want the cell phone company to know where we have been – what areas of town we shop in or travel to? To whom might this information be given or sold without our knowledge? Do we want an image of us collected without our consent as we walked down the street keyed to a specific location and passed on for someone's use? Some will argue that the better good of society is served by such monitoring and that if we are not committing some criminal act, it should be of no concern. In the type of society in which we live, we may not fear the authorities – but we may have some concerns about our movements being tracked by those wishing to take advantage of us or do us harm.

Release from the Tyranny of our Geography

When all of these above-mentioned implications are brought together, we can conclude that in the final analysis, the most important implication is that GPS helps to release us and our economies from the tyranny of location and our geography. This is perhaps the greatest and most important implication of all in the GeoEconomy.

GPS helps to release us and our economies from the tyranny of location and our geography.

Conclusion

For the first time in human history the GPS provides us with an instrument that can instantly provide a measurement of our geographic location with an accuracy of a few metres. GPS is an enabling technology. As important as obtaining a position measurement may be, it is the plethora of diverse technologies and applications that have been and are being developed with its use that has made GPS technology so important in the GeoEconomy. The other four technologies to be discussed all now depend on the availability of GPS. As it is incorporated into ever more of our personal devices such as cell phones and laptop computers, it will probably be ever more a service that is taken for granted perhaps to the extent that we assume there is phone service wherever we go.

3.2.3. Satellite and Airborne Remote Sensing

3.2.3.1. What is it?

We spend a great deal of time and effort in this section of this book describing remote sensing – what it is, and how it is done, and why. Why do we spend so much time on this one topic? Some might say it is because of our background – we both earned Ph.D.s in this field. While that bias may enter into it, we believe that remote sensing is rapidly growing in visibility and importance to the individual, to companies, to local communities as well as to state and national governments. Yet while it is more visible, it is not generally well understood – but it should be.

Remote sensing is important:
it should be better understood by more people.

Remote sensing is not just a technology that does some "really neat" things. It has become one of the technologies on which our society has become dependent. Its impact on society is pervasive, profound, and like many essential services, an aspect of infrastructure of which the public is largely unaware. Remote sensing extends human senses by orders of magnitude. For example, the public takes it for granted that there is weather forecasting and is largely unaware that this capability is dependent on a global network of satellite and ground-based sensors continuously collecting data to estimate sea surface temperatures, temperature profiles through the atmosphere, storm locations and intensity, wind direction and intensity, and a plethora of other factors. Using this data, sophisticated computer models and human experts produce continually updated information on current and anticipated weather conditions. This information is used by the transportation industry to improve the routing of aircraft and ships to avoid storms, and by municipalities to plan for snow clearing, flooding, hurricanes, and other events that will impact the demand on services. While we often make jokes about the accuracy of weather prediction, it is much more precise than it was thirty years ago and much of that precision can be attributed to remote sensing.

While fundamental components of our infrastructure such as the weather service are important, we are largely unaware of and don't think much about them. But the impact of such behind-the-scenes components is immediately apparent when they suddenly fail. Imagine for a moment the impact on airlines if there was suddenly no weather information and pilots would not know in advance if they were flying into dangerous weather conditions, or if the airport where they were scheduled to land was shut down by a blizzard.

Remote sensing is important: it should be better understood by more people.

Remote sensing is the canary in the coal mine: it is remote sensing that has provided the unequivocal proof of the changes that are taking place on our planet – and what has taken place on other planets too.

In addition to the weather example noted above, remote sensing contributes to the mapping and assessment of changes on the surface of the earth. It helps us monitor the environment and assess disaster relief requirements. It contributes to both transparency of decision making and our collective security. Some liken it to the "canary in the coal mine" for our species – it is remote sensing that has provided the unequivocal proof of the changes that are taking place on our planet – and what has taken place on other planets too. For that reason alone, we believe it important for a wider audience to better understand what remote sensing can do – and what it cannot. The importance of remote sensing was clearly underscored in a presentation by Nobel Laureate Jonathan Overpeck at the 2010 American Society for Photogrammetry and Remote Sensing (ASPRS) Conference held in San Diego. He said that "remote sensing has been very important in identifying the problem (of climate change) and will be even more important in the future to help solve the problem." An interesting response to his speech by Dr. Marguerite Madden, a former President of the ASPRS, was that "the last seat in the lifeboat should be reserved for a student of geography."

Satellite Image - Montreal Olympic Stadium. This commercial satellite image has a spatial resolution of 0.61 m. Source: DigitalGlobe

We have already mentioned satellite imagery and how it is well entrenched in today's computer and Internet savvy society through the efforts of Google and Microsoft – as well as more traditional popular sources of geographic knowledge such as *National Geographic.* Indeed, one could argue that *National Geographic* was the only major popular – and regular – source of remote sensing information, imagery and knowledge until several entrepreneurs began to distribute image maps in the mid-late 1980s, and a few coffee table books were produced.[52]

To us it is an interesting curiosity that there are likely more people aware of satellite imagery than of its lower-tech precursor, aerial photography. Again, as perverse as it may seem, this heightened awareness of satellite imagery compared to aerial photography is especially so in developing countries. There the information content of aerial photography has often meant that it has been considered to be sensitive information, with military value much greater than that of the less detailed satellite imagery.

"The last seat in the lifeboat should be reserved for a student of geography."

For some decades satellite imagery has graced the front pages of newspapers in Thailand, Malaysia, and India, for example. Yet for these countries and much of the rest of the world aerial photography has been, and is still considered to be, something that cannot or should not be distributed for the entire country. This is because of the detail such imagery contains in areas judged to be of military significance. Even in Canada aerial photography acquired before the 1950s was often stamped with the word "SECRET" or "CONFIDENTIAL." In virtually all countries there still remain "no-fly zones" over sensitive military installations or potential targets.[53] The implications of these limits on distribution are further discussed in Section 3.4.2 and 3.4.3.7.

We consider airborne and space-borne imagery of many different types together since they have been taken to be part of the same technology – remote sensing. The term remote sensing was popularized in the *National Geographic*[54] and by the late Dr. Robert "Bob" Colwell of the University of California Berkeley – a leader in the field who was a friend and mentor to the authors of this book.

Remote sensing, which came alive with the beginnings of the space program of the United States, has been defined in a number of ways:

- "The measurement or acquisition of information of some property of an object or phenomenon by a recording device that is not in physical or intimate contact with the object or phenomenon under study. The technique employs such devices as the camera, laser and radio frequency receivers, radar systems, sonar, and ... (various geophysical instruments)."[55]

- "Deriving data or information about the Earth's surface or the atmosphere by observing reflected or emitted electromagnetic radiation. The oldest remote sensing instrument is the photographic camera."[56]
- "The science, technology, and art of obtaining information about objects from a distance."[57]

Remote sensing is based on a very simple premise – things that are different may look different on imagery acquired from above.

Remote sensing is based on a very simple premise – things that are different may look different on imagery acquired from above. Expressed more scientifically, things that are different reflect light or emit electromagnetic radiation in a different way from other things (and hence they "look" different).

If you want to monitor or map these different things, you record and measure the reflected or emitted radiation in parts or bands of the electromagnetic spectrum in which they appear different. A simple camera records light reflected from the scene in front of the lens in what we call natural color – the color seen by the human eye – in the blue, green and red bands of the spectrum. Some parts or bands of the non-visible electromagnetic spectrum can be used to assess the health of vegetation (near infrared), while others can measure the relative roughness of terrain (radar), or heat (thermal infrared) emitted from a building – or an enemy tank. Entire books have been written on how we can measure and use reflected and emitted energy to measure the properties of the world around us.

Imaging Building Heat Loss. The colors in this thermal infrared image of buildings on a residential street on a cold night range from purples and blues (cool) to oranges, reds, and white (warm). Window areas are white indicating maximum heat loss. Source: Argon ST Inc.

Image Resolution

One of the terms often used to describe the nature of remote sensing is spectral resolution, which "refers to the number and width of spectral bands recorded for an image."[58] One can think of this as a measure of the amount of color one can see in an image. Another term is radiometric resolution which is a measure of the "number of values available to record the brightness levels in an image."[59]

There are several other terms that are also useful to describe remote sensing data: spatial resolution, temporal resolution and whether the sensor is active or passive.

There are several other terms that are also useful to describe remote sensing data: spatial resolution, temporal resolution and whether the sensor is active or passive.

Spatial resolution is a measure of the spatial detail recorded by the system. While spatial resolution is sometimes considered to be a measure of the smallest feature discernible on an image, it is more technically correct to say that it is the smallest area for which reflected or emitted radiation is recorded. The dimension of this smallest area, as measured on the ground, is often referred to as the pixel size.

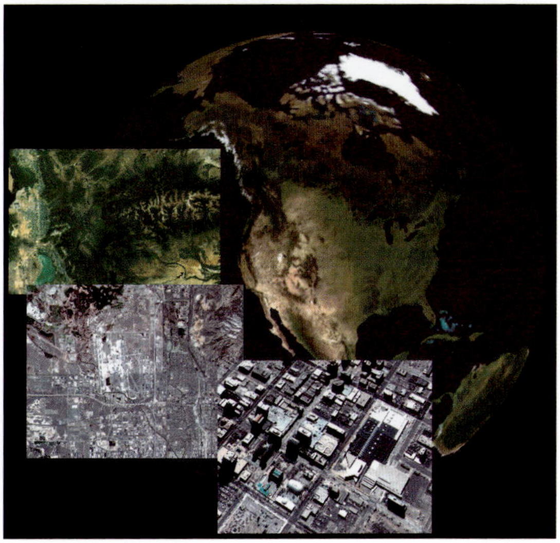

A Range of Satellite Sensors. Satellite imaging sensors differ in their spatial resolution, the number and narrowness of spectral bands they detect, the areas they can image, and the frequency the same area can be imaged repeatedly. Examples of low (AVHRR), medium (Landsat) and high (QuickBird) spatial resolution sensors are shown. Source: NOAA, USGS, DigitalGlobe

The first civilian remote sensing satellite, launched in 1972, had a spatial resolution on the order of 80 metres – the pixel (on the ground) was approximately 80 metres x 60 metres. The most recent civilian satellite has a resolution of less than one metre. If we consider that we are dealing with a satellite image with a spatial resolution of one metre, it may be possible to see (or discern) that there is a something on the lawn in front of a house that is not green grass – but it may not be possible to tell if it is a table, a piece of plywood, or something else.

Temporal resolution refers to the frequency with which data (usually satellite) are obtained. Most satellites return to fly over a given point on the surface of the earth on a known schedule. Obviously certain uses of data require more frequent acquisition of data than others. Exploration geologists, for example, may want imagery acquired at a certain time of year to ensure that vegetation is either present as an indicator of some geologic anomaly of interest, or absent – but the geologist may not care if the data were acquired last year – or ten years ago. On the other hand those searching for icebergs in shipping lanes will want data no more than a few hours "old."

The final descriptor of remote sensing is of the system – not of the data. Systems are either passive or active. Passive systems use reflected light or energy – as does a camera or thermal sensing device. Since a camera is a passive system – it does not carry with it its own source of illumination – it cannot "see" in the dark. Since heat is emitted day and night, thermal sensors, while passive, can "work" at night. An active device carries with it its own source of energy to illuminate the terrain. A radar sensor, for example, beams a radar pulse to the ground and the sensor measures what of that energy returns to the sensor. Likewise, a LiDAR (Light Detection and Ranging) sensor uses a pulse of light (a laser) and measures the return.

For each sensor discussed below we give a short introduction to the technology and a summary of applications. It should be noted that this is simply a summary – for each sensor one could literally write a good sized book – as one could for each application area.[60]

In many ways it goes back to the old adage that a picture is worth a thousand words.

Remote sensing is important in the GeoEconomy for a number of reasons. It is:

1. *A source of information about the world around us;*
2. *A way of allowing the average person to see and understand geo-relationships – including those that are changing over time related to climate change for example; and*
3. *At the heart of Google Earth, Microsoft's Virtual Earth, and similar programs and data sets – such as NASA's Visible Earth.*[61]

3.2.3.2. What is it used for?

This section is divided into two subsections: passive sensors and their application, and active sensors and their application.

Passive Sensors and Their Application

We begin our discussion of passive sensors with the simple camera and relatively straightforward black and white film, color film and the slightly more sophisticated color infrared film. We then add digital cameras and a discussion of the implications of the new imagery types, followed by thermal sensors, passive satellite sensors, and hyperspectral systems.

Film Cameras

In our opinion a camera was the first of many remote sensing tools, and obviously a photograph is remote sensing data, although some of our colleagues in the field would argue with us – to some the only true remote sensing imagery comes from a satellite. For a layman, a photograph is easy to understand and relatively easy to use to extract information. The photograph here shows trees in the foreground and a cropped field in the background. A person interpreting this image will use color, shape, context, and other features of the image to arrive at this conclusion.

This image illustrates the basic principles of image interpretation. To understand even a simple landscape photo such as this, one uses color, shape, contrast, context, and other characteristics of the scene to deduce information.For example, the orange and yellow colours in the foreground suggest the trees are deciduous and the photograph was taken in early autumn. Source: S. Aronoff

Obviously, those who specialize in this type of forest, or cropping system, or geology will be able to interpret this photograph related to their own professions or geographic area of knowledge, using many of the same features of the image. In effect, because of their background, they will know what questions can be "asked" of the image and "how to ask them." However, as we move beyond the common photo to many of the newer remote sensing systems discussed below such as radars, LiDARs, and hyperspectral systems, their use requires at least some sophisticated processing, and often highly specialized knowledge to extract useful information.

> ***Recently the tools and systems used to extract information from imagery have become far more sophisticated – just as the imagery has become more sophisticated – and yet today it is far easier for individuals to access imagery of ever increasing resolution and coverage than ever before.***

What has been happening recently is that tools and systems used to extract information from imagery have become far more sophisticated – just as the imagery has become more sophisticated – and yet one of the paradoxes of the explosion of technology is that today it is far easier for the educated individual to use imagery of ever increasing sophistication than ever before.

As noted in Chapter 2, the history of the use of photography[62] to study and collect information about the earth parallels that of the far more recent GPS in many ways. It began in the military – with photographs taken of battle sites, first from observation balloons in the 1800s, and then by aircraft in World War I.

The first imagery was panchromatic – everything was portrayed in shades from white to black. Color film is simply as its name suggests – it allows one to see the terrain photographed in natural color using sunlight.

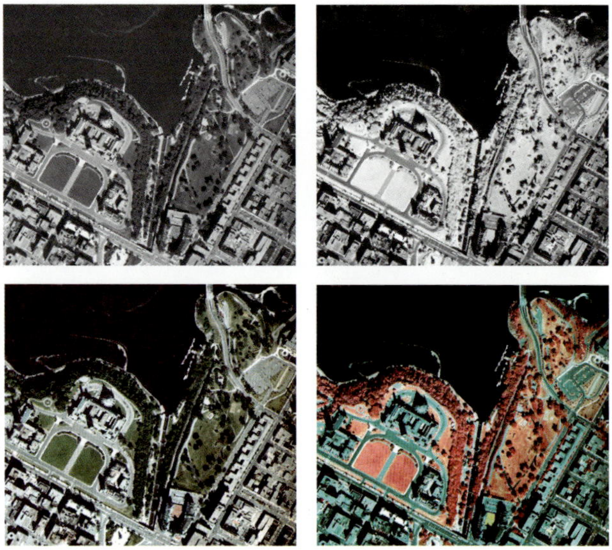

Photographic Films Used in Aerial Photography. The photographic film chosen for aerial photography depends on the application, the cost, and the preferences of the analysts. The examples show (top left to right) panchromatic and near-infrared black and white photos and (bottom left to right) natural color, and false color infrared photographs of Parliament Hill, Ottawa, Canada. Source: Natural Resources Canada

Color infrared (CIR) film senses reflected light in the near infrared, red and green areas of the spectrum and was originally called camouflage detection film (since it could be used to distinguish between dead branches or camouflage paint and living vegetation). The human eye cannot "see" near infrared light. CIR has found use in the study of vegetation condition and environmental change. In such imagery healthy growing vegetation appears a vibrant red, while non vegetated areas appear blue, green, or white; and water bodies typically appear black. In what we hope was a tongue in cheek comment in the original *Manual of Photographic Interpretation*[63] was the statement that such colorful imagery would serve well as a colorful lamp shade! While color infrared (CIR) has not been widely used, it has been applied in studies of vegetation damage from pollution, wind-drift of herbicides onto neighbouring fields, winter damage assessment to crops, forest insect damage, and the like.

Digital Cameras

Today photography using film is rapidly moving from routine application into the history books. Like the film cameras in use at home, aerial cameras using film are being replaced by digital cameras.

Digital cameras offer many advantages including:

- Reduction of time to deliver a product (photo-processing is not required);
- Full compatibility with computers and digital processing (photos do not have to be scanned to generate the digital imagery needed for computer-based analysis and use); and
- Lens distortions and other factors related to consistency are easier to overcome (digital corrections can be directly applied).

Digital cameras can provide the same black and white (or panchromatic) imagery, color and color infrared renditions with which the mapping and environmental monitoring communities have become comfortable.

Today photography using film is rapidly moving from routine application into the history books. Like the film cameras in use at home, aerial cameras using film are being replaced by digital cameras.

To put this into perspective for the casual reader, consider the rapid rise in the use of the popular and seemingly ubiquitous digital cameras we buy for personal use that produce images of over 3000 x 2000 pixels or picture elements. The camera, along with the software to do basic processing on common home computers, now sells for under $300. The first systems, with much less capability, sold for thousands of dollars. When we look at the simple analysis systems that are sold with these cameras, the changes in technology over the years are truly remarkable.

Image Handling

Compare the current availability of inexpensive systems and computers to the world's first commercially sold image analysis system – the General Electric Image 100 bought by the Canada Centre for Remote Sensing in 1974 for $500,000. That system occupied a room the size of a large living room and was able to display an image that was a mere 512 x 512 pixels – and in today's dollars that system cost several million dollars.[64] The digital revolution that has hit personal photography has also hit the aerial camera business – with similar (or greater) impacts.

Today the world is flooded with aerial (and satellite) digital imagery, much of it viewable over the Internet at no charge. This low or no-cost access to imagery has greatly changed the business models used in the industry, as well as how most of us access the data.

Now the world is flooded with aerial (and satellite) digital imagery, much of it viewable over the Internet at no charge. This low or no-cost access to imagery has greatly changed the business models used in the industry. These policy changes are discussed more fully in Section 3.4 below. Earth observation imagery has changed over the past thirty years from a specialized information source to a part of our geospatial infrastructure accessible to all of us. As with other forms of infrastructure, those with the skills, technology, and creativity to use that infrastructure can gain a competitive advantage. One of the technologies critical to our ability to use the geospatial infrastructure we have today was the development of the geographic information system discussed in Section 3.2.5.

Imagery has been used as a backdrop for other information as well as a source of information. The latter is growing in importance.

In general one can say that remote sensing, whether from aerial cameras or passive satellite sensors (discussed below), is used in one of several ways. In the past it has often been used as a backdrop for other information – such as the use of a satellite image as a background for a map – or a geographic information system (See Section 3.2.5.) Imagery is also often used as a source of information – the image data are interpreted in some way to extract a specific set of useful (or in some cases, not-so-useful) information. This interpretation can be approached in different ways depending on the data.

In some cases image data are quickly reviewed (either automatically or by image interpreters) looking for specific features – such as ice that might affect shipping, a specific crop type, a wetland, or a terrorist training base. Large volumes of data can be "consumed" and relatively straightforward interpretation approaches are used. In other cases there is a more intensive interpretation of the entire image or group of images covering a specific area – to map surface geology, land use, to do a forest inventory, or to do more complex assessments of the military situation. These more complex analyses are typically done by specialists in image interpretation related to a specific topic – forestry, agriculture, military intelligence, urban planning, coastal zone management, etc. One of the interesting characteristics of such use is that the same imagery can be used for multiple purposes by different specialists. This is in effect what we see with some of the imagery available on the Internet – and the value that the imagery has – it can be used for multiple purposes by people with very different requirements, and very different backgrounds.

Imagery has been used to study almost everything that can be seen from the air – and many things that cannot.

Imagery from aerial cameras has been used for a tremendous range of applications related to the natural and built environments. Imagery has been used to study almost everything that can be seen from the air – and many things that cannot – things whose presence can only be inferred from what is on the ground – such as the number of cattle likely housed in a barn, or where there may be oil under the ground. The number of topics of application is limited only by the ingenuity of the people needing the information. Usually those needing the information understand what they are interested in looks like from above. There are two basic elements – the spatial relationships between the feature of interest and its surroundings, and how the feature looks different on the image from other things, using clues related to its color, shape, context, patterns (man-made or natural), texture (or roughness of the feature as it appears on the image), and size.

Remote sensing offers a 'bird's-eye view' that allows a person to easily discern features and patterns difficult to identify from the ground. Source: Optech Incorporated

The following table contains a number of oft-cited applications.

Some Applications of Remote Sensing Including Aerial Photography[1]	
• Engineering (for pipeline, road, transmission line, and rail line planning and construction) • City and rural planning and management (land use studies, watershed management, subdivision planning, etc) • Environmental studies (wetlands, coastal zone changes, environmental impact assessments, environmental changes over time, etc) • Agricultural studies (crop areas, crop condition, vineyard management, crop scouting, planning of soil sampling, etc)	• Forestry (inventory and management) • Geological mapping and prospecting (geomorphology, structure, lineaments, etc) • Post-disaster assessments (building and infrastructure damage, flood extent, etc. • Land and property tax data base and change detection (to see un-reported new additions on buildings, etc) • Industrial espionage (what are competitors doing on their oil exploration leases, what sort of chemical processing is company Y's competitor doing at their new facility, etc)
[1]This list is only a sample. Many books have been written on the range of applications of remote sensing in various areas from natural resource management to oceans, earth sciences and human settlements.	

There are, of course, many military applications – from monitoring military bases and troop movements to targeting.[65] In recent years aerial cameras have not been the preferred tools for military applications. Close-to-the-target support is now more often provided by Un-manned Aerial Vehicles (UAVs), while more distant imaging is provided by satellite.

Thermal Sensors

Thermal sensors are used from both aircraft and spacecraft. As implied, these sensors sense heat radiated from the surface of the earth – be it warmer water, heat leakage from buildings, forest fires, or other natural or man-made sources of heat.

The application of heat sensing has been both broad and interesting – ranging from identifying pollution sources and the edge of forest fires to finding marijuana grow operations.

Smoke Penetration of Thermal Imagery. Thermal infrared image (top) and normal color image (bottom) of a forest fire. The hottest areas are clearly shown as white and red colors in the thermal infrared image whereas in the normal color image the ground is obscured by smoke. Source: NASA and the Ames Research Centre

Again, their development comes from military requirements. The principles of thermal sensing are somewhat more difficult to understand than visual interpretation of images acquired with a camera,[66] but for our purposes here we need not go into the details. It is sufficient to say that thermal properties can be sensed.

Applications range from assessing the precise location of forest fires (through smoke), to heat-loss in buildings or underground pipes, and the location of marijuana grow-operations – which tend to use (and emit) a great deal of energy in the form of heat. An interesting side-bar resulting from the latter is the question of the legality of the state using sensing devices to determine if a crime is being committed. In Canada, at least, the government has the legal right to use imagery in such cases.[67] Thermal imaging has also been used in southern Ontario to identify areas where it would be feasible to grow sensitive grapes from which one can produce high quality wines.[68] Another application of thermal sensing has been for finding and/or tracking discharges of hot water or other effluents from factories. Several cases of successful prosecution can be cited after environmental miscreants have been found through the use of thermal sensing at night.

Passive Satellite Sensors

It was the development of space photography (or more accurately space imagery – since much of what is done is not actually done by what is commonly known as photography) that revolutionized remote sensing reconnaissance. Although satellite images could not capture details as fine as those collected by aircraft, satellite images provide broad coverage over difficult and sometimes hostile terrain. As noted previously, in its first successful Corona Mission lasting only a few days, the US collected imagery over more of the Soviet Union than all of its previous aerial reconnaissance missions combined.[69]

Ballistic missile installations were large and easily seen from space. During the Cuban missile crisis in 1962, then highly-secret space photography showed that the Soviet Union had far fewer ballistic missiles and was much less of a threat than it proclaimed. This strengthened the US position leading to a resolution of the crisis. At the time of the crisis, space photography was top secret and the resolution of the images was about 10 metres. It is worth noting that it was not until 1986 – twenty-four years later – that a civilian satellite (SPOT) achieved such spatial resolution.

Early Military Satellite Photography. This 1962 American satellite photograph shows the intercontinental ballistic missile base at Yurya, Russia. The rapid and broad coverage of American photo reconnaissance satellites and distinctive layout of this type of missile base enabled analysts to determine that Soviet missile strength had been substantially over-estimated. Source: Central Intelligence Agency Record Group #263

Today earth observation satellites constantly provide publicly available imagery of the earth, some with resolutions better than 1 metre. While there are some restrictions on the access to high spatial resolution data (some nations are not allowed access to data by the suppliers and/or the laws of the countries in which they are based), data are generally widely available.

To understand the development of applications of remote sensing from space it is useful to know something of the history of the field and how it developed. Images of earth had been obtained from space for some time before the USA launched the first of the Landsat series of monitoring satellites in 1972. The first Landsat had two sensors on board. The primary sensor was to be the wide band panchromatic Return Beam Vidicon Camera (RBV) – with an effective 80 metre resolution. However, the system did not work as expected and the Multi Spectral Scanner (MSS) became the work-horse of the early Landsat series. The MSS was onboard Landsats 1 through 5 and acquired images of the Earth nearly continuously from July 1972 to October 1992, with an 18-day repeat cycle for Landsats 1 through 3 and a 16-day repeat cycle for Landsats 4 and 5. The Landsat MSS recorded information in four generalized spectral bands (blue, green, red, and near-infrared) with a spatial resolution of 79 metres, and approximate scene size of 170 kilometres north-south by 185 kilometres east-west. While imagery was obtained every 16 or 18 days, in fact much of the data in many parts of the world were so covered by cloud as to render the data completely useless.

For those of us involved in the early days these were exciting times, although in many ways the early Landsat data was a solution looking for a problem. The capabilities of the early data were overhyped to the point that some potential users were turned off satellite data for the next decade or more. With the launch in 1982 of the higher spatial and spectral resolution Landsat Thematic Mapper (30 metre resolution in six of the seven bands from the blue to the short wave infrared), a new era saw wider use of data for far broader applications, including forest management and updating of forest inventories.

Satellite images provide broad and usually low cost coverage over difficult and sometimes hostile terrain. Applications range from military reconnaissance to environmental monitoring – and everything in between.

In 1986 a France-led consortium launched the SPOT satellite, with even higher spatial resolution in both multispectral and panchromatic data modes. Since then there have been so many satellites launched by the likes of Japan, India, and many other countries (or groups of countries) that keeping track of them, how to order data and what they can be used for has become almost a full-time occupation. Indeed, in a 2006[70] study it was forecast that 139 imaging satellites worth $16.3 billion would be built by 2016 with 97 by 2011! Perhaps one of the most interesting aspects when one considers the number of satellites is that many of them are designed for a single purpose – monitoring the oceans, or ice, or crops, or some other element of the environment.

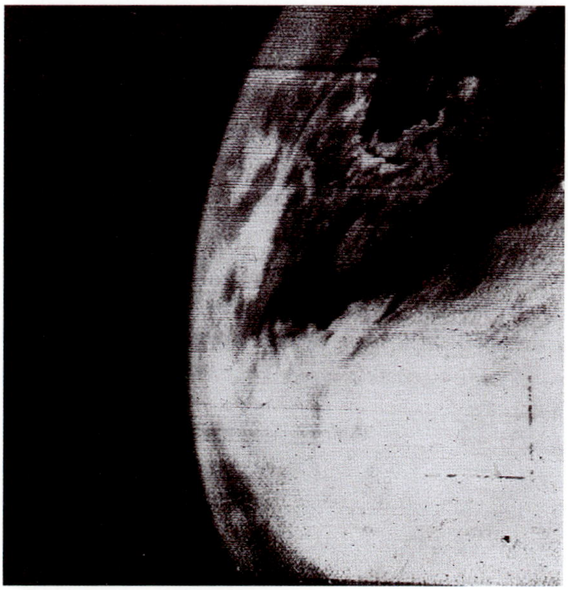

First Space Television Image 1960. In April 1960 the satellite TIROS-1 transmitted the first electronic image from space. Source: NASA Earth Observatory

Like many other areas of science, remote sensing has come a long way in the last forty years. What is of interest to us in looking at applications is the quantity of data now available and how dramatically the data have changed. With spatial resolutions now below one metre (approaching .5 metres), today's high resolution satellites have a resolution that is more than 13,000 times that of the first Landsat – and even 2,500 times that of the game-changing

Thematic Mapper sensor. There are now many governments that support earth resource satellite image acquisition programs including Canada, France, Germany, India, Japan, and Thailand, as well as the international European Space Agency. A number of commercial ventures now operate earth resource imaging satellites including DigitalGlobe, GeoEye, and RapidEye.

With the 1972 Landsat resolution of 60 x 80 metres a common USA football field (120 x 53.5 yards – or 109.7 x 48.9 metres) would be completely covered by no more than 2 pixels. With the mid-1980s SPOT 10 metre data, the field would be covered by 55 pixels. The most recent satellite data would require almost 15,000 pixels to cover the same area.

With the 1972 Landsat resolution of 60 x 80 metres a common USA football field (120 x 53.5 yards – or 109.7 x 48.9 metres) would be completely covered by no more than 2 pixels. With the mid-1980s SPOT 10 metre data, the field would be covered by 55 pixels. The most recent satellite data would require almost 15,000 pixels to cover the same area. This dramatic change has had important implications for the data's use when taken into consideration with data policy, security and privacy issues, awareness, and available associated technologies. Rudimentary systems used to interpret the data that once cost millions of dollars can now be had with far greater capabilities for hundreds of dollars. Even the most advanced analysis systems today can cost less than $4000,[71] although a large system with significant throughput capabilities to analyze large quantities of data in a near automated form can still cost upwards of $500,000.

THEOS Satellite Image of Downtown Washington, D.C. Acquired by THEOS, the Thai Earth Observing System which is owned by the Government of Thailand and operated by the GeoInformation and Space Technology Agency (GISTDA) of Thailand. © 2008 GISTDA

Passive satellite imagery is generally used in applications that require larger area coverage, in remote or hard to access areas, or where (generally) lower budgets are available for data collection. Applications range from military reconnaissance to environmental conservation – and everything in between. Oft cited are regional crop monitoring (yield and condition), forest depletions (from fire or clear-cut harvesting), geological and geomorphologic mapping, water-related studies (such as sediment loading), and generalized land cover and land use mapping. As imagery acquired by

Satellite Image of the Inauguration of US President Obama, January 20, 2009. Source: GeoEye Satellite Imagery

satellite becomes more and more sophisticated there is a growing convergence between what satellite data and airborne data are (or can be) used for.

Hyperspectral Sensors

Hyperspectral imagery has been available for some years from various airborne platforms including the CASI sold commercially by Itres Research in Calgary, Canada, as well as a few custom-built systems used by NASA and other government agencies in the USA, Europe, China, and elsewhere.

The difference between hyperspectral remote sensing data and the more common sorts of data available from digital cameras and satellites such as the Landsat, SPOT and other systems is that, as mentioned above, these other systems have broad spectral bands – and usually not many of them. A hyperspectral system has many more bands (as many as several hundred in research systems we have seen), and much finer spectral (think color) resolution – ten or more times that of the more common systems. The value of these additional bands and the finer spectral resolution is that one can obtain much more precise information about the light reflected by an object on the ground. In theory, this should yield better information about everything from the mineral content of bare rock to the condition of vegetation. It also means that much more sophisticated analysis is required to understand the physics of what is being seen, as well as to extract useful information.

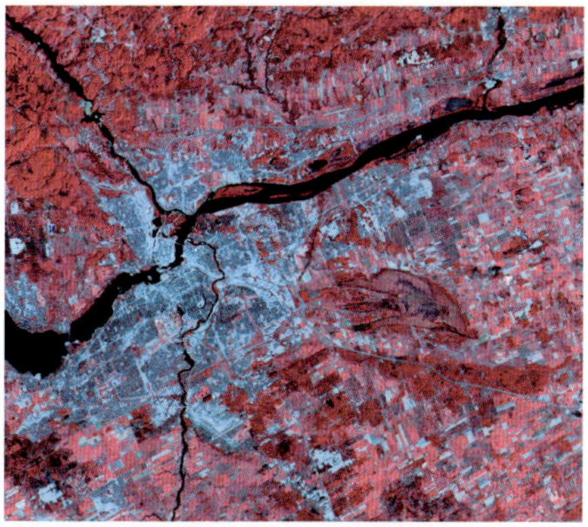

Landsat 5 Satellite Image of Ottawa. Source: Natural Resources Canada

A hyperspectral system has many more bands (256 in research systems), and much finer spectral (think color) resolution – ten or more times the more common systems.

Airborne hyperspectral data was first used in mineral exploration. It has been used to find diamond bearing rock in Africa, for oil and gas exploration, as well as to identify areas for further study for mineral exploration. Itres Research, a Calgary Canada based company has sold more of its hyperspectral sensors than all others combined.[72] They provide a long list of applications that, in general, go above and beyond those that one typically sees for more broad-band sensing systems.

In agriculture, Itres notes that hyperspectral data can be used to provide structure information about the health of the vegetation, extent of infestation or stress damage, map soil characteristics, compliance monitoring (farming practices) and the assessment of crop damage due to storms and drought. On the environment side, the data can be used in a range of mapping applications from vegetation stress and disease mapping to wetlands management and ecosystem status and trends. The detailed spectral information is especially useful in contamination and spill assessment as well as for planning pipeline or other routes. Itres also says that "mapping and understanding changes in ocean color can assist in the management of fish stocks and other aquatic life, help define harvest quotas, monitor the water quality and allow for the identification of human and natural water pollution such as oil and algal blooms, which are dangerous to fish farms and other shell fish industries." In forestry, hyperspectral data have been used in yield inventory, stem counting, and disease management. There is also a strong belief that hyperspectral data will be useful for forest chemistry – determining the chemical composition of trees.

Hyperspectral Image, Fish Lake, Nevada. This color image was generated from 3 of the 120 narrow spectral bands simultaneously captured by the scanner. This type of imagery is commonly used in mineral exploration and geologic mapping. Source: HyVista Corporation Pty Ltd., Australia

While this long list of applications suggests a potentially growing and vibrant business, hyperspectral data application is an area that is not yet well understood beyond mineral exploration and some vegetation and marine applications.

Active Sensors and Their Application

Introduction

After World War II, the Cold War provided the impetus to develop other types of sensors that generated images of the landscape, using techniques other than conventional photography. This era brought us radar (which enabled imaging at night and through cloud cover) and thermal infrared sensors (which could produce images at night detecting the temperature of objects such as the heat given off by military vehicles). Specialized high-altitude aircraft were developed by the United States, such as the U-2 and SR-71, capable of flying higher

Hyperspectral Scanner Imagery for Agriculture. From the 228 spectral bands in the data set for this image a natural color image (left) and various enhancements (right) can be produced as needed to suit a specific application – in this case the assessment of agricultural fields of different crops. Source: ITRES Research Ltd., Calgary, Alberta

or faster than pursuit aircraft or missiles could reach at the time.

There are two active remote sensing systems – Radar sensors and LiDAR – and both are discussed below. They each carry with them their own illumination and sensor to record the return from the ground – or in the case of some LiDAR systems, simultaneously from the sea-surface and the sea bottom to yield water depth maps. Because of the power requirements and complexity of both sending out a pulse and recording the return, these systems tend to be relatively heavier and more complex than the passive systems – and have a far different set of operating parameters and capabilities.

Radar Sensor Systems

Radar sensors are the most complex sensors to both build and use. Unlike the cameras referred to above that rely on the sun illuminating the earth, a radar sensor carries its own source of illumination or "light." In effect, the radar sensor bounces a radar wave off the earth's surface and then records what comes back to the sensor. That information is then used to form an image. Originally developed for military purposes, the first radar imagery was declassified in a series of workshops run across North America in the early 1970s. Those early systems have led to far more sophisticated systems today, and a far greater range of applications – both civilian and military.

> ***Radar sensors are the most complex sensors to both build and use. Unlike cameras and electronic sensor systems that rely on the sun illuminating the earth, a radar sensor carries its own source of illumination or "light."***

The basic principles of radar sensors, the principles of radar interpretation and even how the data are used, are all easily understood through simple geometry and elementary physics. However, putting these principles into application and understanding and using the mathematics and physics lead to an extremely complex task.[73] To put the task into a lay perspective, radar satellites flying hundreds of kilometres above the earth's surface with ground resolution (i.e. the minimum area for which a return is received) of 10 to 30 metres can be used to detect subsidence of the land (near a volcano, over an active oil field, resulting from an earthquake) of but a few millimetres. How this is so is usually quite perplexing to a layman – but we can assure the reader that the physics explains everything!

Rise and Fall of Santa Ana Basin, California. Satellite radar interferometry measured 5 cm of subsidence in the Santa Ana Basin between May and September 1999 using ERS-1 image data with a 90 m spatial resolution. Repeating color bands show the radar -derived differences in elevation. The background image is generated from digital elevation data. Source: USGS

As radar data become more sophisticated, there is a broadening of what radar data can be used for – moving inexorably into areas once thought solely the purview of optical satellite sensors or airborne data.

The first radar data was from what is called a real aperture system – the length of the radar antenna dictated the spatial resolution. A major advance was the development of what is called synthetic aperture radar (SAR). In a SAR system the sensor in effect tricks the system into regarding the sensor's length of movement, during a very short time period along the path of the aircraft or spacecraft, as the length of the radar's aperture – hence the name "synthetic aperture." The synthetic aperture leads to much higher (or greater) spatial resolution, yielding more detail.

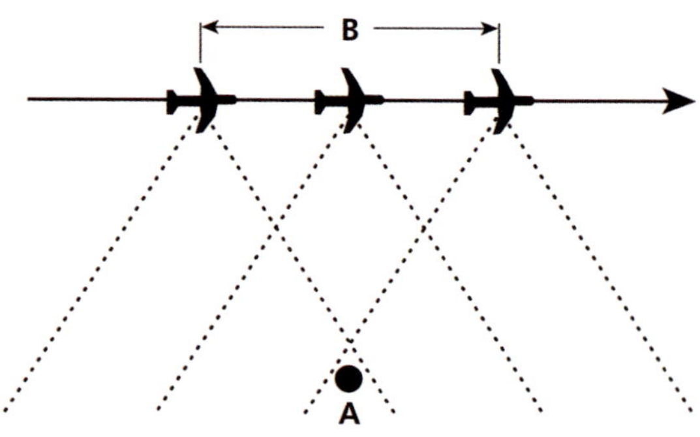

Operating Principle of a Synthetic Aperture Radar. By processing the returns from a target (A) for the entire time it is within the sensor's field of view, a short antenna can operate as if it was many times longer (B) than its actual length, thus providing improved spatial resolution. Source: Natural Resources Canada

Much of the early known (i.e. accessible to civilians) research in such systems was done at the Willow Run Laboratory in Ypsilanti, Michigan. In 1972, the Willow Run Laboratory was renamed the Environmental Research Institute of Michigan (ERIM). It was from ERIM that the Canada Centre for Remote Sensing (CCRS) bought a SAR system, and from which it hired Dr. Keith Raney one of the world's leading experts in this technology. That SAR system, flown on the CCRS

Convair-580 aircraft, was in effect the test facility for both the European ERS-1 satellite and for the future RADARSAT-1 and RADARSAT-2 systems built by Canada. Canada's interest in Radar data was tied to two major areas of application: the need for Canada to monitor ice during the development of northern off-shore oil fields and for northern shipping, and the need to monitor Canada's northern territory to maintain sovereignty. For a time CCRS was regarded as a world centre of expertise in radar remote sensing, until budget cuts in the mid 1990s virtually dismembered the program.

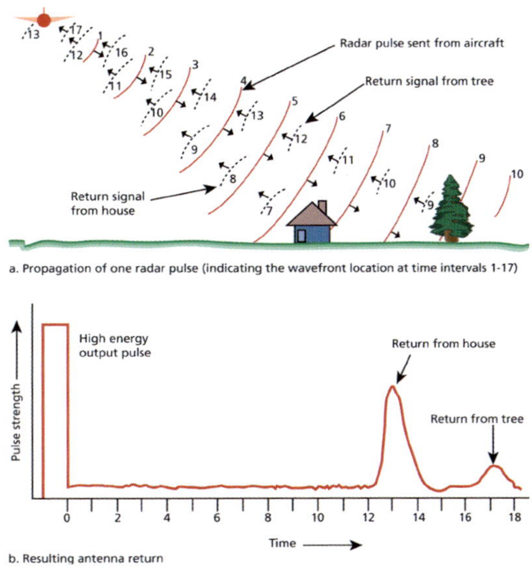

Operating Principle of a Side Looking Airborne Scanner. The upper diagram shows the travel of a single radar pulse towards the ground and the return signal. The lower figure is a graph of the signal received by the radar antenna for that pulse over time. Source: Image in Remote Sensing and Image Interpretation. 2nd ed. By T.M. Lillesand and R.W. Kiefer. © 1987 John Wiley and Sons, Inc. Reprinted with permission of John Wiley and Sons, Inc.

The first widely available commercial SAR data came from airborne systems that in effect grew out of the ice monitoring requirement. One of the early pioneers in this area, Intermap Mapping Technologies Corporation,[74] has been providing data for many years from its SAR systems – first for ice monitoring in the northern reaches of Canada where it is dark for half of the year and later in tropical environments that are perpetually cloudy. SAR data has been used to make 3-D maps for general mapping, for flood planning and modeling, and to assist in oil exploration in tropical areas. More recently the technology has been used to provide precise information on changing elevations (related to land subsidence or uplift – both natural and man-induced), among other applications.

Radar satellite data are considered to be militarily and strategically important.

The importance of radar related to sovereignty and for military applications has received widespread public scrutiny in Canada with the proposed sale of MDA's space assets, including RADARSAT-2, to the US-based military contractor Alliant Techsystems for $1.325 billion in cash. According to Canadian law, the Government of Canada has the right to review the sale of any company worth more than $200 million. Under the Investment Canada Act a sale can be rejected, as this one was, if it does not provide a net benefit to Canada. This was the first time in the twenty three years that the Act has been in place (and 1587 sales), that a sale was rejected.

The first civilian radar satellite, and the first designed specifically for the collection of information about the oceans, was the USA's Seasat launched on June 28, 1978. Seasat was managed by NASA JPL and had on-board a number of sensors including a radar altimeter to measure spacecraft height above the ocean surface, a microwave scatterometer to measure wind speed and direction, a scanning multichannel microwave radiometer to measure sea surface temperature, a visible and infrared radiometer to identify cloud, land and water features, and a Synthetic Aperture Radar (SAR) L-band, HH polarization, fixed look angle to monitor the surface wave field and polar sea ice conditions. This sort of information is essential in weather forecasting and shipping, to name but two applications. The SAR operated for 105 days until the electrical system failed. While the mission was short, it whetted the appetite for SAR data around the world.

Seasat Image of the Rocky Mountains in British Columbia, Canada. Source: Natural Resources Canada

The failure of Seasat so soon into the program underlined the complexity of the systems and the need for robust, and consequently expensive, engineering compared to optical satellites. Seasat also showed the difficulty of obtaining, processing and using SAR data from space. The complexities are such that, unlike optical satellites that can (and have been) built in university labs in less developed countries, SAR equipped satellites remain the purview of more advanced space-faring nations.

The next SAR mission (not technically a satellite) was the first deployment of the Shuttle Imaging Radar by NASA in November 1981. While the next free-flying (as opposed to shuttle-based) SAR-equipped satellite that was said to be civilian was the Russian Almaz 1 launched in March 1991, imagery from it was neither easily accessible nor widely used. Almaz was based on military systems that had flown many years earlier.[75] The next civilian SAR system that followed was the European Space Agency's (ESA) first remote sensing satellite, ERS-1, with a C-Band SAR launched on July 1991, and ERS-2 in April 1995. These systems were especially useful for monitoring ice for shipping and oil exploration in northern waters. Japan put a joint optical/L-Band in space in February 1992, while the first fully operational SAR satellite was Canada's C-Band RADARSAT-1, launched in November 1995. With RADARSAT there was significant growth in the use of data for everything from the standard ice mapping to monitoring shipping and rice crop assessment.

The February 2000 Shuttle Radar Topography Mission (SRTM) obtained elevation data as NASA has reported "on a near-global scale to generate the most complete high-resolution digital topographic database of Earth. SRTM consisted of a specially modified radar system that flew onboard the Space Shuttle Endeavour during an 11-day mission."[76] For some areas of the globe the SRTM provides the most up-to-date and complete 3-D view of the terrain. Underlining that the military is never far away when it comes to even civilian remote sensing, the SRTM is led by the National Geospatial-Intelligence Agency (NGA) and the National Aeronautics and Space Administration (NASA).

In 2002 ESA launched Envisat, said by ESA to be the largest Earth Observation spacecraft ever built: it had to be to carry the ten instruments on board. Envisat's largest sensor is the C-Band Advanced Synthetic Aperture Radar (ASAR). The Medium Resolution Imaging Spectrometer (MERIS) is an "imaging spectrometer that measures the solar radiation reflected by the Earth, at a ground spatial resolution of 300m, with 15 spectral bands in visible and near infra-red, and programmable in width and position. MERIS allows global coverage of the Earth every 3 days. The primary mission of MERIS is the measurement of sea color in oceans and coastal areas:"[77] the imagery gives an indication of the ocean's health and productivity.

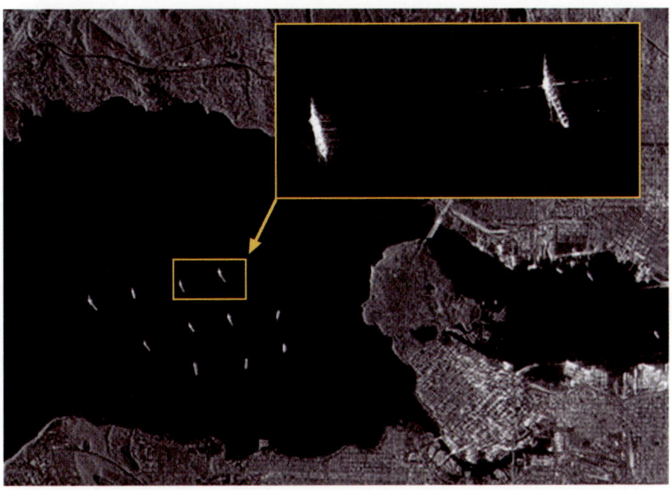

Radarsat-2 Satellite Radar Image – Vancouver. Radar imagery with resolutions as high as 3 m for commercial use and 1.5 m for some government clients see use in a range of reconnaissance, mapping, and emergency response applications. Source: MacDonald-Dettweiler Corporation

Germany joined the SAR club with the launch of the aptly named X-Band SAR satellite TerraSAR-X in June of 2007. Following an operational readiness review by the German space agency (DLR) and Infoterra GmbH, the company responsible for commercial marketing, the system was commissioned (officially approved to routinely deliver imagery) in January 2008. TerraSAR-X can function in three different basic modes at resolutions between 1 and 16 metres. The sensor can provide single or dual polarised data – leading to a wider range of potential applications. On an experimental basis, additionally quad polarisation and along track interferometry (to measure minor changes in elevation from one date to another) are possible.[78]

One of the more recent additions to the SAR systems is RADARSAT-2, launched in December 2007. This system is the most sophisticated of all of the civilian SAR systems launched to date – and some have said that it may even be more sophisticated than most of the military SAR systems currently operating. Another recent SAR system was launched by India in April 2009. That system, built by Israel Aerospace Industries, is said to have military applications and an even better spatial resolution than RADARSAT-2.[79]

RADARSAT-2 offers both flexibility and speed of delivery. The spatial resolution can vary from 3 to 100 metres, the sensor can look both left and right of the spacecraft flight (reducing the revisit time to a specific location), and it can operate in a number of different polarization modes. The latter can be explained by understanding how the sensor sends out the radar pulse and how it receives the return. The pulse can be transmitted either horizontal to the plane of the spacecraft orbit or vertical. Similarly, the return can be received either horizontally or vertically. If it is sent horizontally and received horizontally that is referred to as HH. Other modes are then VV, HV, and VH. Objects will show up differently depending upon the polarizations used. The dual-polarization and quad-polarization modes allow the simultaneous acquisition of multiple polarizations on transmit and receive. In the quad-polarized mode, four different polarization channels are acquired, which is in contrast to the single channel for RADARSAT-1. Data can be available within 4 hours of reception in Canada and, with what is called accelerated processing speed, it is possible to produce up to eight RADARSAT-2 products in less than 90 minutes from when they are received at a receiving station.

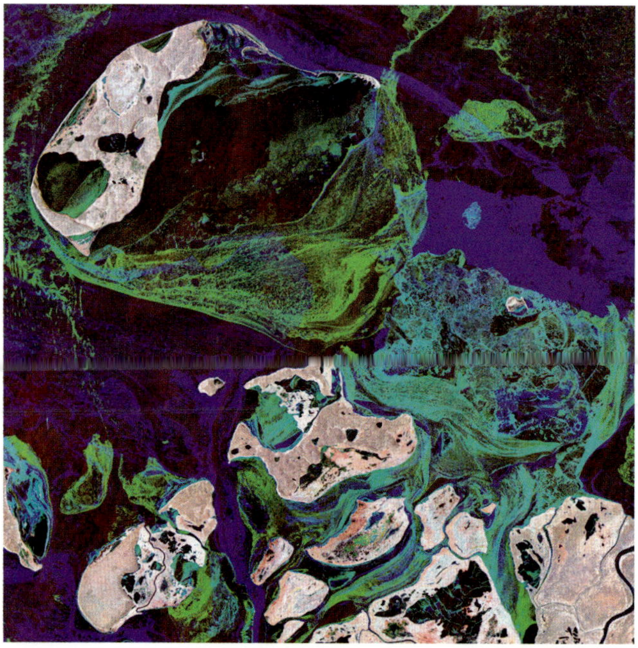

TerraSAR-X Multi-Date Radar Satellite Image - Mackenzie Delta, Northwest Territories, Canada. This image was generated from three radar images acquired on different dates and displayed as red, green, and blue. Source: Infoterra GmbH

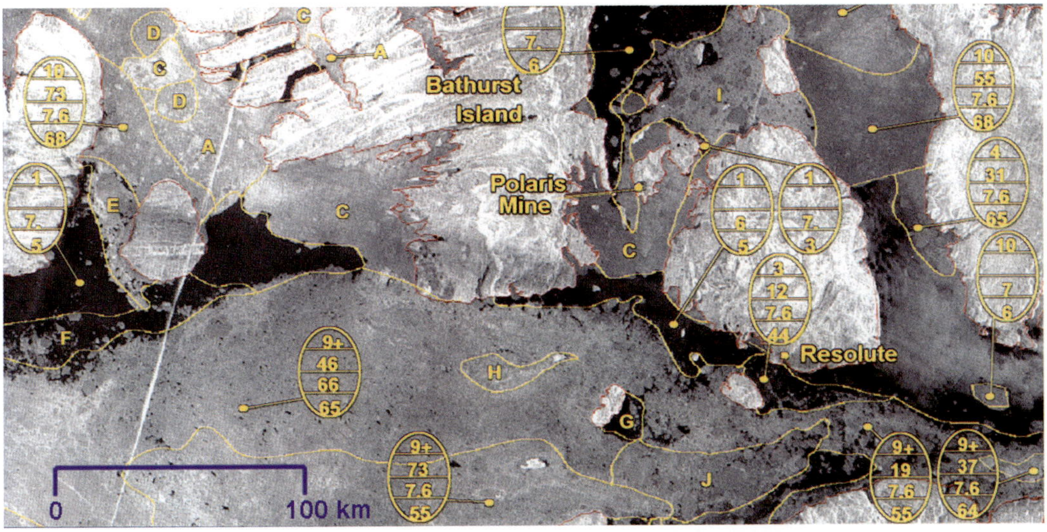

Ice Condition Map - Resolute, Northern Canada. Satellite radar imagery from Radarsat-1 was used to produce this ice condition map. The data ellipses indicate estimated ice thickness and coverage conditions valuable to plan ship routing. Source: Natural Resources Canada

The first radar systems such as RADARSAT-1 and ERS-1 proved useful for a number of specialized data collection tasks – many of which are simply not possible with optical sensors, and none of which can be done by optical sensors in cloudy regions or at night-time. They include:

- Monitoring ice conditions affecting shipping;
- Monitoring waves – an important input into weather prediction and modeling;
- Coastal zone and shoreline mapping;
- Some elements of wetland mapping (including rice cultivation);
- Ship detection on the open ocean (for security and sovereignty concerns);
- Oil detection and movement on water (both natural seepage and oil spills);
- Changes in elevation caused by ground water or hydrocarbon removal or seismic activity;
- Flood mapping; and
- For certain applications related to mineral and hydrocarbon exploration, such as linear feature mapping.

While it is likely too soon to provide a definitive list of all applications of the new systems such as TerraSAR-X and RADARSAT-2, research suggests a far broader range of applications will be possible than for previous radar systems. The use of airborne simulations and ground study have suggested what parameters should be used to differentiate between different features – be they crops, natural vegetation, ships, ice, floods, etc.[80] The following table, derived from information on the TerraSAR-X web site,[81] provides an indication of radar data use.

TerraSAR-X Potential Applications

Agriculture

Mapping of acreage / detection of small fields

Mapping of field heterogeneities

Crop type recognition

Growth stage assessment

Crop parameter assessment

Hydrology

Snow and ice

Snow extent

Snow parameters

Glacier mass balance

Glacier ice motion

Soil moisture

Groundwater resources

Wetlands extent monitoring

Land Environment

Land cover mapping

Mapping of broad land cover and land use classes

Change detection analysis

Forestry

Tree type identification

Mapping of disturbances caused by fire, insect attacks, deforestation

Forest parameter assessment

Oceanography

Wind and waves

Wind speed and direction

Wave direction and height

Ocean dynamics

Internal waves

Currents

Ship detection

Ship and wake detection

Ship speed estimation

Oil slicks

Urban areas

Extent of settlements

High precision settlement mapping

Characterization of urban areas

Delineation of infrastructure

Traffic monitoring

Topographic mapping

Repeat-pass interferometry

Radargrammetry

Bathymetry

Water depth (general level)

Geology

Geological mapping

Features and structures

Lithology

Disaster management

Landslides

Volcanoes

Earthquakes

Disaster management

Risk and vulnerability assessment

Monitoring of hazard prone zones

Planning and management of rescue operations

Damage assessment

Coastal Areas

Coastal protection and change monitoring

Delineation of coastline

Mapping and monitoring of major land cover types, e.g. mangroves

Sea ice

Iceberg detection

Sea ice extent

Sea ice type mapping

Sea ice draft

Source: http://www.dlr.de/tsx/main/applications_en.htm

One of the other interesting aspects of radar data is the way in which the data are made available. The European Space Agency tends to regard its systems as being for research. For that reason cost of access to data is low, as is the general level of accessibility. TerraSAR-X is a hybrid system. Researchers whose research is approved can obtain data for a low price, while all other access is regarded as commercial and at much higher prices. RADARSAT data are, by and large, only available commercially. The commercial involvement in both RADARSAT, and to a lesser extent TerraSAR-X, has been contentious in some quarters given the amount of the system cost that has been born by taxpayers in the respective countries – and the costs that are still being charged to governments by what some contend are the state-approved monopolies running them. This issue is dealt with in more detail in Section 3.4.2 below.

Laser sensors on-board aircraft (LiDARs) have revolutionized how elevations are mapped – both for mountains and valleys on shore, and in shallow water. A LiDAR system sends out narrow pulses or beams of light and then measures the return – and the time it takes to return.

LiDAR Sensor Systems

Lasers were first developed for remote sensing in the 1970s. Laser radar or LiDAR (Light Detection and Ranging) works much like the radars discussed above. While a radar system sends out a radio wave and measures the return and time to return, a LiDAR system sends out narrow pulses or beams of light and then measures the return – and the time it takes to return. One could think of it as a pulsing laser pointer – except in this case what is reflected back is measured – a laser pointer with attitude is how we might describe it. Each pulse illuminates or covers a narrow swath of territory. The movement of the airplane forward leads to a build-up of information as one swath is joined to the next. A major complexity is in joining the succeeding swaths one to the other.

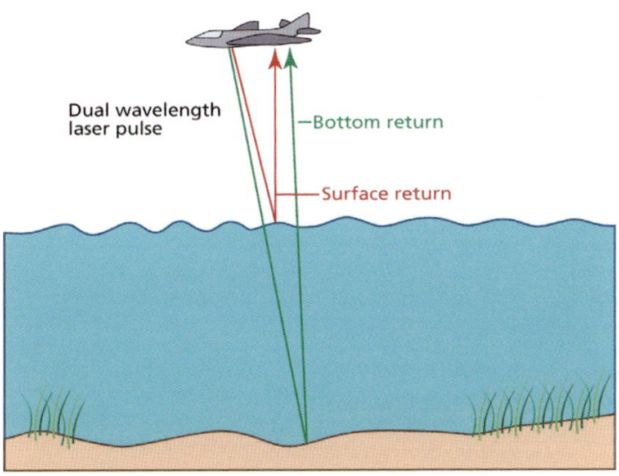

Hydrographic LiDAR. The SHOALS hydrographic LiDAR provided by Optech uses two lasers operating simultaneously to calculate water depth. The beam of green light reflects off the bottom (maximum penetration depth depends on water clarity) and the near-infrared beam reflects off the surface. Source: S. Aronoff

The first civilian widely-used commercial laser systems were based on research at York University in Toronto and were subsequently commercialized by Optech.[82] Optech (and others) sell systems that can measure the depth of water (laser bathymetry), and the height of terrain. Since the measurements are largely done automatically, and are based on total area coverage (not a sample), mapping accuracies for LiDAR bathymetry (water depth mapping) in particular are better than for competing ship-board systems in shallow and relatively clear water.[83] The US Navy has accepted this technology for its mapping purposes. LiDAR technology has also come into its own over the last few years for terrain mapping and digital elevation models – technical words for mapping the height of land.

By early 2008 it was estimated that over 240 LiDAR systems had been deployed worldwide,[84] and approximately one-half of those have come from Optech.[85] While it cannot be said that LiDAR systems have completely replaced traditional imagery for mapping of heights, they have certainly dramatically altered the market for the derivation of height information used for a wide range of purposes. At the same time LiDARs provide useful information related to trees and other ground cover. LiDARs are often used alongside aerial cameras – the LiDAR to provide height, while the imagery provides thematic information.

The applications of LiDAR summarized in the following table give an indication of the importance of accurate mapping of heights – be it the height of the trees on the land for forest inventory and volume estimation, the underlying height of the land (for engineering and construction, cell-phone tower location, or flood planning to name but a few uses), or depth of the water for navigation and security. LiDAR can also be used to differentiate objects based on differences in heights and multiple returns can be used to measure vegetation canopies for forest height and biomass measures.

<table>
<tr><th colspan="2">LiDAR Applications</th></tr>
<tr><td>Urban
• Wireless communications
• Emergency route planning
• Signal propagation
• Other GIS applications

Shorelines and Bathymetry
• Coastal engineering
• Flood management
• Erosion monitoring
• Water depth (bathymetry)</td><td>Power-lines/Pipelines/Corridor Mapping
• Tree locations
• Tower locations
• Catenary models
• Vegetative critical distance analysis

Vegetation
• Forestry applications
• Volume calculations
• Removal of non-ground features
• DTM creation (under the canopy)</td></tr>
<tr><td colspan="2">Source: http://www.optech.ca/altmapps.htm</td></tr>
</table>

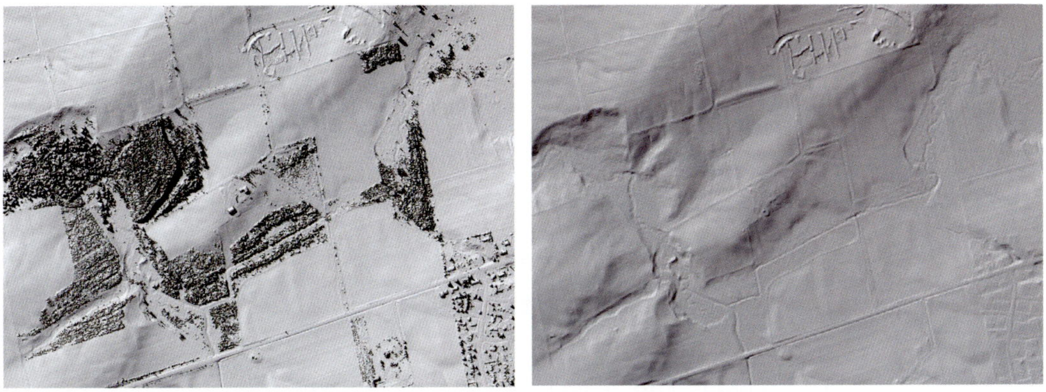

LiDAR Terrain Data. LiDAR data are displayed here as a shaded relief image (left). Sophisiticated processing of the data can be used to generate a digital terrain model with the surface features, in this case the vegetation, removed (right). Source : Ambercore/Terrapoint

While LiDAR does not produce as wide a range of information as do some of the other sensors discussed here, it has revolutionized the mapping industry. The technology has made it much easier and faster to obtain information about the height of land and depth of water – which in turn leads to a different way of thinking about mapping – and the ability to do it faster – and better. Some have said that once all of the heights have been measured, the only changes will be in glaciers, coastal zones (including deltas), areas of seismic activity, deserts, forests, and urban areas. Thus, these experts say, there will be less need for LiDAR data. Such regions, of course, represent those that are of most interest and most important to our continued survival!

3.2.3.3. What are the implications?

Overview

We believe that the implications associated with remote sensing can be discussed in several ways and on several levels. One could look at the implications by type of application. For example, remote sensing imagery can be used for geo-monitoring – monitoring of changes at a specific location. It can also form part of a national, state or local geo-information infrastructure. Remote sensing has also been used for geo-intelligence – both for business intelligence and for military intelligence. Related to its use in intelligence is its use in fostering transparency of decision making related to the environment, planning, and resource management.

Such application-oriented categories may be more useful for those who have a specific topic of interest and who know what they want to use the imagery for. Topics might include the earth sciences, natural resource management and environmental monitoring, its application to the marine environment, or to human settlements. Each of these has, in fact, been the subject of a detailed manual as part of the American Society for Photogrammetry and Remote Sensing's *3rd Edition of the Manual of Remote Sensing.*[86] Those who wish to gain an in depth understanding of one or more aspects of the field might consult one, or more, of those volumes.

The highest and best use of remote sensing is to improve the quality and security of all human life everywhere.

The last two sentences of the volume on human settlement applications of remote sensing edited by Ridd and Hipple bear repeating to a broader non-specialist audience:

> "The contents of this volume have demonstrated that remote sensing technology has applications for human settlements of every size and type, located in every kind of environment. Thus we close with the assertion that the highest and best use of remote sensing is to improve the quality and security of all human life everywhere."[87]

This statement leads us to another way of looking at implications. We can look at them from the perspective of the individual and the various communities that individuals join to create – companies, communities (from mega-cities to rural villages in developing countries), states (or provinces) and nations. Of course, in sum, we are eventually looking at the implications for us in the ultimate collective – as a species.

The implications of remote sensing to individuals and collections of individuals in companies, local communities and especially to state and national governments are significant. While we naturally focus on governments, we also devote a significant amount of the discussion in this section to individuals, industry and municipalities. This is because in these areas imagery use is not as well established and the implications are perhaps not as obvious nor as well understood.

The implications associated with the use of imagery are many, complex, and often inter-related. Some examples include:

- Concerns of privacy – of people and companies;
- Empowerment of the individual person or company in the face of powerful governments and big business – which once controlled all access to image data;
- Monitoring for security and sovereignty related to nation states;
- Monitoring and assessment of the situation in what some call rogue states by the international community;
- Access to and use of imagery by rogue states and terrorists;

- Monitoring the state of the environment – from local to global scales; and
- Monitoring associated with the collective good – be it the identification of illegal use of buildings (marijuana "grow-ops," for example), theft of irrigation water, identification of those not paying taxes, illegal release of contaminants in the environment, or ships illegally dumping on the high seas.

Many more implications can be identified. As imagery is used more widely, and as more sources become available, we can expect that there will be implications not yet envisioned.

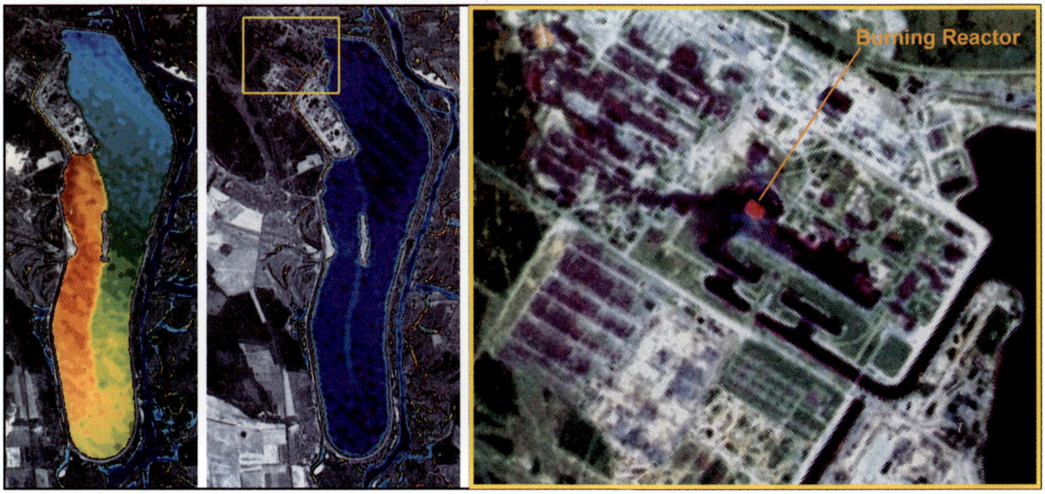

Satellite Image Documentation of Chernobyl Nuclear Accident 1986. Landsat satellite images of the cooling pond next to the nuclear power plant at Chernobyl in the Soviet Union document its status before (left) and after (center) the explosion that stopped its operation. Thermal infrared data was colour coded blue to red to indicate cool to warm water temperatures. Before the explosion the cooling pond shows the normal decrease in water temperature from the outlet around the pond to the inlet for the cooling water. After the accident the pond is uniformly cool indicating no input of warm water. The composite image (right) of imagery from the Landsat and SPOT sensors shows a more detailed view of the reactor site. The very hot reactor core shows as a red dot. Source: USGS (left and center), USGS and CNES/SPOT Image Corporation (right)

As individuals we used to be concerned that "Big Brother is watching us." While Big Brother can and does watch us, in many more ways than most people realize, we can now better understand how that is done – and turn-about is fair play. With our access to today's technologies we can now all watch Big Brother – and each other. This concept of using imagery to watch Big Brother started with the 1986 Chernobyl crisis in the Soviet Union. The heat released by the accident was obvious on imagery made widely and publicly available. The authorities could not hide the impact – any more than North Korea could hide the evidence of their development of a nuclear capability.[88] These uses represent a subtle shift from the time of the 1962 Cuban missile crisis when it was the government of the United States using imagery it had acquired to put forth its case. Now almost anyone can obtain imagery.

As with GPS, literally hundreds of uses for imagery have been identified – and many more are being developed as this is being written. And, as with GPS, privacy advocates (and some governments!) have expressed concern about too much imagery being available. Several countries have been accused of restricting the quality of data available in Google Earth – and not all of these so accused have been less developed or undemocratic countries. As noted above, some of these concerns have already been tested in court. In Canada, as in the USA, society has typically won when cases are argued on the basis of the good of society over the rights of someone or some entity engaged in what would be considered to be criminal or illegal activity.

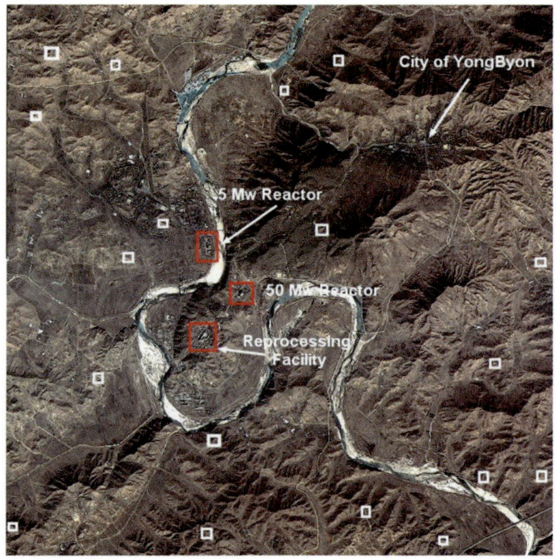

Satellite Image of Yongbyon Nuclear Facility, North Korea. Source: DigitalGlobe

We now focus more specifically on the implications for the following communities:

- Individuals;
- Industry;
- Municipalities; and
- State and national governments.

The Individual

Introduction

At the individual level, to date much of the use of imagery has been simply to "see my house" or to "look at my summer home." But with imagery being more widely available, and with more and more high school, university and college courses requiring the study of imagery, and more game software using sophisticated imagery, use of it by individuals is growing apace. That use will continue to not only grow but grow exponentially, as more and more people become comfortable with imagery and aware of its many uses. The implications of remote sensing technology for the individual are outlined below:

The Implications of Remote Sensing Technology for the Individual

1. *It allows the individual to better understand his or her environment – both good and bad;*
2. *It empowers the individual; and*
3. *It provides the ability to monitor and evaluate changes to the environment.*

Better Understanding of the Environment

Today we all have access to imagery types and sources that were not available ten to fifteen years ago to any but those working in the so-called dark world of military image intelligence. With better understanding of imagery, there has also been a growing realization that there are historical sources of imagery – including satellite imagery for the entire world[89] and aerial photography in most of the developed world – that can complement new image data or other data sources. The implication is that imagery provides us with a better understanding of our environment and our place in it.

Imagery will play an increasingly important role in the lives of individuals as society better understands the usefulness of imagery.

Empowers the Individual

Access to imagery can empower the individual or small groups of individuals – it can help us to think globally and both act and monitor the results locally. We can expand the quote by Ridd and Hipple to say that "remote sensing technology has applications for" (not just) "human settlements of every size and type, located in every kind of environment," but for every type of environment located everywhere.

Monitor and Evaluate Changes Important to Individuals

A number of questions relevant to individuals (as well as to communities, businesses, and even state and national governments), can be answered by a set of historical imagery acquired over many years. For example imagery can be used to trace the history of a piece of property. The questions one can ask about the property are simple – and the responses can have serious implications. The questions: In the past was this land ever a land fill or garbage dump? Were fuel and/or chemicals stored at this location? What was manufactured at this site? Many pieces of land that would appear to be valuable but which remain vacant in the centre of major cities in North America will be found to have interesting histories if they are explored through the use of historical aerial photography. Simply stated, soil contamination has precluded their use as anything other than a parking lot.

We can think of many more questions that need answers that might be asked of imagery taken ten, twenty, fifty, or even seventy years ago. Imagery can tell us what was upriver from the beautiful waterfront property we are looking at buying, and can help us decide where we might want to take soil samples or look for information on the ground. Truly, the imagination of the user is the only limitation.

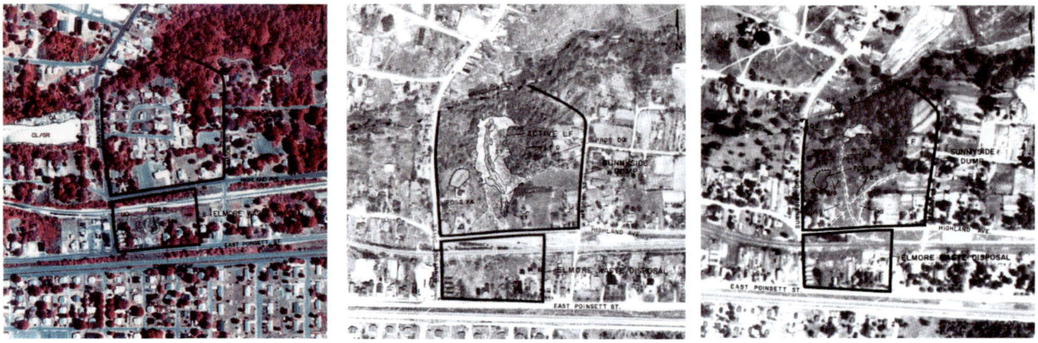

Aerial Photography Documents Redevelopment of Waste Disposal Site. Aerial photographs show that the site of a housing development in the 1981 false color infrared image (left) had been an active landfill site in the 1955 (center), and 1944 (right) airphotos. Source: US Environmental Protection Agency

Today, for a relatively small amount of money, groups of concerned citizens can obtain new imagery specifically acquired for their own use. Farmers can use imagery in the management of their farms – and see problems on or from neighbouring farms that might impact them – from herbicide drift to vegetation stressed by pests and disease. Groups of farmers can acquire imagery to assess if farmers upstream are taking irrigation water to which they are not entitled. One vegetation-related application that is often spoken of – especially by those who are unfamiliar with the use of imagery – is "pre-visual stress detection." The concept suggests that one can use imagery to identify crop stress before you can actually see it in the field by the naked eye. However, to date no-one has been able to demonstrate that any imagery source can actually do pre-visual stress detection.

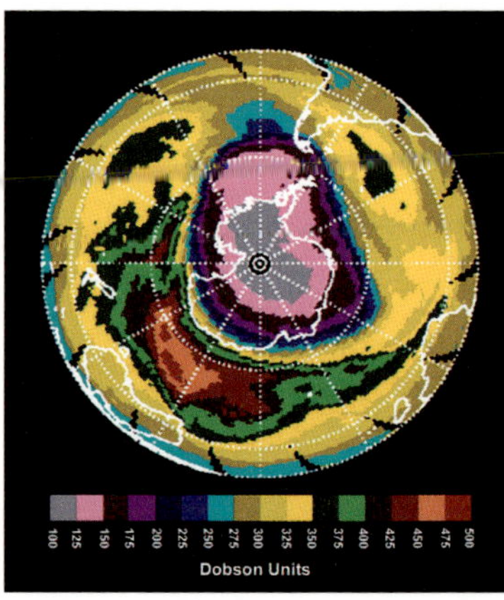

Ozone Concentration Over Antarctica – September 1998. A satellite sensor maps the ozone concentration over Antarctica. The data have been coded to show low concentrations as grey, pink, and purple. The very low ozone concentration over Antarctica that became known as the "ozone hole" prompted international action to ban the use of certain chemicals in consumer products. Source: Ozone Processing Team, NASA Goddard Space Flight Center

We can, however, see damage to the environment – be it damaged vegetation or increased turbidity or algal growth in the water. We can also see ecosystems important to us and overlay development plans on them. In a recent case, one of the authors used a series of images to see if a small piece of land his church was buying was contaminated. (It wasn't.) The other author developed methods to monitor the environmental conditions associated with the early development of the Alberta Oil Sands.

Jumping from the individual to the ultimate collective – our species and the world in which we live – remote sensing has also been of fundamental importance. If we go back to the concern of ozone depletion and more recently changes to the world's ecosystems brought about by man, it was startling remote sensing imagery from space that provided the first clues. The ozone hole, the tremendous decrease in the Aral Sea, the advancing desert in Africa, and the decrease in the size of wetlands and glaciers world-wide – all were recorded by remote sensing. As noted above, remote sensing has provided the canary in the coal-mine: the warning that all is not well with our environment. By the same token, imagery can also provide an indication that things are improving – that the ozone hole is shrinking, that the vegetation is recovering, that the water is less turbid – and this can lead to positive re-enforcement for positive actions.

Retreat of Larsen Ice Shelf, Antarctica. Satellite radar images from the European Space Agency's ERS-1 and ERS-2 sensors have documented the 100 km retreat of the Larsen B ice shelf between 1992 and 2002. Source: ESA

Industry

Introduction

The implications of the use of imagery for and by industry – especially for those dealing with natural resources or resources that are based on the land or sea – are significant. The implications of remote sensing for industry are noted below;:

The Implications of Remote Sensing for Industry

1. ***Business practices for some industries can be streamlined, making them efficient, more competitive and more profitable;***
2. ***Companies can demonstrate compliance with various government regulations – or its use can be forced on them to demonstrate compliance;***
3. ***Business practices are, or can be, more transparent; and***
4. ***There is a greater risk of, and opportunity for, industrial espionage.***

Streamlining Business Practices

Streamlining business practices can include the aforementioned use in agriculture, finding diamonds in Africa, locating the best areas to explore for oil in Papua New Guinea, or identifying the safest shipping route through ice infested waters.

Perhaps the broadest penetration of imagery into an industry other than mineral and oil exploration is in forestry. Imagery has been used world-wide for decades to better manage forests and to assess forest depletions and forest stock. For privately owned forests this is especially important, for the forest stock is the capital that provides the value of the company. For bankers and those financing the development of mills and factories based on the forest – be it public or private – it is the forest stock that will fuel the factory. If the yield of the forest is sustainable over time, it may lead to an adequate return on investment and the project may prove to be economically worthwhile. If the forest stocks are insufficient to support the mill, it should not be built. Examples of forest applications can be cited in Scandinavia, British Columbia, across the USA, Thailand, India, China, and Brazil.

The implications of the use of imagery for and by industry are significant in resource management, mineral and oil exploration, transportation and industrial espionage.

Imagery has been a significant tool for companies doing all manner of other more unexpected things – from entrepreneurs identifying where there may be valuable wild mushrooms to harvest, to lawn maintenance companies assessing where there might be problems in specific neighbourhoods to further their marketing. Commercial services targeting the agricultural industry at the farm level can be found in both the USA and Canada. In the Napa Valley of California imagery is routinely used by wineries to better manage their vineyards, by identifying areas where there are problems or management issues, and to help plan pre-harvest sampling that leads to increasing the amount of higher value "reserve" wines. A recent study we did found that five hundred dollars invested in imagery and its analysis can result in a payoff off of over $50,000 for a 40 acre vineyard.[90] As imagery becomes less expensive, and as demands on our agricultural productive capacity grow, we can expect imagery to be even more widely used.

Demonstrate Compliance

Imagery can also be used to demonstrate compliance with various statutes dealing with the environment, and for production management. What better way to illustrate that you are not damaging the surrounding ecosystem than by showing imagery that supports such claims! Of course, if there are problems, the imagery may be used to identify the extent and areas where field visits and ground sampling are required. Any company facing requirements for environmental reporting,[91] or concerned with the potential impacts of its operations on the surrounding region would be well advised to consider the acquisition of imagery before there is a problem. In this sense, imagery can in some respects be regarded as a relatively inexpensive form of insurance against future liabilities. As a picture is worth a thousand words – so too is an image worth a thousand words – and a thousand words written by a lawyer (especially one suing you!) can be very expensive indeed!

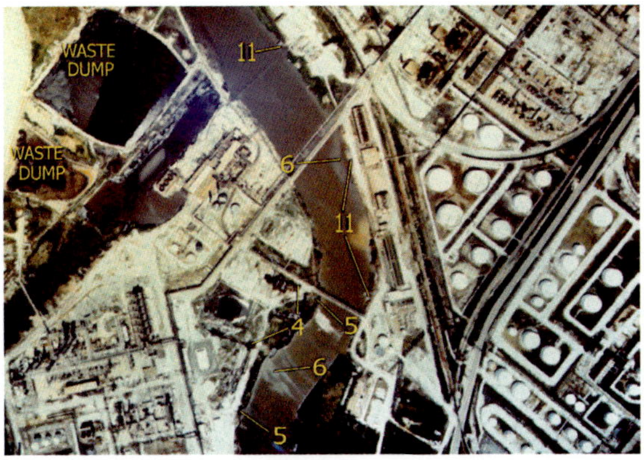

Aerial Photography for Industrial Site Inspection. The overhead view provided by this aerial photograph assisted the field inspection of this oil storage containment site. Key to features noted by analysts: 4 – product spillage outside containment, 5- point of entry of old spill, 6-oil slick and/or sheen on water, 11- wastewater outfall. Source: US Environmental Protection Agency

In some jurisdictions where there are requirements for companies to exercise due diligence[92] in environmental matters, image acquisition and use is increasingly becoming part of customary practice. This in itself provides an interesting dilemma that will increasingly be faced by industry and governments. If a company does not monitor their facilities with imagery they will not be seen to be exercising due diligence. If they DO monitor their facilities, find problems, but do nothing about them, they too are not exercising due diligence. As one executive once told us, "we don't want to know."

One suggestion has been that law firms should acquire the imagery and have it interpreted. If problems are found the company is protected by lawyer-client privilege. While perhaps a legal solution, we question the morality of such an approach – we do not believe that it would be acceptable practice for those of us who are Certified Mapping Scientists.[93]

Compliance monitoring has also been a significant application. In the 1970s, the US Environmental protection Agency (EPA) developed a method to use color aerial photography to monitor compliance with laws dealing with oil storage and chemical processing. It proved to be much more efficient to acquire imagery and assess it in an office for potential problems than it was to actually visit each site and walk the storage areas. From imagery they were able to assess where there may be problems. Only those areas where there appeared to be problems were then visited. The number of fines and speed with which problem sites were identified, visited (and revisited) increased dramatically.

Transparent Business Practices

While imagery can be used for compliance monitoring, its wide availability also can assure us that business practices in some industries are transparent. Whether this is an environmental group monitoring the Alberta Oil Sands, or a neighbourhood monitoring the construction of a pipeline or highway through their community, in many countries of the world industry must assume that someone is watching them.

Industrial Espionage

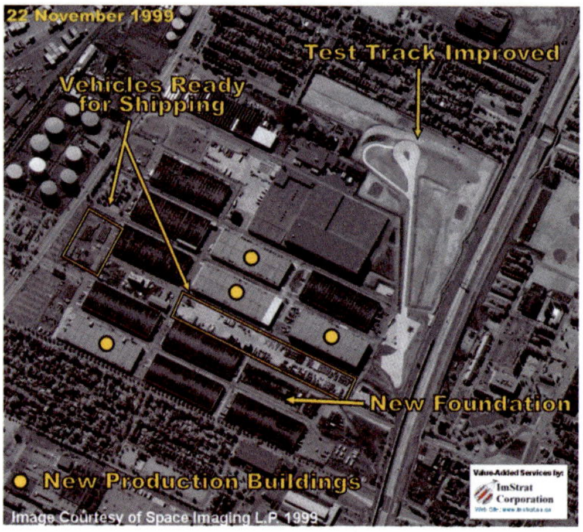

Business Intelligence from Satellite Image of Vehicle Production Plant. An experienced analyst comparing this image with earlier ones noted features indicating the business was significantly expanding. Source: ImStrat Corporation © 2003

Industrial espionage is another, but not often talked about, application of remote sensing. A major use of early 10 metre SPOT satellite data in Canada was to monitor who was doing what on oil leases in Alberta. Oil industry experts were monitoring activity and, from an analysis of that activity, could determine the status of exploration on neighbouring properties – which in turn was valuable to companies deciding whether or not to renew their leases. In such analyses time was of the essence. Analysis of the data was done within a day or two of the satellite having acquired the imagery. Others have monitored competitors' factories[94] using, in effect, the same approach that the military uses to analyze (and target) installations in enemy territory.[95]

Municipalities

Introduction

City governments (like state and national governments), have long used imagery to make traditional maps. That has all changed, as is discussed further in the next section, and that has had some interesting implication for municipalities and their management.

While municipalities have long used imagery in engineering applications, many now use imagery in innovative ways that have implications for how their cities are managed.

There are four implications of remote sensing for municipalities.

The Implications of Remote Sensing for Municipalities

1. *A new paradigm for data collection is possible, has been evolving, and will impact how data are both obtained and distributed;*
2. *They can better monitor and manage their jurisdictions – including their parks, waterfronts, and environmentally sensitive areas;*
3. *They can perform several security and tax related activities; and*
4. *They can simulate, model and then visualize changes to the environment caused by construction, new developments, and disasters.*

A New Paradigm for Data Collection Impacts How Data Are Obtained and Distributed

Imagery is now being used to "feed" geospatial data bases, not just make maps. An interesting approach advocated in Canada[96] is that "Data should be collected only once, closest to the source and in the most efficient way possible. Data should be integrated both horizontally and vertically across jurisdictions, to reduce duplication of effort and to promote the concept of collect data once – use many times." The implication of such a policy is that municipalities, as the level of government closest to the source, should be the basic collector of geospatial data. Another factor related to this issue is how data are collected and distributed, and who pays for what. This has led, in some areas, to groups of municipalities banding together to buy data on an on-going contractual basis, rather than as a "one-shot" effort as was often the case in the past. Many would suggest such data should be shared with citizens and perhaps even cost-shared with higher levels of government and that the data should be distributed free of charge.

Better Monitor and Manage Their Jurisdictions

Municipalities are now beginning to use imagery in many innovative ways that have implications for how their cities can be run more efficiently and transparently and how the environment is monitored – also more transparently. In planning, images are used to map land use and land cover and the changes in both. Images can also be used to monitor and manage parklands, wetlands, or water sources – especially in larger cities or those with widely dispersed holdings – including reservoirs whose surroundings must be kept pristine. Areas of unhealthy vegetation can be identified and this information can be acted upon. In still other cases, simple aerial photography serves as a backdrop for tax rolls, or other information.

Tax and Security Related Activities

There has been increased use of imaging for policing at several levels. For example, new airborne imagery can be used to identify building additions or improvements such as swimming pools, garden sheds and the like that have not been properly "permitted" – or which are not being taxed. It has also been reported that municipalities in California have used imagery to warn homeowners whether brush is too close to their homes and thus presents a fire risk.[97] As noted previously, thermal imagery can be used to identify buildings that may be illegal "Grow-Ops" used to grow marijuana. When Ryerson was working as a scientist with the responsibility of answering questions from the public and other government agencies, a question that was asked at least once a year by police agencies across Canada was "can we find a grave of a possible murder victim using heat-seeking imagery or some satellite imagery?" The answer then, as now, was an emphatic "no." One might, however, use ground penetrating radar to locate where there may be some inconsistency in the soil.

Simulate, Model and Visualize Changes to the Environment

Images are also often used to show view-sheds – what the view would be like with another building or development (or removal of trees) superimposed on the city or one of its neighbourhoods. Indeed, for some developments, three dimensional models are built with imagery draped over a 3-D representation that has come from LiDAR data – such models make it easier to understand and visualize what the result will be of a specific decision. In a more dramatic application, imagery can be used to show what the potential impact of flooding would be on a municipality and its surroundings.

State and National Governments

Introduction

State and national governments have a much longer history of imagery use than do individuals, industry, and municipalities. However, that use and the implications associated with it are changing and changing rapidly. Imagery is no longer simply used to make static maps that usually go out of date before they are updated – and long before they are out of print. In general, however, governments have yet to take full advantage of imagery – and the many other advanced geospatial tools we discuss in this book. One of the objectives of this book is to help governments at all levels better understand how

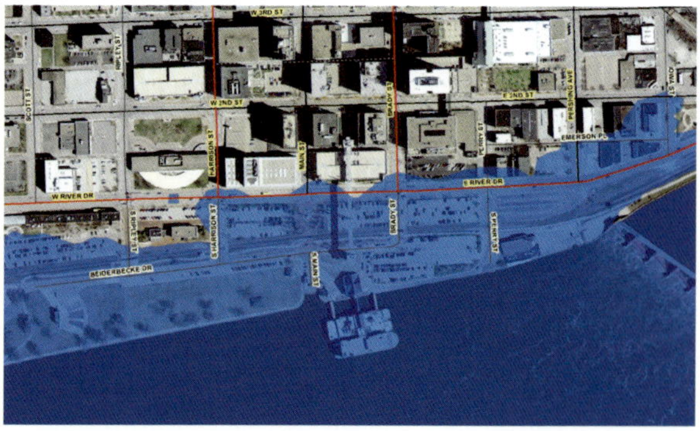

Flood risk mapping. Davenport, Iowa. Combining imagery with digital elevation, a GIS was used to model areas that would be flooded by a 6.3 m (19 ft) rise in the river level. Source: City of Davenport

they can best take advantage of geospatial technologies and the resulting data. One of the ways of doing that for government is to discuss the implications of using (or not using) remote sensing imagery, and the information derived from it.

> ***State and national governments have a long history of imagery use. Today that use is broadening dramatically. Today not only is Big Brother watching us, we can watch Big Brother – and Big Brother knows this. This has led to more transparency in government.***

The implications of remote sensing for governments of remote sensing are interesting in their diversity and impact.

> ***The Implications of Remote Sensing Technology for State and National Governments***
>
> 1. *By its nature the use of remote sensing leads to increased demands for more remote sensing imagery;*
> 2. *Better decisions can be made and grow to be expected by the population that the governments serve;*
> 3. *The ability to monitor its citizens and, by the same token, allows the citizens to monitor its governments;*
> 4. *Increased security is provided by governments;*
> 5. *Resources and the environment are monitored more effectively, leading to improved quality of life for the citizenry; and*
> 6. *There is a paradox in how governments speak about, acquire and use remote sensing: governments NEED the data, but seem unwilling to PAY for it.*

Increased Interest in and Demand for Remote Sensing

Governments are being forced to use more imagery. This has come about because of the improved access to, and use of, the tremendous volume of imagery now being made available. Everyone has imagery and everyone expects governments to be using that imagery. This new interest in imagery is driven in turn, at least in part, by the likes of Google Earth and Microsoft's Virtual Earth. One can argue that in this sense, both Google and Microsoft have had a positive impact on government in terms of both transparency and improved decision-making.

Better Decisions Made – And Expected

The more imagery that is acquired, and the more readily available it becomes, the more it is seen. The more it is seen, and as it becomes better understood as a source of information, the more it is used by those responsible for protecting and managing resources, including our environment. The more it is used, the more good decisions are made based on accurate and reliable information. This is so whether one is considering the identification of areas to be harvested in the forest, protection of sensitive recharge areas for aquifers providing ground water, or the monitoring of protected wetlands near developments. The list of applications is almost limitless. As the use of the imagery becomes more common, so too does the public's expectation that better decisions will continue to be made, putting even more pressure on governments to acquire and use the imagery. To those of us in the field of remote sensing this becomes a happy circle of increasing pressure on us and our technology to continue to perform well.

Monitor the Population – and Government

Obviously, if governments can monitor what we are doing in our forests, on our land and near or on our water, we can also watch our "Big Brother" governments and industry. At the same time there is a growing realization on the part of governments everywhere that they had better be watching themselves too. This self monitoring with imagery was at one time forced by those arguing aboriginal land claims or who were on one side or other of a major development issue – such as pipelines or oil sand development. Inevitably if one side began to use imagery, the government side soon followed, and vice versa! As monitoring moves into more hands in government and the private sector, there are also implications related to privacy issues. The Privacy Commissioner for Canada, for example, has held seminars and discussions on this topic in which the authors have participated.

Increased Security

There are implications for national security associated with remote sensing. The rationale behind Canada's interest in a remote sensing program (and the interest of virtually every other nation that has embarked on a program to deliver imagery) was clearly stated many years ago: Canada wanted to know at least as much about its own resources as did anyone else. The bottom-line (and hot button) issues were, and still are, security and sovereignty for most countries – both economic and political. The debate that accompanied the planned sale of MDA's space assets to an American military contractor, including the sophisticated RADARSAT-2 satellite designed and conceived by MDA but largely paid for by the Canadian tax payer, underlines the growing importance attached to imagery. That Canada said "no" to the sale may have shocked some in industry, but the decision was consistent with security and sovereignty policies developed over the past thirty-five years. Imagery is, as has been stated above, an important element in how many countries define their national identity.

More Effective Monitoring of the Environment and Resources

Additional implications are associated with other issues important in many societies today. As we begin to encounter more and more concerns about food shortages, water availability, and rising oceans, it is clear that imagery that can monitor small changes over time will grow in importance. Imagery can help assess crop conditions, discover where water can be found, determine where ground water can best be re-charged in the dry regions of the world, and show how much the sea is actually rising. Today's imagery from satellite is even being used to track and monitor refugee movement, as well as the transportation bottlenecks in getting food to them.

The Paradox in Government Actions Concerning Remote Sensing

We can see a paradox in the support for imagery. While recognized as useful and important in political speech after political speech (note the Earth Observation Summit called by President Bush or the 1994 Beijing Declaration agreeing to the importance of space imagery in development[98]), the political level rarely puts their money on the table to support imagery acquisition unless there are direct and pressing military concerns, or some significant hot-button issue that makes an overwhelming case for imagery. As public demand for imagery and the benefits imagery brings increase, we can expect there to be more targeted and substantive support, not just the more general support for geospatial cooperation or general infrastructure, as was President Clinton's support referred to in Section 3.2.1.

> ***The political level rarely put their money on the table to support imagery acquisition unless there are direct and pressing military concerns or some significant hot-button issue that makes an overwhelming case for imagery. This political short-sightedness is a problem in almost every country in the world.***

Part of the lack of support may relate to the fact that one of the drivers of image acquisition in the past has been topographic mapping. Topographic maps depended upon imagery. That form of mapping, which once underpinned government geospatial activity, has been largely rendered obsolete. How and why this has happened, and the unfortunate consequences for governments and their citizens, is explored in the next section.

3.2.4. Mapping On-demand and Traditional Mapping

3.2.4.1. What is it?

There are four basic types of maps to which most map-literate people have been exposed. These are: (1) continental, country or state maps showing administrative (i.e. country or state) boundaries, location of large cities, and major rivers; (2) road maps; (3) topographic maps; and (4) image maps. These are explained below, culminating in some observations about map accuracy and the changes in mapping from what it was, to what it is today.

Continental, Country or State Maps

Continental, country or state maps are usually small scale maps (i.e. large areas shown on a small piece of paper) that tend not to change very dramatically over time. However, they do change. Names of places can change (such as Bombay becoming Mumbai on maps in the English speaking world), as can national boundaries, as states dissolve or merge (Czechoslovakia becoming the Czech Republic and Slovakia, for example). For our discussion here, continental or regional maps are not especially important – although for those countries with border disputes, or should you venture into a country with a map showing a place name or a sensitive border not acceptable in that country, such maps may become very important indeed!

Road Maps

Road maps are far more interesting for our discussion. Most of us still carry road maps in our cars – be it for the city or for the state or province where we live or frequently travel. It wasn't many years ago that if you were going from one place to another over unfamiliar roads, your first stop would be the American Automobile Association or AAA (and its Canadian affiliate the CAA) to have custom-made maps prepared for the route. These maps were printed beforehand and stored as linear segments from one place to another. Your guide would start with your town and each succeeding map in your custom-prepared map guide would take you towards your destination from one point (or town) to another. Together they would suggest the best route to get from where you were to where you wanted to go. I remember the first time I ever saw a highlighter used – it was on a map my father had the CAA prepare from our hometown to Ottawa returning through Algonquin Park. Today of course these maps are online, and the basic principle is the same whether you use MapQuest[99] or the AAA.[100] And they still use a highlighter – only now it is electronic.

Of interest is that the copyright of the map shown at the AAA site is "© AAA, Navteq, TCS, and ESRI." That these different groups are listed as holding the copyright suggests that the application involves both GPS and GIS. This involvement of the AAA, ESRI, TCS, Navteq, and by extension Nokia (which has purchased Navteq[101]), again underlines the convergence that we highlight in our discussion of the GeoEconomy and converging technologies in Section 4.2.

Today you can call up MapQuest on line, enter a beginning address and an ending address and it will produce both word directions and a map – door to door. In effect, what MapQuest (or the AAA) does for us is create a map "on-demand" – a map where we want it, for the purpose we want, and when we want it. And the map is disposable. Obviously this sort of system requires precise addressing, and road information. In some areas addresses are not registered as consecutive numbers along a street – they may be based on the age of the building or perhaps some other naming convention. In some areas road maps are not up to date or accurate. In such areas, it becomes much harder to develop an application such as MapQuest.

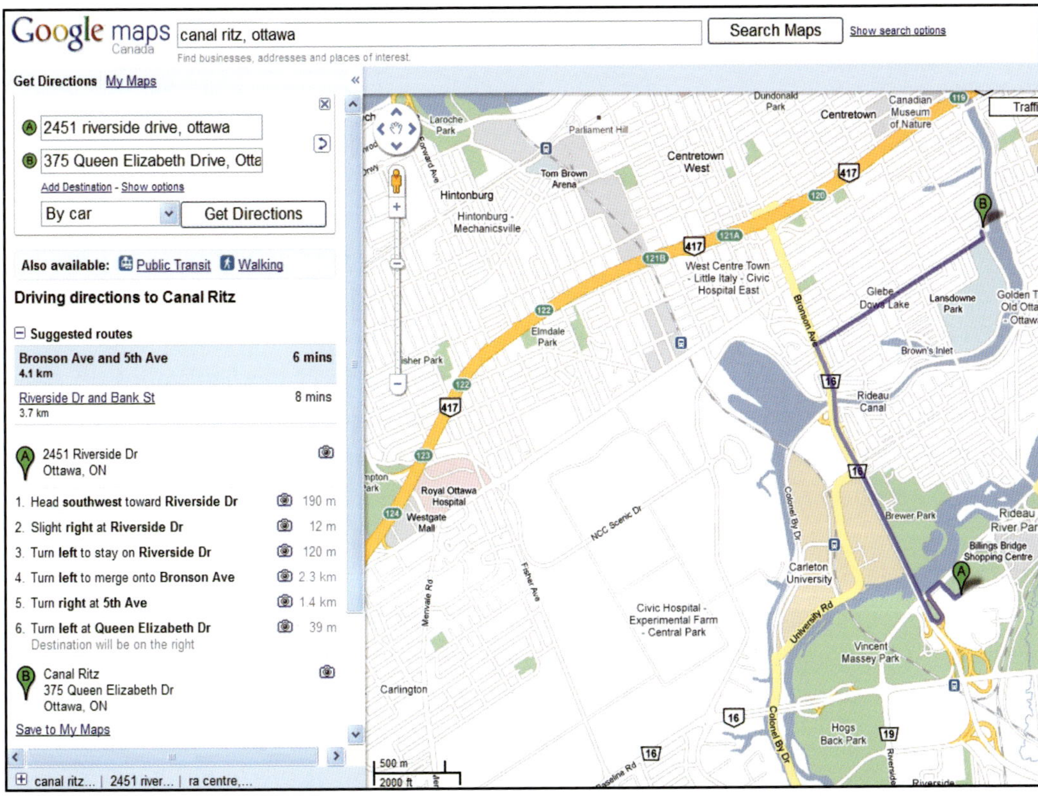

Online Travel Directions. There are now many free online sites where the public can generate a customized map with recommended route and driving directions. Source: Google, Inc.

While these systems are fine to plan how to get from point to point by car, road maps are used for far more than planning a visit to Aunt Minnie in Sheboygan. They are used by delivery companies to plan pick-ups and deliveries,[102] trucking companies to plan long haul routes as well as pick-ups and deliveries, and road maintenance crews – all to minimize travel time and fuel consumption. They are also used for emergency planning (such as evacuation routes for fires – be they in Canada, California, or Australia) and for emergency vehicle routing. When one considers the latter, the matter of getting updates for new roads as well as precise and consistent addressing of all buildings takes on special importance.

Depending on the jurisdiction, updates are done in a number of different ways. In some cases new road information is entered into a data base as a matter of course – when the road is approved for opening, its coordinates are placed in the appropriate road data base as part of the "road layer." In others, the road is entered into the data base as part of the planning process – and shown as an "actual" road when ready for use. While citizens have rarely been regarded as a source of information to update maps, today there are experiments in a number of countries and companies aimed at allowing citizens to update maps – so called "neo-geography," "Web 2.0," "crowd-sourcing," or "wiki-mapping." While rarely accepted as workable, the idea of having a large number of citizens involved in making maps is not new. In the 1930s Sir Laurence Dudley Stamp used school children and volunteers to map land use in England – work credited with helping England avoid starvation during WW II.[103] The major change is that today we have the technology to easily accept input from citizens.

One can (and should) ask the question: if we can get road maps "on-demand" from MapQuest or the AAA, why not for other types of maps? In 1999 the national agency responsible for mapping in the USA (the United States Geological Survey) answered that question. They announced that it would begin to use mapping on-demand for some of the less frequently sought after maps.[104] But the maps so discussed were really not so much specially made customized maps, as they were a different delivery mechanism for the same old (and some might say tired) products. But why? Why not be more flexible? To answer this sort of question, we have to go back in time – over twenty centuries – and look at the topographic map and how it developed.

Topographic Maps

The topographic map is THE father of all maps and to understand how mapping has developed as it has one must understand how and why the topographic map developed.

With the exception of the source data (aerial photography), printing techniques, materials used, and the small matter of the recent use of computers and digital photogrammetric devices, traditional topographic mapping has not changed substantially in many aspects since the first Chinese maps were made in the Han Dynasty some 2200 years ago. We are not sure if this is a comment on the advanced state of Chinese mapping at the time, or the slow development of the mapping field as it waited for the right technologies to arrive on the scene – or perhaps both.

> ***The topographic map is THE father of all modern maps and to understand how mapping has developed as it has one must understand how and why the topographic map developed as it has.***

Jin[105] described the Han Dynasty "topographical map" as being 1:180,000 (a sensible scale given the units of measurement of the time), and depicting "river systems, landforms, residential areas and roads. The rivers are shown in great detail and the relatively larger rivers are noted with their names. The mountains are reflected in closed curves which reveal fairly clearly the shape, run and size of the mountains. In addition, the map offers a detailed and reliable description of more than 80 residential areas, including eight county towns shown in rectangular symbols and the rest in circular symbols. The roads are of two grades indicated in solid and dotted lines respectively."

In addition Jin reported that there was a "garrison map" showing the location of rivers (field-green in color), mountains (black) and military information (red) – such as the location of nine different army units, beacons, outposts, headquarters, etc. It also shows villages with numbers of households, immigrant settlements, and roads with distances.

For much of history a topographic map has been made following carefully prescribed approaches by the cartographer or map maker who assembled data from a variety of sources (largely based on aerial photography after the availability of that data source) to make a map. That map was then printed and became the data base used by everyone, from the military officers planning exercises, to engineers doing initial site plans for roads, urban and regional planners doing zoning and suburban planning, and individuals doing everything from hiking to planning bird watching.

In many countries in the world topographic mapping was (and still is) seen to be of such strategic importance that it is carried out within the purview of a military mapping agency. Both the US and Canada have separate civilian and military mapping agencies. In Great Britain the national mapping agency's name shows its military heritage: it is called the Ordnance Survey.[106]

Undertaking a large scale topographic mapping program was (and still is) expensive – so perhaps it is good that it is often done within the military as part of their security and reconnaissance budget allocations.[107] High cost and the range of skills, number of people, and the time involved, together led to the need for a prescriptive approach with detailed and often complex rules so that products retained consistency no matter where, when, or by whom the maps were created. Unlike the more specialized maps for exploration, topographic maps were made by large teams of anonymous cartographers.

Topographic maps were designed by mapping specialists who by nature tend to be meticulous – and, we would argue, reluctant to change. That reluctance to change was at one time a positive characteristic – since national and state mapping programs wanted to ensure consistency of their outputs over time and across the region being mapped. Resistance to change has since turned, at least in part, into a liability.

The topographic maps so produced neither were, nor are, the most "user-friendly" of documents. Indeed, most of us who have studied geography (or who were in Boy Scouts or Girl Guides) did some training in "map-reading" so that we could make sense of the squiggly lines and colors that seemed to bear little resemblance to what nature painted on the geographic palette that we saw in the "real world." But (and this is an important "but") every one of the thousands of topographic maps made in almost any country on the planet held many similarities, and topographic maps within a specific country were (and are) generally consistent from one end of the country to the other. This consistency came from the application of standards – and standards are even more important in the digital age than in the hard-copy paper-map age. In this sense, now that we have entered the digital age of mapping, we should thank the traditional topographic map makers for insisting on consistent standards.

Why are standards so important? This relates back to what maps are supposed to be. They have been considered in old geography texts as a second level abstraction of the real-world – a photograph (such as from remote sensing) is the first level abstraction.[108] The traditional view of a map as a snapshot at one time – usually the time at which aerial photography is taken – seems to have been at the core of topographic mapping, and hence national level map making agencies, for many decades. Furthermore, topographic maps were rigid in their boundaries from map to map. In effect the country was regarded as a series of tiles – and for each one of these tiles a map was produced. Certainly the map maker adopted new technologies – but typically applied these to produce more of what they had always produced – a map sheet – a snapshot of a specific and limited location.

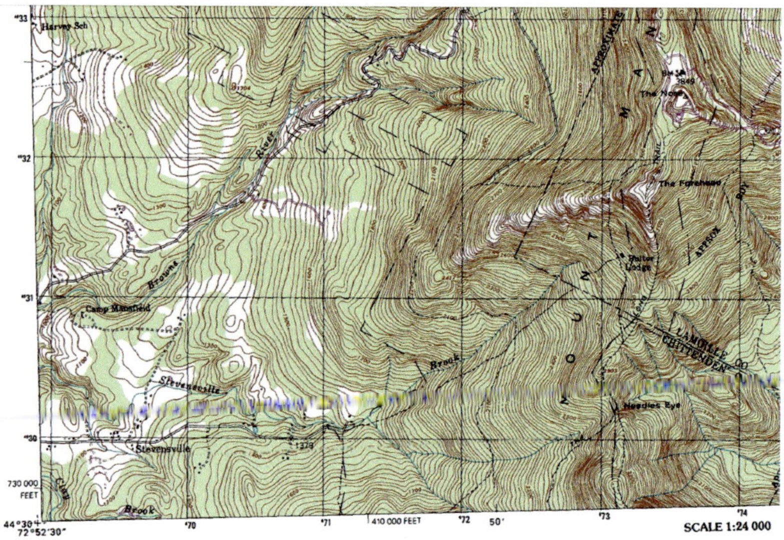

Portion of a Topographic Map of the Mt. Mansfield, Vermont Area. Source: USGS

In many countries warehouses exist that contain nothing but neat piles of maps – many of which will never be used for their intended purpose.

To make matters worse (or better depending on your point of view), the mapmakers typically wanted to cover the entire country with exactly the same scale of map, regardless whether or not the area is populated, of interest for mineral or petroleum exploration, or has some other strategic value. In Canada this led to a long term plan to create over 13,000

1:50,000 scale maps. (1:50,000 scale means that one inch on the map represents 50,000 inches on the ground.) This desire to "fill-in-the blanks" seems to have been common whether in Canada, the USA – or Mozambique. Canada has yet to finish mapping the entire country at 1:50,000 – nor, in our opinion, should it. Canada (and every other country) should concentrate geospatial data collection efforts only where the effort is warranted.

Today, using the new geospatial technologies now available, new maps can be very quickly generated. So when more detailed mapping is required of an area, a new topographic map can be generated in a matter of weeks – not the years that traditional methods required. Now a just-in-time approach can be used for new mapping.

In some jurisdictions attention has been paid to the differences between settled regions and those regions with no population – leading to somewhat more detailed maps being made of the former. In other regions, maps in large numbers are produced for all areas – leading to warehouses that contain nothing but neat piles of maps – many of which will never be used for their intended purpose.[109] Where this problem is most severe is often in countries that can afford it least – developing countries. In many developing countries the same "wall-to-wall fill-in-the-blanks" mapping paradigm has been followed. There has been wall-to-wall coverage, often using inappropriate technologies to produce maps of areas that have seen little change in centuries – areas that need not be re-mapped. Nowhere has that approach to "wall-to-wall fill-in-the-blanks" mapping been more noticeably (and wrongly) applied than in Mozambique, as explained below.

> ***A study we did in Mozambique identified a need for new maps in support of land mine clearing. While there was insufficient budget available to re-map the entire country, there was sufficient funding available to re-map the populated areas and those affected by land mines. One of the complicating factors associated with land mines in Mozambique was that mines were often laid near schools or the community well. When the local village site became dangerous, the populace would often pick up and move the village to a new location. In some cases where roads (usually unimproved tracks) were mined they were no longer used, and new tracks developed. The result was there were new roads, and some existing roads that were no longer being used. While villages retained their old names, they would often be found some kilometres from their original site. These problems rendered existing maps useless. We therefore recommended that new maps be made for the populated areas.***

> ***However, following the advice of a mapping organization, an aid agency decided to scan existing maps to provide uniform digital mapping of the entire country. All of the errors in village names, locations, and the now abandoned roads were, of course, captured from these old maps. The resulting product simply changed the map format from paper to digital – it didn't improve the geospatial information in the maps. For the purposes of reconstructing the country, the limited budget would have delivered far greater benefits if priority had been given to improving data quality by producing new digital mapping of the populated areas, rather than delivering complete digital map coverage of the country based on old maps. This is especially so given that these digital maps contained major errors in the most important regions, those where the people live and work.***

Another limitation of topographic maps is that while they are far more accurate than, for example, road maps of the state or province, they are far too inaccurate in most cases to be used to portray the special case of urban land ownership. Yet another form of map base must be used to map land ownership – and that base is usually dependent on land survey – not photogrammetric mapping (i.e. mapping with air photos). And, with highly different land values and parcel sizes depending upon location, the accuracy and detail required will vary greatly between land ownership maps in downtown New York compared to rural Montana, for example. Of course, if one finds oil or diamonds in rural Montana – or Mozambique – the land values and consequent need for precision will change dramatically.

In addition, newer technologies contribute to the confusion about what is an acceptable level of detail in mapping. For example LiDAR systems produce more detailed and accurate elevation maps than traditional aerial camera surveys (and do so faster). But LiDAR systems do not produce the full range of information that can be obtained from aerial photography. For some applications where more detailed elevation information is required, LiDAR is ideal, and yet for others where broader information is required, LiDAR may be limited. With the demand for specialized maps that are used once (and then often discarded) increasing, the right choice must be made about what technology to use where, and when, otherwise maps will be produced that do not meet the user's needs.

The need for different levels of detail in different areas resulted in the sensible, if sometimes difficult to implement, approach to data collection suggested in Canada by the Interagency Committee on Geomatics (as noted above):[110] "collect data once, closest to source." The implication is that governments should work together to collect the most detailed information needed and then generalize that information if it is required at a more general level: in other words, municipalities should share their detailed information with higher levels of government, and everyone should share in the cost of acquisition. This approach, while sensible, is hard to implement because of the many different data policy models found in the many different political jurisdictions and the many different entities that would have to be involved. Much has been written on data policy – some of it by the present authors on studies done in a number of different countries.[111]

A basic rule that Canada has tried to follow in data creation is useful: collect data once closest to the source of data. While difficult to implement across several levels of government, this approach reduces overlap, duplication, and saves money.

Image Maps

Alternatives to traditional topographic maps for more general map use began to surface with the advent of higher resolution Landsat Thematic Mapper satellite data in the 1980s. In the case of image maps, the images have been (in some cases, but not all) corrected to accurately represent the geometry of the area. An image map may or may not show things correctly placed to the same level of accuracy as a topographic map – it may be the same, better, or worse.

It surprised us not at all that in unpublished research we did in the mid-1980s we found that topographic map sales declined for any area for which the company Worldsat International[112] produced an image map. People wanted a first-level abstraction from the real world – not a cartographer's rendition or interpretation that was inherently hard for the lay-person to understand. Some cartographers have seemed loath to accept this decision – as the piles of maps gathering dust in warehouses will attest.

Map Accuracy

It is useful to address the potential (and in some cases real) differences between a map, an image map and a representation. First let us consider accuracy. A map with a certain scale and a certain level of accuracy can be assumed to have placed things, or show heights of the topography, in the right position within certain known limits of error. (Interestingly enough, the original accuracy standards of topographic maps were related in part to the thickness of a line that could be drawn by hand on a map at a certain scale, as were the minimum areas depicted on maps of land use and land capability made in the 1960s and 1970s.)

Maps, image maps, and representations are different and should be seen as such.

With the many changes in technology associated with the development of maps, there is now a wide range of map types, each with differences in the accuracy of the location of things depicted on them. There are the standard topographic maps with which many school children are familiar. The Canadian Centre for Topographic Information defines a topographic map as "a detailed and accurate graphic representation of cultural and natural features on the ground." They further state that the information that can be found on a topographic map "identifies numerous cultural and natural ground features which can be grouped into the following categories:

- CULTURE: roads, buildings, urban development, boundaries, railways, power transmission lines;
- WATER: lakes, rivers, streams, swamps, rapids;
- RELIEF: mountains, valleys, slopes, depressions;
- VEGETATION: wooded and cleared areas, vineyards and orchards; and
- TOPONYMY: place names, water feature names, highway names."[113]

Some topographic maps are now produced with image backgrounds. If the imagery is current, it can provide updated information without the expense and time delay of digitizing and adding information to the map database. For example, a municipality looking for structures that have been built without a permit (and which are thus not taxed) could use a map displayed over recent orthoimagery – imagery that has been geometrically corrected to accurately fit a map, i.e. without terrain and other distortions.[114] The added structures not yet being taxed can be easily spotted visually without the expense of digitizing.

Another element that must be considered in map accuracy is that the geometry of the image or map may not have been properly accounted for, or it may have been changed in the processing, or may vary by map type. For example, a road map will not have the same positional accuracy as a topographic map – nor does it need the same accuracy. Country or continental maps showing boundaries of countries or the names of features may be or may not be accurate in several ways. There may be errors associated with positioning <u>and</u> names. For example, in maps from India one can expect to see the boundary with Pakistan in a slightly different position than a map from Pakistan might depict it. A map from Iran will show the Persian Gulf, while a map from Saudi Arabia will no doubt call the same body of water the Arabian Gulf.

We should also mention the difference between a map and a representation – and use a rather humorous anecdote to do so. Dr. Bob O'Neil and Peter Paul and their team at the Canada Centre for Remote Sensing convinced the appropriate authorities to place a modern map of Canada, a depiction of an older map, as well as an image of RADARSAT-1 and a ground receiving station for satellite data, on the back of the $100 bank note of Canada. This was to commemorate the 100th birthday of the National Atlas of Canada of which O'Neil was in charge. It was noted by the team that at the scale used, both Vancouver Island and Prince Edward Island would not show as separate from the mainland. But EVERYONE knows that they are islands. Using the miracle of modern mapping the islands were moved out from the shore. As Director General of CCRS, I received howls of protest from geographers and mappers. "Did we not know that the islands were improperly placed? "Yes," Dr. O'Neil replied, "but this is not a map of Canada, it is a REPRESENTATION."

Fundamentally the person making a map, although adhering to standards, must still make many choices in creating the final map product. The choices are made to best suit the intended uses. In this case a strictly accurate map would not have suited the intended use because certain key geographic features would not have been discernible. Even in traditional topographic mapping there are rules allowing the geographic position of features to be shifted so they can be clearly distinguished. For example, in traditional mapping a railway line running next to a road had to be drawn slightly out of position so both the road and railway could be clearly printed on the map. Today, the correct position of features can be stored digitally even if they are in close proximity. However, as in the case of the map on the $100 bill, when a map is generated adjustments may be needed for the product to suit its intended use.[115]

This is a useful concept to keep in mind. Are we dealing with reality, a map of that reality (as interpreted by a cartographer), a representation, or some other means of depicting the data as it relates in terms of position to other features? How is accuracy determined? How important is accuracy? For a road map it only matters that distances are more or less correctly shown, that the roads link up and cross one another as they do in reality, and that the towns and village and thruway off-ramps are in the right order. Specific small bends in the road can be (and usually are) ignored. Small villages are left off the map, and in some cases so are larger towns – so that other towns can be identified. We have heard it said by some cartographers that "road maps aren't really maps." Perhaps not: but if properly constructed they are a useful spatial representation of the road network, and that is all that matters.

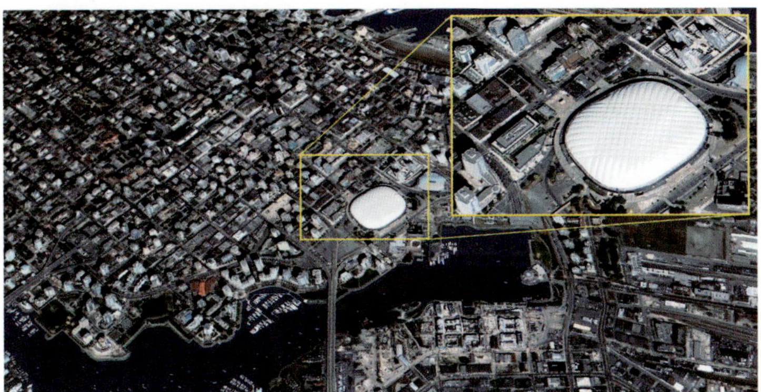

Satellite Image Map – Vancouver, B.C. Satellite imagery can provide surprising detail of urban areas as seen in this image of downtown Vancouver. Source: Robert Stacey, Worldsat International

What this discussion shows is that there is sometimes a mismatch between the specific information that one is intending to communicate with a map and the unintended assumptions that the user might make. When dealing with maps people tend to make general assumptions about map accuracy that might not apply – usually assuming that the maps are more accurate than they really are. The problem here is that although two maps may look similar, their properties might be very different depending on how they were made and there isn't an easy way to find this out unless it is explained. Most maps simply don't come with this explanation. By default users tend to assume that the map has the characteristics of maps with which they are familiar. Unfortunately, this assumption can infer far more accuracy to the information on a map than it actually warrants. Those who buy products or services are cautioned: "buyer beware." We caution those who use maps: "user beware."

Canadian Currency Highlights Importance of Geospatial Information. This image is the back of the Canadian $100 bank note. It was designed to celebrate the 100th anniversary of the founding of the National Atlas of Canada which was created at least in part to help promote the settlement of the western part of the country. The bottom left shows an old map of Eastern Canada and the tools that were used to make such maps, while the right shows how maps are made today with satellite data sent electronically to receiving stations, as depicted. See the text to understand why this is not a map. Source: Used and altered with permission from the Bank of Canada.

One clue to judging the suitability of a map for your purpose is to consider the intended use for which it was created. A road map is created primarily for drivers to find their way to a destination. It would probably give reasonable measures of distance along roads, accurate enough for the traveller because that is its intended use. But it is unlikely it would provide accurate measurements of areas or shapes of lakes, since it was not designed for that purpose. You don't use road maps for boating, any more than you would use nautical charts to plan your cross country drive.

Now that we have you thoroughly confused about topographic mapping, image maps, and the rest, we can tell you that with today's technology, topographic mapping and the traditional paper maps don't matter anymore in many ways.

Changes in Mapping: Some Concluding Observations

Now that we have you thoroughly confused about topographic mapping, image maps, and the rest, we can tell you that with today's technology, topographic mapping and the traditional paper maps don't matter anymore in many ways. The shift can be seen in the 5.5 x 8.5 inch pads of note paper that have been handed out by the United States Geological Survey to advertise their new National Map initiative.[116] The front and back of these is shown on the next page – the front has the USGS identifier along with the words "National Map" at the top and at the bottom the web address and above that a note that suggests that the "National Map" is "The Nation's Topographic Map for the 21st Century." This is all printed on the back of old topographic maps – in this case from rural Kentucky. A similar cost saving use of old maps has also been done in Canada.

The making of maps has changed dramatically – and, we would say, forever. The following factors and technologies have exerted a tremendous influence:

- We can now acquire digital imagery from digital cameras, cutting out the photographic processing step in mapping – thereby avoiding the necessity of scanning film to take what was an analogue product (a film-based photograph) into a digital environment – a side benefit is that this makes modern mapping more accessible in developing countries;[117]
- While digital image and map files are very large, modern computer power is equal to the task of handling and manipulating these very large files;
- Geospatial-specific data bases that can efficiently handle these large volumes of data with a position identifier have been developed by the likes of Oracle;
- Algorithms and standards have been developed to overlay disparate data layers one on the other in these data bases – much as the data base forerunner topographic map has printed on it blue rivers, brown contour lines, red residential areas, black roads, etc. In the data base one can overlay various data digitally;
- With the data in these sophisticated data bases, one can bring spatial and other data together for further analysis. This is where the true value of the data comes in;
- The Internet allows us to send large amounts of data from place to place;
- Technologies have been developed by CubeWerx[118] that allow for data to be overlaid and combined on the Internet – meaning that the data used never have to be brought to one location. This opens up a number of possibilities – including the maintenance and updating of specific data bases by their owners – not by mapping agencies;
- Printing technologies now allow the production of print-on-demand hard copy maps. In effect, users can select what information they want, how they may want

the data generalized (i.e. presented using less detailed categories), if at all, at what scale they would like the map produced – and then produce it. If small enough areas are to be printed out, a simple desk top ink-jet printer can be used; and

- Data base technologies, modern GIS systems, and related technologies mean that one need not necessarily print a hard copy – data layers can be brought into a system and used within that system.

The biggest change in mapping today is that a map is in effect a window into a geospatial data base. That window can be viewing one or more data layers; can be up-dated as new information is added; and can roam over the data base as required to make a map of whatever layers are of interest.

The biggest change in mapping today is that a map is no more and no less than a window on a geospatial data base. That window can be on one or more data layers; can be up-dated as new information is added; and can roam over the data base as required to make a map of whatever layers are of interest – it need not be restricted to a specific map sheet or "tile" arbitrarily selected by a mapping scientist 50 or more years ago. (As a frequent user of such maps over a long career doing field work, a basic truism of field work with maps is that the area of interest is always where two map sheets join – or worse, where four of them join!)

Surplus Maps Reused as Notepads by the United States Geological Survey. Traditional production of paper maps sometimes resulted in large inventories of out-dated high quality paper products printed on only one side. Some mapping agencies have reused the paper, producing products such as notepads (front and back shown) instead of sending tons of paper for recycling. Source: USGS

A great deal of attention has been paid recently to what makes up the fundamental layers in a geospatial data base – this is discussed in more detail in Section 3.2.6.

Two final and important points must be made as we finish the discussion on making customized maps. First, if you begin with poor data, you will end up with poor maps – "garbage in leads to garbage out." Second, just because a customized map has been made especially for you, does not necessarily mean that it will meet your needs.

3.2.4.2. What is it used for?

Traditional uses for maps or types of maps have been with us for centuries and have been well catalogued. We list a few in the following text box. But what is most interesting is the range of new applications that open up with mapping on-demand. Those who may wish to use information from a geospatial data base are no longer restricted to whatever product the cartographer in the national (or state) mapping agency has decided to supply. The implications of this new reality are discussed in the following section.

Traditional Uses for Maps[1]

Agriculture Irrigation catchment areas **Land Environment** Land cover mapping Mapping of broad land cover and land use classes **Forestry** Mapping of disturbances caused by fire, insect attacks, deforestation Forest access roads **Hydrology** Snow extent Potential flood zone Ground water recharge areas	**Urban and Populated Areas** Topographic mapping Land ownership Delineation of infrastructure Roads Public transit routes Planning Land Ownership **Bathymetry** Water depth/hydrographic charts **Disasters** Response Mitigation	**Geology** Geological mapping Lithology **Coastal Areas** Delineation of coastline Mapping and monitoring of major land cover types, e.g. mangroves **Recreation and Tourism** Canoe routes Bicycling paths Hiking trails Tourist attraction maps Boating routes **Sea ice** Sea ice extent (info from satellites) Sea ice type mapping

[1]To provide a basis for comparison we have used the same categories as for the Table showing TerraSAR-X applications in Section 3.2.3.2 and have added a few additional ones. Maps may be of the several different types discussed above.

On-demand services have already become a replacement for road maps in many people's travel plans – including many of our own family trips! In many government agencies on-demand services now provide a full range of geospatial information to the many traditional users of maps. But the true significance of mapping on-demand is that it has opened up the potential to go far beyond traditional mapping in the breadth and depth of its application. It can provide information that is both current and specially designed to meet specific needs of a far greater range of users, be it for the individual, an interest group, industry or any one of several levels of government. In the future, the only warehouses holding map data will be on computers – geospatial data warehouses.

There is another important factor in what mapping on-demand maps are used for. It does not necessarily result in the production of a traditional map: in its simplest form it allows one to obtain and display information selected from a geospatial data base. The data may simply flow from the mapping agency data base into the user's computer where it may be used or merged with other information.

Studies we have done suggest that few of the national mapping agencies in developed countries are actually keeping their data current. Many of them "talk a good game," but few seem to have had the ability to impress upon their political masters that geospatial information is one of the essential building blocks of a modern society, and as such, updating the data warrants regular significant expenditure.

For example, to minimize travel time to a company's golf tournament one might want to superimpose the location of all of the public golf courses nearby, as well as the company's offices and employee's residences, on a road map. By applying a simple minimum distance travel algorithm one could identify the golf course that will minimize travel for staff to the tournament. Such an application would only require that we see the result – we need not necessarily print an actual map. Indeed, the company CubeWerx,[119] which specializes in software and systems to move geospatial data using the Internet, has developed software that allows one to bring together information on the Internet, combine or overlay it elsewhere and only bring back the result of the analysis – without ever actually possessing the basic information. This sort of technology leads to people not buying information, but rather renting it. This, of course, has profound implications for things such as copyright and information ownership.[120]

It is also argued that since maps are made only when needed, they use the latest information available. Of course, this is true and a benefit only if the data bases actually are updated and do contain current information. Unfortunately, most geospatial data bases do not contain up-to-date information. Yet people tend to assume that if a map has just been produced to suit their needs, then the information it contains must also be current and sufficiently up to date for their needs. It may not be.

Some have also said that mapping on-demand will allow the mapping specialists to focus more attention on updating and maintaining their data bases, and this in turn should lead to better data. Again, this is true only where any savings are actually spent on improving or updating data. Studies we have done in our consulting firm suggest that few of the national mapping agencies in developed countries are actually keeping their data current. Many of them "talk a good game," but few seem to have had the ability to impress upon their political masters that geospatial information is one of the essential building blocks of a modern society, and as such warrants regular (and significant) expenditure.

Perhaps this book will help the mapping agencies in government make that argument to their political masters.

3.2.4.3. What are the implications?

For such a simple concept, mapping on-demand has already turned the mapping industry on its head – and for the foreseeable future will continue to both wreak havoc on, and create opportunities for that industry and those who use its products and services.

> ***Mapping on-demand has already turned the mapping industry on its head – and for the foreseeable future will continue to both wreak havoc on, and create opportunities for that industry and those who use its products and services.***

People usually like to talk about opportunities that come with great change – and leave the threats that accompany change as a minor afterthought. We hope to provide a more balanced view. While the opportunities associated with mapping on-demand are tremendous – so too are the threats and problems that come with these opportunities. In short, there are a number of important implications associated with mapping on demand.

> ***Implications of Mapping on Demand***
>
> 1. *Democratization of access to geospatial information – mapping on demand leads to easier access to geospatial data, potentially lower map accuracy, greater risk of inappropriate use, and a greater possibility for litigation;*
> 2. *The potential for broader application of geospatial data and, as a consequence, better decision making;*
> 3. *Reduced costs for map storage and map printing facilities; and*
> 4. *Raised expectations of the public.*

Democratization of Access to Geospatial Information and Potential for Inappropriate Use

The increasingly computer and Internet literate society we live in provides a ready market for on-demand mapping. Indeed, on-demand mapping is one of the technological pillars of the GeoEconomy, and on-demand mapping is well positioned to bring geospatial data to a much broader range of users than did more traditional mapping. This is both a blessing and a terrible curse.

> ***On-demand mapping is one of the technological pillars of the GeoEconomy and is well positioned to bring geospatial data to a much broader range of users than did more traditional mapping.***

The problem is that while our society is increasingly computer and Internet literate, we are not geospatially literate or geo-literate. One of our fundamental observations as we began this book was that over time modern society has, in general, lost its geo-awareness. But geo-literacy is even more than geo-awareness – it involves knowledge, understanding and application of complex geometry and the measurement of the effects of the forces that literally shape our earth – the arcane field of geodesy. This complex knowledge, which together we call geo-literacy, has resided in the hands of the selected few we entrusted with the making of our maps, completion of our land surveys, and measurement of the geoide – the Earth. Over the last century our increasingly sophisticated use of geographic data has rested on the scientific and engineering knowledge of a special and very small segment of the geo-literati.

> ***Over time modern society has lost its geo-awareness. We are no longer geo-literate and few of us have the background to take advantage of the opportunities new geodata and technologies are providing.***

The corollary of the development of this very small geo-literate subclass is that most of society is basically geo-illiterate, but for those directly involved in map creation, geodesy, and land survey. This "rest of society" knows little about map or geo-positional accuracy. It knows little or nothing about how to project the curved surface of the earth on a flat piece of paper (or flat computer screen). It knows nothing about the complex field of map generalization that allows us to take detailed maps and combine them with more generalized maps.[121] It knows nothing about map standards – either from the Open Geospatial Consortium (OGC www.opengeospatial.org) or from International Standards Organization (ISO) Technical Committee (TC) 211 – which at the time of writing has published 36 different standards, and is in the process of developing another 22.[122] There are even standards about how one describes the standards! Indeed, there are so many standards, that we believe it accurate to say that no-one individual among the geo-literati understands them all.

The implications are significant. Consider that with on-demand mapping the user or client must decide what data sets will be used, what mapping accuracies will be selected, and how they will be overlaid and integrated – in effect this untrained individual is being asked to make decisions about standards of which he or she knows nothing – and should be expected to know nothing. Who should supply this expertise? Need it be supplied?

For the user who is not geo-literate, the following paragraph is a warning – with especially interesting implications in an increasingly litigious society. What happens if a map created by a non-specialist leads to erroneous conclusions? Who is at fault? Who gets sued?

While many types of data are provided in formats and in ways that allow them to be accurately overlaid one on the other using guidelines such as those established by the Open Geospatial Consortium, not all types of data can be easily overlaid on all other types of data or give a valid result.[123] How will the user select the right data, the right tools, and use data in the right way? Will such help come from mapping organizations in government? Our response is "not likely." For the geo-literate mapping specialist this would appear to represent an opportunity. For the user who is not geo-literate, this is a problem – with especially interesting implications in an increasingly litigious society. What happens if a map created by a non-specialist leads to erroneous conclusions because it was not suitable for the application? Who is at fault? This leads to the next, and perhaps most important question. Who gets sued?

Broader Application and Better Decision Making

There are of course some positive implications tied to the wider and faster creation of maps or map-like data – assuming that they are sufficiently accurate to meet the needs of those who are creating them or paying for their creation. We believe that on-demand mapping will lead to rapid growth in the range and number of applications of geospatial data – as there has already been for new sources of imagery. We expect that this growth will have an impact at many levels. Individuals will be able to create maps for their own unique needs – be it for house hunting, planning holidays, or assessing the changes in environment that may be occurring near their homes. Companies that want to assess markets using geospatial information in unique ways will be able to do so more easily – especially using tools of the sort developed by companies like MapInfo – now Pitney-Bowes Software Inc. Municipalities will more easily be able to create maps for their employees and political masters' use, as well as for the public at large.

This impact will be even more significant in national and state mapping agencies – although perhaps not so much in the range of application, but rather in how geospatial data are delivered. Paper maps will no longer go out of date so quickly since they will be printed on-demand (where they are printed at all) and as noted above, inventories of maps will largely become a thing of the past. We expect that the growth in use of maps and mapping-on-demand will be especially important in developing countries.

We expect that the growth in the use of maps and mapping-on-demand will be especially important in developing countries.

•••

Traditional mapping used to be done using heavy optical/mechanical photogrammetric devices that were usually so heavy that they had to be located on the first floor or in the basement of a building. The input was photographic film. Both the instruments and film were extremely sensitive to damage from dust – which in many developing countries is almost impossible to keep out – especially on the first floor. The equipment (both photogrammetric devices and scanners that eventually were used to convert the film to a digital file) were difficult to maintain. Typically maintenance was done only by highly specialized staff usually based in Europe or North America.

•••

Today, systems use digital information from digital cameras and widely available personal computers. Personal computers are, of course less sensitive to dust, are lighter and can be located on higher floors where they are more easily shielded from dust. In addition, personal computers are easily serviced almost anywhere, as Ryerson learned from personal experience in Maputo, Mozambique.[124]

While some might also believe that a modern society's apparently growing need for near instantaneous access to up-to-date information is important, there appears to be a disconnect between this view and what state and national governments will agree to finance. Simply put, governments seem willing to support the construction of specialized satellites for monitoring, but they are loath to spend the money to turn the data from these satellites into information that can be accessed by governments, industry, and the population at large. Yet it is the provision and use of that information that will lead to a more efficient country and more transparent government.

Governments seem willing to support the construction of specialized satellites for monitoring, but they are loath to spend the money to turn the data from these satellites into information that can be accessed by governments, industry, and the population at large. Yet it is the provision and use of that information that will lead to a more efficient country and more transparent government.

The implications are clear. Those jurisdictions that provide sufficiently accurate and up-to-date information will be more efficient, will make better decisions, make them faster, and ultimately, be more competitive in the GeoEconomy. Those who do not provide such data and do not use it wisely will become increasingly less efficient, and less competitive. What is especially sad is that those who are not geo-aware, and who do not become geo-competitive will likely not even know it and will not know why they are falling behind. We can cite a number of examples in retail of this happening. Indeed, we suggest that the American retail juggernaut that has crushed local competition in head-to-head battles around the world has done so in large part through the intelligent use of geo-information. American companies such as Pizza Pizza, McDonalds, WalMart, Toys R Us, and others have long had access to low cost geographic information at home in the USA. With the low cost, it was easy to obtain and experiment with the data. This led to them realizing its value for such things as optimizing warehouse locations, distribution networks and delivery routes to reduce cost. While this knowledge about the value of geo-information was being gained along with experience on its application, many international competitors only had access to expensive data (if any) and thus had not developed the understanding of its value or experience in its use. When these geo-aware American companies looked at international markets where there was limited or high cost geographic data, they either bought what was available even at a high cost, or they created their own data. Local competitors often ended up in the unenviable position of knowing less about their local market than did the foreign competition. The local competitors didn't have a chance.

When these geo-aware American companies looked at international markets where there was limited or high cost geographic data, they either bought what was available even at a high cost, or they created their own data.

Local competitors often ended up in the unenviable position of knowing less about their local market than did the foreign competition. The local competitors didn't have a chance.

Reduced Costs for Map Storage and Map Printing Facilities

The reasons for a mapping agency to use mapping on-demand are clear. Mapping on-demand reduces inventory (and hence storage costs), provides a more flexible product better tailored to the needs of the user (assuming that the user knows what to ask for), makes it easier to update information, and the product is not tied to artificial boundaries. Furthermore, mapping on-demand reduces the need for access to massive printers of the sort needed to produce large runs of highly complex colored maps – many of which may never be sold.[125]

Raised expectations of the public.

With the advent of Google Earth, Microsoft's Virtual Earth, MapQuest, and other similar on-line services, the expectations of the public have been raised. In some cases these expectations are met, in others they are not. When they are not met, there can be disappointment and even frustration. And of course, who gets blamed? The map-maker of course!

As a case in point, recently, one of the authors (Ryerson) used one of the Internet direction services to locate the street address in a new unfamiliar neighbourhood. As is usually the case, there are alternative views – one can select map (which in this case showed a number of streets) – or aerial view. When the aerial view yielded a nice colored aerial photograph that showed none of the development, only a nice cornfield, there was a certain amount of frustration. "Surely, there must be more recent imagery." Even many of us in the geo-business (who should know better) have succumbed to the higher expectations of the general public! Of course, since the source of this imagery is the local city, the lack of up-to-date imagery can ultimately be laid at the feet of we the public who elect the politicians who have decided that more up-to-date imagery is not needed!

3.2.5. GIS and Data Integration

3.2.5.1. What is it?

Geographic Information Systems (GISs) are computer hardware and software designed to handle the input, storage, analysis, and output of geo-referenced information. Virtually any data in digital form or that can be converted to digital form, both geo-referenced and not geo-referenced, can be input and stored in a GIS. Data as diverse as property records, transportation networks, topographic information, natural resource data, competitive business intelligence and market analyses, and disposition of military forces are routinely handled using GIS technologies.

Geographic Information Systems (GISs) are computer hardware and software designed to handle the input, storage, analysis, and output of geo-referenced information.

The development of GIS technology was outlined previously in Section 2.9.3. Geographic information systems were originally developed because conventional database systems were not able to effectively handle the spatial component of information, or perform analytical operations unique to this type of data. For example, identifying features within an area or in proximity to a site (location of schools near a toxic waste site for example), overlaying maps of an area showing different features or conditions (such as land capability for agriculture vs. zoning maps), generating perspective views, and identifying obstacles within a line of sight are examples of spatial analyses provided by GIS that cannot be effectively implemented in a non-spatial information system.

The sophisticated data handling and analysis methods developed for GIS are now being incorporated into conventional enterprise database solutions by all major suppliers of these tools. As a result, the capability to use geo-spatial information is becoming more commonly available within the information infrastructures of both industry and government, and in turn is more accessible to the broader public. Tomlinson (2003) often considered to be one of the fathers of GIS, has written on the process of implementing a GIS within an organization.

3.2.5.2. What is it used for?

Introduction

GISs provide the means to search, view, and analyze a wide range of information pertaining to specific geographic areas. Perhaps the most important capability of a GIS is its ability to integrate different data sets to produce new information. For example, data of the vegetation types, soil classes, and slope characteristics can be combined to produce a map of terrain units suited for certain types of development in land-use planning.

> ***Geographic information systems provide the means to search, view, and analyze a wide range of information pertaining to specific geographic areas. Perhaps the most important capability of a GIS is its ability to integrate different data sets to produce new information. The old saying "garbage-in-garbage-out" is perhaps no truer than in a GIS. When used well with good data a GIS can be powerful – when used well with poor data incorrect results could be costly.***

GIS technology has enabled a more integrated approach to managing human activities within both natural and man-made landscapes. In so doing, GIS offers the opportunity to improve decision-making through a more complete assessment of available information, the ability to model multiple scenarios, and develop simulations to predict outcomes. In short, a GIS provides the means to optimize the use of the land and resources. Of course, this assumes that the tool is used skilfully and that the data used are accurate. Garbage-in and garbage-out is nowhere truer than in a GIS. In short, when used well with good data, the technology can help us use human and physical resources more efficiently, by doing more with less, and by better anticipating outcomes.

Much as we ask questions of imagery, GIS analyses may include queries of tabular data linked to location. These include, for example, land record information, and the representation of information in the form of maps and reports. A GIS can be used to model future outcomes such as urban expansion or the spread of a forest fire, and the development of prescriptive actions such as changes to police service areas to improve response times, or the planning of reforestation projects. Some of the many examples of GIS applications are listed in the following table.

Examples of GIS Application Areas and Example Applications

Application Areas	**Examples**
Resource Inventory	Forest inventory, habitat mapping, coastal resources
Land Records Management	Maintain accurate information on property ownership, boundaries, valuation, zoning, and other characteristics
Network Analysis	Provide optimal routing for delivery vehicles
Business Geographics	Market Analysis to optimize store locations
Incident Mapping	Record location and characteristics of crime events
Corridor Selection	Optimize transportation route design
Logistics Planning	Provide emergency response support to identify available equipment and personnel and task groups
Facility Management	Track facility condition, respond to changing facility needs, coordinate facility maintenance, plan facility de-commissioning
Topographic Analysis	Assessment of terrain trafficability, viewshed analysis
Demographic Analysis	Project change in population age distribution to assist planning school locations

Engineering Design	Transportation project design, building site selection
Watershed Analysis	Manage watershed land use to preserve water quality
Municipal Workload Planning	Coordinating municipal maintenance service to reduce road closures
Transportation Modeling	Predicting traffic volumes, designing evacuation routes
Resource Exploration	Identifying promising exploration sites from multiple data sets
Military Planning	Situation awareness of battlefield conditions
Public Health Management	Monitoring disease spread, defining hospital service areas, projecting future health service facility needs
Change Detection	Using remotely sensed data, assess urban expansion, reforestation, changes to coastal regions, climate change

GIS functions can be usefully grouped into Input, Visualization/Output, Storage, and Analysis capabilities, and GIS applications can be categorized as Descriptive, Prescriptive, and Predictive.

GIS Functions

Input

A GIS is used for accepting and combining all inputs that are geo-referenced – no matter their source. A GIS can accept as input virtually any data in digital form or data that can be converted into digital form, that is in some way referenced or "tied to" a location. This includes conventional geographic information as commonly portrayed in map form, aerial photography, satellite imagery, and other forms of remotely sensed data displaying areas of the Earth. It also includes data typically presented as tables of information or statistics that can be given a geographic location, such as census data. As well, other less conventional geo-referenced data such as video clips, photographs, or audio clips pertaining to geographic locations can also be included in a GIS database. For example, a user could ask a question of the GIS about a region and obtain materials as diverse as audio recordings of interviews with aboriginal elders, photographic evidence of road surface damage, imagery from satellites, and expert discussion of findings. Even without these non-traditional information sets tied to location, the possibilities now for the application of GIS are staggering. This is especially so when one considers that a study in the USA by Business Week Research[126] suggested that 80% of all data in industry data bases in the USA have a location attached to them.

A study in the USA by Business Week Research suggested that 80% of all data in industry data bases in the USA have a location attached to them.

Visualization/Output

A GIS is used to provide geographic information in virtually whatever format is most effective to communicate with the intended audience. Maps, tabular data, and reports can be produced in the form of digital files or hardcopy output. Multimedia output may include audio with video and still images showing, for example, the appearance of a new housing development superimposed on the terrain with a backdrop of the hills in the distance, the expected growth of a forest area over time, the spread of flood waters or a forest fire, or what has become known as a "fly-through." A "fly-through" yields a 3-d perspective view of an area before and after development, as if one was flying over it in a slow (or fast) moving aeroplane. The debt the technology owes to the military is obvious: "fly-throughs" come from the same technology used in flight simulators, and military mission planning – be it for fighter aircraft, cruise missiles, or other means of applying force through airpower. Of course, the technology is now widely used in video games.

Data Storage

A GIS can be used to store data locally on the system supporting the GIS software, or access remotely stored data. If not stored locally, the data may be maintained in a centralized database that multiple systems access using an intranet or public network such as the Internet. With the networking of databases held by different jurisdictions, it has become common to perform GIS analyses using data accessed from different databases, housed in different physical locations and maintained by different jurisdictions. This all depends, of course, on having common standards for such data. With the Open Geospatial Consortium, such standards are becoming more the norm than the exception that they were just a few years ago. For example, census data obtained from one database might be combined with health services data to assess the capacity and type of health facilities that will be needed, based on projections of population growth and demographic makeup.

GIS technology offers a fundamentally different paradigm for working with data.

Analysis

The main strength of a GIS is its analysis capability.

GIS technology offers a fundamentally different paradigm for working with data. Rather than being restricted to a snapshot of conditions at a particular point in time, a GIS can be used to provide a window into a constantly changing data set, as has been discussed in the section dealing with On-demand Mapping. Information can be extracted from that data set and presented in many different ways, tailored to a specific application. Generating information products such as a map, a tabular summary of information, or a visualization of the appearance of a new housing development in its proposed location, are relatively fast and inexpensive to produce so long as the data are well-maintained.[127] This ability to rapidly perform analyses has changed the way many projects can be done. Complex projects can be progressively refined by altering parameters and assessing the results.

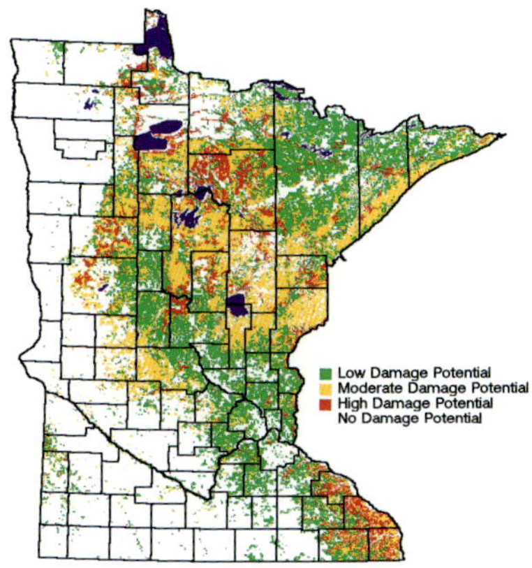

Predictive Model of Gypsy moth Infestation. A GIS was used to combine vegetation information and professional judgement to generate this model of susceptibility when Gypsy Moth populations were expanding towards but had not yet reached Minnesota. Source: Department of Natural Resources © 1997 State of Minnesota. Reprinted with Permission

For example, a proposed housing development will affect the local region in many ways. There will be economic factors, environmental issues, and impacts on local infrastructure such as traffic volumes, and the viability of the project will be affected by the demographic makeup of the local population. Modifications to the project will impact several factors ultimately affecting the project's commercial viability, adherence to local regulations, and acceptability to community and other stakeholders. By modeling these factors within a GIS environment, planners can progressively refine design alternatives, incorporating the concerns of different stakeholder groups and identifying the trade-offs of different proposals.

GIS Applications

A useful categorization of GIS applications is: descriptive, prescriptive, and/or predictive. Depending on the scope of an application it may include components of one or more of these types.

- Descriptive applications are used to present and analyze information pertaining to an area at one or more points in time for which data have been collected. These applications describe the situation. For example the tabulation of property ownership and values in a municipal area, maps of road networks, forest cover, topography, or a fly-through visualization of the Grand Canyon.

- Prescriptive applications are used to help develop a plan of action. For example, a municipality can optimize infrastructure maintenance activities by prioritizing and scheduling the repairs to roads, sewer, water, and other facilities to minimize both disruption and costs. Emergency responders can be presented with optimized routes to the emergency site, based on current traffic conditions, known road closures, and one way streets.

- Predictive applications are used to simulate conditions that would occur at some time in the future or in the past, based on the available information and models of processes that would produce changes. For example, the spread of flooding through a municipality and its impact on services for a specific flood event could be modeled to help develop emergency response plans. As another example, one could model the impact of resource development such as mining or forestry on water quality and sediment loading in a watershed over the proposed development period, to help develop procedures that mitigate impacts.

One can see how the three categories of GIS work together if we take one example, such as the study of the West Nile virus. Data from insect traps describe the current occurrence of the virus. Maps of vegetation types in a municipal district can be used to prescribe areas most likely to benefit from spraying to control mosquitoes and from draining of standing water. Information about progress of the disease northward, trends in summer temperatures and rainfall can be used to develop a predictive model for planning future activities of the virus – for example, identifying the regions where the virus is most likely to spread the following year, so mitigation measures such as reducing the area of standing water or applying insecticides to critical areas can be implemented.

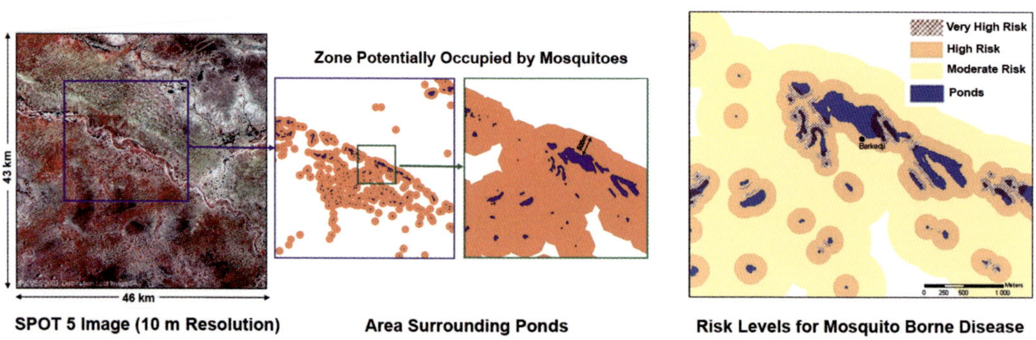

Satellite Imagery for Human Disease Management. SPOT satellite imagery was analysed to identify new ponds, an indicator of greater risk of exposure to mosquito borne diseases. Spatial modelling was used to produce a map showing zones of higher risk of exposure that public health services could then use to prioritize mitigation efforts. Source: Yves M. Tourre Copyright: Medias-France product, CNES 2003. Distribution Spot Image SA.

GIS technology can be applied by anyone from individuals to the largest of governments. A GIS implementation can range from stand-alone systems operating on PCs, to enterprise wide systems with multiple users accessing a central database, to networks of users accessing multiple databases. Miniaturized applications are available in handheld mobile devices that enable technicians to assess features at a location, enter their observations, and have the data automatically coded with location information using GPS technology. Data may be stored for later download or input directly by wireless transmission to a centralized database.

There are numerous vendors of GIS technology and services from which to choose. While some vendors developed proprietary formats, the drive for data and systems from multiple vendors to be able to work together has fostered interoperability among systems and the ability to read and write common data formats. The drive for interoperability has been encouraged by the Open Geospatial Consortium (OGC). This organization has developed standards for geo-processing capabilities developed by consensus with vendors and users. The specifications enable the systems of different vendors to exchange data and instructions facilitating their use in a wide range of information systems. The OGC also makes available public domain software to promote the use of geographic information and interoperability.[128]

3.2.5.3. What are the implications?

Introduction

The practical applications of GIS technology developed at an exponential rate during the 1990s and that development continues today. Annual sales of GIS-related technology and services have been estimated to be as high as $4 billion or more. This, of course, does not include the cost of using the systems (staff, data creation), nor does it reflect the potential benefits from the use of the systems. GIS technology has been incorporated into virtually every type of computer-based information system: handheld data collection units, computers, laptop and desktop systems, enterprise systems, and distributed networks.

The application of geographic information has advanced beyond the limited range of dedicated GIS software packages and is now included in mainstream enterprise database software and web-based consumer information sites. It is safe to say that someone somewhere has already incorporated GIS technology, in some form, into almost every application that can make use of geographic information. GIS technology has become ubiquitous within the information infrastructure of business and government, and increasingly is used by the individual consumer. GIS methods are often embedded in what is generally perceived to be conventional information queries and display. For example, websites such as Google Maps offer free access to users who want to find the location of a street address and driving directions to the local eatery, and display it as a map or over an air-photo or satellite image background.

Over the past 40 years, geospatial information technologies (be they identifiable GIS software or GIS functionality incorporated into a variety of information services) have grown to become part of the information infrastructure used by business and government of both developed and, increasingly, developing nations. Among developing nations, the technology is being adopted as it becomes more competitively priced and skilled personnel become available to implement and maintain it. The implications for GIS technology, like remote sensing, are somewhat different for each of individuals, industry, and governments.

Implications for the Individual

Introduction

The implications of GIS technology and its application by individuals is less well studied and thus perhaps not as well understood as for business and governments, for it is only recently that individuals have begun to have access to GIS. The growth in availability and ease of use of GIS technology have four important implications for the individual.

Implications for the Individual of Use of GIS Technology

1. ***With access to GIS the individual is empowered to take advantage of available geospatial data, leading to more complete participation in the GeoEconomy;***
2. ***The individual expects to have access to the required geospatial data when, as and where he or she requires it;***
3. ***Increasingly individuals will play a role in both creating and updating geospatial information in geospatial data bases accessed via GIS technology; and***
4. ***Privacy and access to information will continue to be a concern for many individuals and this will create challenging trade-offs between the individual's right to privacy and society's need of information for the general public good.***

Empowered to Participate in the GeoEconomy

The first, and perhaps most important, implication associated with the individual and GIS technology is the fact that it allows the individual to begin to take charge of his or her participation in the GeoEconomy. It has been said that the computer made us all numbers, and in giving us all numbers made us all individuals. Allowing users to tailor data products to their needs not only makes use of the data more effective, it makes the data resource itself that much more valuable. This is so because of the much wider range of benefits that can be derived from it. This in turn should justify increased expenditure to maintain the quality of the data – its accuracy and the frequency of updates. This of course assumes that governments have bought into the notion that geospatial data are an important part of a national infrastructure.

In the past many government websites provided access to a national atlas that consisted of a set of static maps representing the resources, and physical and cultural geography of the nation. It is now common for this information to be presented as an online map service where the user chooses the area of interest, selects the data to be displayed, and decides how the information will be represented. In addition, one can bring in more and different types of data. In effect, the users are able to customize the data to generate a geospatial information product, such as a map, that is tailored to their needs. Such customization is the first step in allowing the individual to participate in the GeoEconomy.

Expectations on Access to Data

The introduction of geospatial technology to online consumer sites has rapidly educated the public in the simple uses of this technology. Free online access to geospatial mapping and images offered by Google Earth and Microsoft's Virtual Earth, covering much of the inhabited urban and rural areas of the Earth, has effectively set a standard for how such data "looks," and the ease with which it can be accessed. The public expects to be able to search and access information from their governments and firms with whom they choose to do business. Not surprisingly, firms that have developed these geospatial display services have found a lucrative business in licensing the application to business and government clients, as a base over which they display the licensee's data for distribution to the licensee's clients.

The power of these geographic modes of information display, particularly the overlay of data on imagery of an area and the ease with which people can understand the information, ensures that communication of information using geospatial visualizations will become ever more prevalent, accepted, and expected by the public. One of the implications of this is that governments are not only wrestling with how to deal with Google and Microsoft, they are playing catch-up to their citizenry's geospatial expectations. Individuals will have to lobby their elected representatives to ensure that geospatial data are adequately supported.

Individuals are now using and asking for access to more and better geospatial information from a number of competing sources. This is done with or without government support. If the government mapping agency won't provide the street network for an area (either because it is recent and they have not had time to update their data or the government wants to restrict that information) then the individual has options to get that information from another provider, including other individuals willing to provide it by means of Internet-based open mapping platforms.

As with remote sensing, the ability to get information from many providers and the potential to provide more information to the public leads to increasing transparency. It is an important trend today that governments are progressively less able to control the information that their citizens are able to access. Witness the futile attempts in 2009 of Iran's government to prevent images and video clips of political demonstrations from reaching the media.

> ***Governments are not only wrestling with how to deal with the geospatial activities of Google and Microsoft, they are playing catch-up to their citizens' geospatial expectations. Individuals will have to lobby their elected representatives to ensure that government-produced geospatial data are adequately supported.***

Technology has provided so many tools for individuals to receive and send information that it is very difficult to control. That said, it also means (as has been stated elsewhere) that if you don't make clever use of data integration tools, your competitors surely will.

Updating of Geospatial Information

An interesting aspect of the use of GIS technology by the individual is that the individual (or, more accurately, many individuals) now have the ability to update geospatial information. Some geospatial information such as street mapping for the public to "find its way" is becoming a commodity. There are a number of options one can choose – for example Mapquest, Google, or Microsoft Virtual Earth. Each has its pros and cons. With this competition there is an incentive for mass market geomatics companies[129] to use the public to improve their products by alerting the companies to streets that have been omitted or incorrectly named. However, there is probably not that much incentive to improve position accuracy much beyond what they have now because their users usually only need directions – not accuracy of a specific position. Some suppliers of mass market data already allow users to directly update information including roads, not just alert them to errors. As this capability to instantaneously update data bases becomes more widely appreciated, it will become expected by users, whether they be the general public or personnel within government or industry.

Concerns of Privacy

Nowhere can geospatial information, the individual, and anything to do with the Internet be discussed without mentioning the issue of privacy. With its ability to organize information spatially, GIS technology can also be a tool for invading the privacy of the individual. For those using or updating data on-line, the issue of how one's privacy is protected should always be considered.

Implications for Business

Introduction

The application of geospatial information for business is commonly termed Business Geographics. Using data on demographics (e.g. from a census), consumer buying preferences derived from customer data, business performance and sales data, and core geographic data sets such as road networks, administrative boundaries, land-use, and projected development plans, businesses can analyze their current and potential future customer base. They can develop customer profiles, divide their market into trade territories, plan locations for future facilities, conduct competitive analyses, and identify emerging business opportunities. Indeed, many retailers use point of sale information and other data to plan the layout of their stores to maximize sales. While this is an area in which there exists one of the juicier urban legends – the so called "parable of diapers and beer"[130] – the reality is that data, geospatial and other, are being mined at an increasing rate by companies seeking to gain competitive advantage. We name only two implications of the use of GIS technology in the business context, but they are important.

The Implications of the Use of GIS Technology in the Business Context

1. Those using GIS technology can gain a competitive advantage; and

2. GIS has opened markets (especially retail) to increased competition.

Gaining a Competitive Advantage

Businesses using GIS gain a competitive advantage from more effective use of resources, identification of opportunities to expand their customer base, and optimize the location of distribution warehouses, retail outlets, and other facilities including store design. In some cases, a GIS can be used to analyze information on where people in certain neighbourhoods shop, or where your clients (or your competitor's clients) tend to live or be based.

Increased Competition

As businesses become more sophisticated in their use of GIS, finer distinctions can be made in, for example, characterizing a given market or a specific competitor and the territory the competitor serves. Armed with such information competition can be better targeted and often more effective. One can expect GIS technology to lead to both increased and smarter competitors. Those not using the technology may well be like those unsuspecting victims of the US-based retail "geo-juggernaut" – they will never know what hit them.

Implications for Government

Introduction

Geospatial information and the technology used to handle it, GIS, have had an impact on virtually all of the functions one would ascribe to government. Unfortunately, while many governments are adopting and effectively using geospatial information and GIS, many more are just beginning to realize the implications of this statement.

Geospatial information and the technology used to handle it, GIS, have had an impact on virtually all of the functions one would ascribe to government. Unfortunately, some governments are just beginning to realize the implications of this statement.

While it seems to us that far too many governments have either not heard or not understood the message, luckily there are some who have and that number is growing quickly – and we hope that this book will help accelerate that growth. It is these geo-aware governments that provide some useful insights on what the government could do for its citizens through the use of geospatial information and technology.

The implications of geospatial information technologies for government are many and important.

The Implications of GIS Technology for Government

1. *The integration and study of geospatial data through the use of GIS technology leads to better decision-making and better delivery of services;*
2. *GIS has proven essential in providing good and transparent government including communication with the public;*
3. *GIS impacts the way data are collected and stored;*
4. *The management and regulation of land and its resources; and*
5. *The assertion and enforcement of security and sovereignty.*

Better Decision Making

At all levels of government, GIS technology has become essential in making better, and more informed, decisions. This is so for everything from the management of government facilities such as parks and recreation areas, to the management of natural resources and vehicles, and on to permitting and regulation. At the municipal level in particular, development planning has come to rely on GIS.

Good and Transparent Governance

GIS has become an accepted part of good and transparent government. For example, the provision of information on development plans to the public is usually accompanied by a series of maps and view-sheds constructed on a GIS. Indeed, many of the major suppliers of GIS technology provide software specifically for this purpose.

Data Collection and Storage

At the municipal level, geospatial technology solutions are typically incorporated into municipal information infrastructures and, increasingly, data are shared with the public either through city portals or via a mass-market geomatics provider. No longer are data collected and used only by government.

Land Management

Municipal records, asset management, permitting, the inventory and management of municipal infrastructure (e.g. roads, parks, sewers, water, garbage collection, and lighting), and tax collection all make use of geo-referenced data and GIS. In developed countries the trend has seen GIS technology adopted by smaller and smaller municipalities as the costs of doing so have decreased. The same phenomenon is beginning to repeat itself in many developing countries.

Security and Sovereignty

National sovereignty depends not only on the ability to demonstrate a presence, set and enforce regulations, and provide services to a region. It is also critical to be able to effectively monitor activities, show relationships (for example between geological features under the ocean and the land), and exert control over the territorial extent of the nation. To do so effectively has increasingly required the use of GIS to integrate many types of geospatial information.

> ***National governments when they have less comprehensive or less current information about their own resources than the governments or business interests with whom they negotiate agreements are at a distinct disadvantage.***

To ensure that national economic policies governing international business safeguard national interests, some governments have come to recognize the importance of developing expertise and baseline data to ensure that the national government has information about its own resources (be they natural resources, intellectual property, or access to information and skilled personnel) that is at least as comprehensive and current as the information other countries have acquired. To manage this information and use it effectively requires a GIS. National governments with less comprehensive or current information about their own resources than the governments or business interests with whom they negotiate agreements are at a distinct disadvantage. They cannot ensure that their national interest is well-served if they are unaware of the full extent of the benefits they are offering.

Implications Conclusion

The recent rapid growth in the adoption of GIS technology within government, business, and applications targeted to the general public has been driven by a wide range of benefits at all levels. One fundamental benefit is associated with the ability to make use of more information to develop solutions or to understand conditions in a way that helps decision makers. CEOs, political leaders, and individuals achieve results suited to their needs more quickly, less expensively, and more accurately when they have access to high quality geospatial tools and data. An added side benefit is that all of this comes with increased clarity of communications. However, in doing so, these technologies also raise the bar. They raise the standard by which such information products are judged. They set the expectation of the information tools that should be provided to get a job done or make a decision. Perhaps most importantly, they set the standard which must be met to compete with the most capable competitors, be they individuals, businesses, or nations.

> ***CEOs, political leaders, and individuals achieve results suited to their needs more quickly, less expensively, and more accurately when they have access to high quality geospatial tools and data. Geospatial technologies have become essential for survival in an increasingly competitive world.***

To modify an old saw, geospatial knowledge is power. In providing the GIS-based capabilities to better generate information more quickly, more accurately, and at lower cost, GIS has become essential for survival in an increasingly competitive world. Anyone not using geospatial information with the power of a modern GIS offers their competitors an opportunity for advantage that sooner or later will be recognized. And in a global economy, with information tools widespread, competitors may arise unexpectedly.

3.2.6. Geospatial Data Infrastructures

3.2.6.1. What are they?

Introduction

Geographic information system technology provides the means to analyze diverse geospatial data sets together to create a wide range of valuable information. However the technology is dependent on the availability of quality data, suitably formatted for GIS use. As GIS technology was adopted by governments, it became clear that, in order to make use of the vast geospatial data holdings spread throughout government departments, it was necessary to have some means of accessing the data over a network, and it was important to institute suitable standards so that data from different sources could be used together.

Geospatial data infrastructures were developed to fulfill this dual need. The concept was then extended to include data sets held by other levels of government, the private sector, international organizations, and even other countries. Substantial government funding was provided for these initiatives because without a geospatial data infrastructure it was difficult, if not impossible, to make use of the geospatial data already collected and being maintained at great expense to the sponsoring government.

A geospatial data infrastructure (also termed a spatial data infrastructure or simply a GSDI) is a computer network-based framework designed to enable digital geospatial data residing in multiple databases in multiple locations to be retrieved and used together in a timely and cost-effective manner. Thus, a geospatial data infrastructure consists of the policies and standards, technology, and human resources needed to enable geographic data residing in multiple locations to be remotely accessed and used together for multiple applications. It is in many ways analogous to a transportation infrastructure.

A transportation infrastructure may consist of a network of roads, railways, waterways, and airports. Although the individual components of this transportation infrastructure are built and maintained by a variety of public and private sector organizations, within any given jurisdiction there are usually well-defined standards for how these facilities are to be built, the signage and safety features to be provided, and a host of other specifications to be met. There is information provided about the components of the infrastructure, such as the number and size of runways at an airport, and standard procedures for making use of the facilities e.g. cars have to be licensed and meet certain safety standards. An organized transportation infrastructure enables a wide range of users to make use of the facilities in diverse ways, from driving to work to running a shipping service using trucks, aircraft, and ships. And to do so safely.

A geospatial data infrastructure (also termed a spatial data infrastructure or simply a GSDI) is a computer network-based framework designed to enable digital geospatial data residing in multiple databases in multiple locations to be retrieved and used together in a timely and cost-effective manner. A GSDI consists of the policies and standards, technology, and human resources needed to enable geographic data residing in multiple locations to be remotely accessed and used together for multiple applications. It is in many ways analogous to a transportation infrastructure.

In a similar manner, a geospatial data infrastructure provides a means for the orderly and effective use of diverse spatial data sets, managed (and maintained) by different organizations and housed in locations where they can be connected to a computer network.

A geospatial data infrastructure is implemented by some form of coordinating body, usually government funded. This body is charged with defining policies and standards for the quality of the data and the way data characteristics (termed metadata) are reported, and ensuring that there are suitable means (usually online access over the Internet or an internal computer network) to search for and retrieve geographic information. The coordinating body also typically plays a major role in encouraging cooperation among data providers and the promotion of data use within the public and private sectors.

In so doing, this infrastructure offers the means to reduce the duplication of effort among data producers, reduce costs of accessing and using geographic data, and increase the benefits derived from the data, by making it more widely available within an organization and also to outside users – be they other organizations or the public.

Typical components of a GSDI are: (1) framework base layers, (2) standards, data and metadata specifications, (3) some form of data clearinghouse to facilitate searching, viewing, and obtaining data, and (4) GSDI support services. These services promote the addition of new data sources, assist users and data providers, and offer data search and viewing tools such as online search engines and data acquisition portals. Each of these components is explained in more detail below.

Framework Base Layers

Framework base layers are data sets used as geographic reference data to which all other geographic data sets are to be registered so that the data can be used together.

> ***What is considered framework data differs among jurisdictions, so the process of defining framework data sets requires consultation among the various stakeholders producing and using the data.***

A useful categorization of data layers commonly used for this purpose is: alignment layers, land feature/land form layers, and conceptual layers:[131]

- Alignment layers provide geometric controls that are used to ensure the position accuracy of the geospatial information in other data layers. A geodetic control point layer, for example, will contain accurate position data obtained for a set of locations using survey methods.

- Land Feature/Form layers include natural or man-made physical features that are easily observed or measured – features typically found on topographic maps such as transportation features (e.g. roads, railways, airports), hydrography (surface water networks), and land elevation. Air photo or satellite imagery layers may also be included for use as background data layers over which to display other data sets

or from which information can be directly extracted. These data layers are useful for many applications, as well as being useful references to register other data sets.
- Conceptual layers represent the organizational framework used to describe and administer land. They include geographic feature names, administrative boundaries such as political boundaries (national, provincial, state, municipal), electoral districts, and land ownership information (termed cadastral data). Conceptual framework layers may also include interpreted information such as ecological zones derived from vegetation and climate data, or policy-related regions such as prioritized economic development zones.

What is considered framework data differs among jurisdictions, so the process of defining framework data sets requires consultation among the various stakeholders producing and using the data. In our work we have found that this discussion is typically a positive and useful activity for all concerned. It usually develops a better understanding of user needs from the point of view of the data providers, as well as a better understanding by users of the challenges of providing the data.[132]

Standards, Data and Metadata Specifications

We are now about to discuss another topic that usually glazes over the eyes of even the most seasoned geospatial professional – standards and metadata. But we ask that you bear with us for a few minutes to consider these important topics.

Geospatial data infrastructure (GSDI) initiatives are generally implemented by governments. To ensure that data produced by different agencies can be used together and produce reliable results, minimum standards are typically developed in consultation with the producers and users of the data. As well, in order to be able to search and access data sets, there must be a standardized form of reporting data characteristics.

Metadata is information that describes the characteristics of a dataset. As part of a GSDI, metadata format and required entries (i.e. data descriptors) are specified so that the data search engine can catalogue each data set. Characteristics

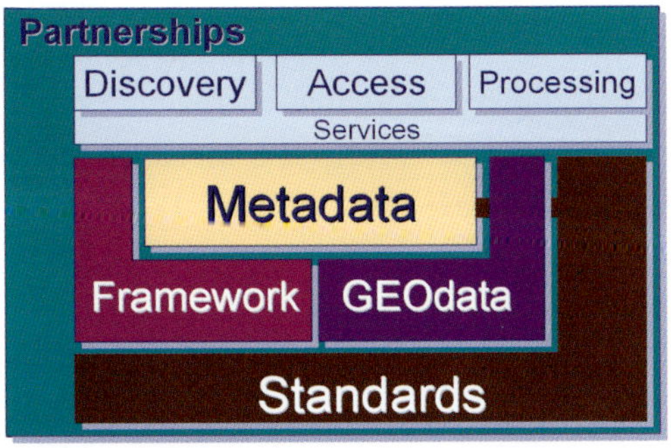

Geospatial Data Infrastructure Conceptual Diagram. A geospatial data infrastructure is an administrative and technological structure that provides widely used baseline information (Framework) and other more specialized data sets (GEOdata) in a format and with a quality level (Standards) that can be easily used by diverse users. Information about the data sets (Metadata) is used by applications (Services) to search for data (Discovery), retrieve relevant information (Access), and generate useful information products (Processing). These functions are provided by diverse organizations that develop formal working relationships (Partnerships) designed to facilitate the flow of information among data users and contributors. Source: D. Nebert 2009

such as the geographic area coverage of the data, the date the data were collected, what the data describe, information about the accuracy of the data, the data source, and method of production are typically included in the metadata file.

Data Clearinghouse for Search and Retrieval

To access geospatial information, a GSDI supports some form of clearinghouse – generally implemented as an online web portal – which is a website providing a single point of access to search for and obtain (via links to the providers) data from any database registered with the GSDI. Commonly, the search tools incorporate some form of online mapping functions to display data. As online mapping capabilities continue to develop, it is becoming ever more practical to not only access data remotely but also to perform geospatial analyses remotely by using online mapping tools, without having to download and store the data or install the software locally. Only the results are downloaded.

Geospatial Data Infrastructure (GSDI) Support Services

A GSDI typically exists as a community of distributed data providers. They provide information to the GSDI that describes their map and data resources. The data providers are typically responsible for the collection, maintenance, and distribution of their own data. They also develop metadata that both describes the data sets in their collection and conforms to the standards of the GSDI. The metadata typically documents the data quality, characteristics, and terms of access.

Organizations that host data accessible via the GSDI typically publicize the availability of their data and provide support to users. In addition, there is a technical and administrative structure responsible for coordinating the activities of a GSDI. This institutional structure takes leadership responsibility for the creation of working groups with data users and providers to identify and address issues in the operation and development of the GSDI.

For example, in Canada, the Canadian Geospatial Data Infrastructure (CGDI) is delivered by the GeoConnections program which promoted the collaboration of data providers throughout the country and developed an online portal, the GeoConnections Discovery Portal (http://geodiscover.cgdi.ca/), from which one can search and access geospatial data sets available in Canada. It also supports the national framework data layers commonly used as a mapping base, to ensure that new mapping efforts will register correctly with the existing data sets. Commonly used framework data layers include road networks, elevation data, administrative boundaries, geographical place names, and satellite imagery. As well, for data sets that contributors have permitted free access, GeoConnections provides repository and distribution services at http://geogratis.cgdi.gc.ca. The National Atlas of Canada, topographic maps, satellite imagery, and framework data layers are some of the national data sets made available free of charge from the GeoConnections Portal. This is done as a way of promoting the use of geospatial data and encouraging the development of, and access, to new data sets throughout the country. CGDI has implemented standards, developed with Canadian and international partners, that are sanctioned by international standards organizations.

In the United States, the Federal Geographic Data Committee (FGDC) established in 1990 was formed as an interagency committee to promote the coordinated development, use, sharing, and dissemination of geospatial data on a national basis. Its activities are administered through the FGDC Secretariat, hosted by the National Geospatial Programs Office (NGPO) of the U.S. Geological Survey. The NGPO also oversees the Geospatial One-Stop site (http://gos2.geodata.gov/wps/portal/gos) – a search facility to locate geospatial data sets. From this portal, metadata for the data sets accessible through the National Spatial Data Infrastructure (NSDI) Clearinghouse Network can be searched and accessed.

Online Search for Geospatial Data. The Geospatial One-Stop web site operated by the U.S. Geological Survey allows users to find data sets that satisfy such criteria as area of interest, time of data collection, and subject. Source: USGS

The Europeans have developed the Inspire GSDI initiative, while many other jurisdictions (from Bermuda to Thailand – and many places in between) are developing GSDI's tailored to their own needs.

In our work on behalf of our clients across a range of geospatial fields, we have found that another key ingredient in a GSDI, or any other technology implementation that crosses the usual departmental mandates, is the presence of a high level advocate – often a minister, prime minister or president. While a law mandating the development and use of an GSDI is perhaps best, an edict such as the Presidential Directive issued by President Clinton in the USA on their GSDI can also be effective.

3.2.6.2. What is it used for?

Virtually all industrialized nations have some form of national level geospatial data infrastructure and many also have local initiatives at the state or provincial level. Useful links to the websites of these GSDIs can be found at the website of the Global Spatial Data Infrastructure Association at http://www.gsdi.org/SDILinks.asp. These GSDI initiatives promote the development, sharing, and distribution of geospatial data nationally and often internationally as well. They provide online access to search engines to acquire datasets available from the original source, for sale or at no charge. These sites are valuable as a single point of access for a wide range of datasets that are immediately accessible and downloadable.

Geospatial data infrastructures are used by government, the private sector, and individuals as a source from which to obtain data needed for their diverse needs. In making the process of finding, comparing, and accessing data more efficient, GSDIs promote the development of geospatial applications, make their use cost-effective, and introduce effective competition among data suppliers. As a consequence, GSDIs can improve the planning and utilization of resources for the benefit of the society's economy, the competitiveness of its industry, the provision of government services, and the overall quality of life.

3.2.6.3. What are the implications?

Introduction

The implications of Geo-Spatial Data Infrastructures or GSDIs are important to the understanding of the GeoEconomy. Indeed, we would argue that GSDIs are becoming the engines of the GeoEconomy.

The Implications of GSDIs

1. *There is relatively easier access to geospatial data (where GSDIs exist and are not constrained by restrictive data policies);*
2. *Collaboration and sharing of geospatial and geo-referenced data is promoted between governments, industry and individuals;*
3. *Because of better collaboration and sharing of data there will be better results and better services – especially of intra and inter-governmental activities;*
4. *The existence of better services tends to increase the expectations within similar jurisdictions of having similar levels of services; and*
5. *Those without a GSDI will find themselves less competitive in the GeoEconomy.*

Geospatial data infrastructures are becoming the engines of the GeoEconomy.

Easier Access

In much of the world, access to geospatial data has become fast, easy, and relatively inexpensive. Supplied online, 24/7, users do not have to acquire and maintain large geospatial datasets. At one time, gaining access to standardized data and the cost of doing so were significant obstacles to the development of geospatial applications. Today the development of GSDIs has often provided a solution to four of the issues that have limited the use of geospatial data:

1. Data access;
2. Standardized quality reporting;
3. Distribution; and
4. Cost.

Increased Collaboration and Sharing of Data

By their nature (and virtually by definition) GSDIs promote collaboration between various agencies that put their information into the data bases and allow that data to be shared on some (usually) equitable basis. In a further element of collaboration, GSDIs usually allow data users to identify sources of expertise that can assist in the selection of suitable data, and support the development of applications that generate new, unique and valuable information products.

Better Services

We have already explained the many ways that increased geospatial data use can lead to the provision of better services – whether by governments at all levels or by industry. And a GSDI is the most effective means to deliver the broadest range of high quality geospatial data to the greatest number of potential users. When combined with the collaborative environment that the establishment of a GSDI demands, improved services are almost sure to result.

Increased Expectations

Standard geo-information products, such as framework data, are valuable, and numerous government agencies are tasked with generating these data sets and promoting their use. Not only is this an opportunity for those with the skills and technology to benefit, it also sets a baseline expectation, pushing the adoption of the technology and use of its outputs. In this vein, one can consider the introduction of freely available online services such as Google Maps and later Google Earth that provide search and display of map and remotely sensed imagery for regions around the globe. These services educated the public to the notion that spatial information, be it the location of the nearest Pizza Outlet or driving directions from their location to a destination, could and should be provided as a map-like display over some form of map or image background, virtually in real-time and at low or no cost.

Expectations have been raised with the coming of the GeoEconomy.

Lower Competitive Position without a GSDI

Organizations that make optimal use of the geospatial data resources and GSDI technology gain a competitive advantage over competitors less competent in geospatial applications and without access to such data. Most data used in business have a geospatial component and, in a competitive environment, making use of relevant geospatial data becomes essential to maintain or grow market share and profitability. In making data search and access more effective, GSDIs have become an important resource to maintain competitiveness in a global market. This competitiveness is not, however, just restricted to the private sector. Governments that are not engaged in developing and using GSDIs will also be less efficient and, no matter the industry sector, such governments will be providing a less healthy environment to nurture a globally competitive industry.

3.3. Understanding New Data and Making Them Valuable

This section discusses the new data that are now available to the geospatially competent user, in terms of:

- What the issues are;
- What new data can be used for; and
- What the implications are of using this new data.

3.3.1. What is the issue?

3.3.1.1. Introduction

The availability of new data is one of the forces driving the increasing opportunities in the GeoEconomy.

The availability of new data is one of the forces driving the increasing opportunities in the GeoEconomy.

This phenomenon is not new. As previously discussed, during the 17th and 18th centuries the development of new data in the form of more accurate and more complete navigational charts enabled those countries and entrepreneurs who possessed these new data, and also had the skills and technology to make use of them, to explore the globe, conquer new lands, and develop major new economic resources. But what was important then, and is if anything more important now, is how geo-data are understood and, as a consequence, made valuable.

As explained in Section 3.2.4 on Mapping on Demand, a number of American companies have gained advantage from their knowledge and use of geo-information. This is, in effect, representative of a new value proposition – geo-information is being seen as having far more value than even just a few short years ago. Added to today's new value proposition has been the continuing development of new geospatial data sources and tools. In the past, as now, even "old data" were made to seem new by the application of new technology and new ways of understanding the data. Today virtually all data collection, be it geo-referenced or not, is collected in digital form and is often stored in digital data bases. That has allowed data to move from being a "thing" to being part of a process. Data are no longer just a snapshot of reality at one time. They can be updated, placed in a data base, and accessed instantaneously. Furthermore, users can access only those data they need – not a pre-selected data set – as is the case with a topographic map whose information content and coverage has been pre-determined. Of course, with the freedom to access data differently comes the freedom to make mistakes.

The whole question of data access is further complicated by the many new data sources now available. These new sources include higher resolution remotely sensed imagery, polarimetric SAR data, LiDAR, and more accurate (and real-time) GPS data, discussed in previous sections. These are but a few of the new geospatial data sources becoming available.

However the mere collection and storage of the data does not in itself provide tangible benefits. It only provides potential. For the benefits to be realized a number of other factors must also be present. For example, the availability of digital satellite imagery is of little benefit unless it is geometrically corrected to provide position or location accuracies sufficient for the intended applications. Furthermore, the fact that the imagery exists is of no benefit if the particular imagery required cannot be found, retrieved, and interpreted at an acceptable cost and within a suitable time frame. Access to the data at an acceptable cost and in a suitable time frame is dictated by a combination of the technology associated with the satellite system and by data policy which controls who gets access to data, when, under what conditions, and at what cost. Data policy is considered in more detail in Section 3.4.2. For emergency response to a natural disaster such as flooding, satellite imagery is of tremendous value – but only if it is available within hours of the event and can be updated regularly as mitigation and recovery measures proceed. In that case, the imagery can be used to plan the deployment of recovery personnel and equipment to focus the greatest effort in areas of greatest need. However, if the imagery only arrives days or weeks after the event, its value is greatly diminished. It cannot be used to guide operations and may only be of use to assess damage and assist in distributing financial assistance.

The factors that affect the opportunities that new data provides in the GeoEconomy can be grouped into the following categories: content, maintenance, and access.

3.3.1.2. Data Content

Content refers to what the data represent. This is the aspect that commonly receives all of the focus.

Content refers to what the data represent.
This is the aspect that commonly receives all the focus.

In preceding sections we have discussed some of the major new geospatial data sources that have recently been developed and the data that they provide. These included:

- Remotely sensed data, including digital aerial photography and optical satellite imagery, produce coverage of small areas up to and including the entire globe at different scales, and some satellites do so on a daily or near-daily basis;
- Weather satellites provide continuous monitoring of the Earth;
- Radar satellites provide all-weather, day and night acquisition capabilities;
- Satellite positioning systems such as the GPS, GLONASS, and Galileo systems enable geographic positioning and provide data used for navigation, vehicle tracking, and related data collection;
- LiDAR systems and interferometric radar provide data used for rapid terrain elevation mapping, and bathymetric systems, both sonar and LiDAR based, are used to map the sea floor topography and characteristics.

Data sets on cultural features such as transportation networks, the urban framework of buildings, parks, services, and other assets are now routinely collected and regularly updated.

Less widely known are the myriad geospatial data sets in operational use in business geographics, for more specialized applications such as demographic statistics, household income, and purchasing behaviour used for strategic business planning. All of those questions asked at your local store "could you please tell me your postal code" are collecting the necessary information for analysis of your buying habits and location.[133] Health services use data on disease occurrence (e.g. the spread of West Nile virus or H1N1) to anticipate demand on health services and plan mitigation measures. Fire services use data on street networks to optimize routing of emergency vehicles, and data on buildings, locations of hazardous materials, and location of fire response assets to increase safety, reduce response times, and improve fire suppression. Police forces use geospatial data sets on crime incidents to anticipate work-loads, develop patrol areas, and optimize allocation of personnel.

There are innumerable additional examples that could be given. Fundamentally, any data that pertains to a geographic location can be collected in a manner that makes the data useful for geospatial applications. Even commonplace information, such as a snapshot photo or an audio recording of a speech, can be tagged with the geographic location to which it refers, and can then later be retrieved as part of the information available for the area.

The principal reason why this plethora of data is now available has been the revolutionary development in technology that has made capture of data and geographic positions fast, inexpensive, and accurate. For example, municipal workers use handheld devices costing only a few hundred dollars, that are GPS equipped mobile computers with GIS functionality, to display maps and images of the city area they are assessing. They can directly enter information about the condition of city property needing repair or report that a building has passed inspection. That information is then later downloaded or sent by wireless communication directly to the municipality's information system. There it is logged, any photographs taken of the property are stored with the report, and it is directed to the tax assessment office to trigger a new building assessment, or to another group to schedule maintenance or another inspection.

The assessment of pavement conditions can be done using specialized vans that collect imagery of the roadways. While they drive, the systems can automatically geo-reference the imagery. Later the imagery is analyzed and repairs scheduled in a way that minimizes traffic disruption and optimizes equipment and personnel time.

Much of the data now available is input by the user of a service. For example, in retail outlets, information about buying patterns for the local area are analyzed from the cash register data. Retail chains can assess differences in buying habits in different areas, relate these to demographics, income, and other geospatial characteristics to help identify potential new store locations, and anticipate the product mix that would be purchased.

The ability to collect high quality geospatial data faster and at lower cost than ever before in human history, provides tremendous opportunities. However, there are other factors that determine whether that potential can be realized, notably data maintenance and data access, as outlined below.

3.3.1.3. Data Maintenance

Introduction

Data are often out of date as soon as they are obtained. For some applications, such as emergency response to a forest fire, the value of data decays rapidly – in a matter of an hour the fire line may have moved significantly and personnel in what was a safe location may be at risk and need to be moved. Other data, such as geologic mapping, may remain valuable for decades or more until additional information collected using other instruments is obtained – or new tools are developed to better use existing data.

Data are often out of date as soon as they are obtained.

The importance of maintaining data suited to its use cannot be overstated. How frequently the data are updated is only one aspect of data maintenance. Other important maintenance issues are data quality, quality assurance and standards, and reporting.

Data Quality

Data quality includes the accuracy of the attribute both with respect to its characteristics and position. If the tree diameter recorded is 20 centimetres, is it 20 centimetres plus or minus 1 centimetre or plus or minus 10 centimetres? If the crime incident reports indicate there have been 5 robbery incidents within the past week in a precinct, will it actually be 5, or because of the reporting methods used could it actually be 3 or 8 because of variation in the way robbery incidents are defined, or because of delays in processing paperwork?

Positional accuracy refers to how closely the reported geographic location matches the actual location. For some applications, a position within 10 metres of the actual location is sufficient. In other applications, such as municipal construction, a 10 metre position error could lead to impinging on a neighbouring property – and legal action.

Quality Assurance and Standards

Data maintenance and the need for quality require an effective quality assurance process to ensure that the data added to a database meet well-defined standards. Standards ensure that the data can be used with some level of confidence that the end result will not be wrong because the data are in error: as noted previously, "garbage in-garbage out." Standards also impose a planning process.

It can be expected that almost all geospatial data collected will at some point be used in a manner and for a purpose not anticipated by those responsible for its collection.

Data collection is expensive. To realize maximum benefit from limited financial and human resources there is a tremendous incentive to collect data once and use it many times. Storing multiple copies of a dataset is also costly, as it increases the updating effort and introduces the risk of two copies of the same dataset being different, because one was changed and not the other. All too often we have encountered this problem in our work. In planning for the many applications for which data may be used, the data collection effort can be optimized. This is done by anticipating the need for compatibility with different analysis systems, and by setting standards that will be suitable for as many applications of the geospatial data as possible. The setting of and adherence to standards multiplies the usability of the data, while only marginally increasing the cost.

Automated methods are used extensively to ensure the integrity and quality of datasets. They are used to search for and highlight potential problems, such as data that may be wrong because the value is outside an expected range. Wherever possible, data validity checks are included in the data collection process so that errors can be immediately recognized and corrected. The more efficient the quality assurance methods, the more reliable the data, and the more quickly data can be made available for use.

It can be expected that almost all geospatial data collected will, at some point, be used in a manner and for a purpose not anticipated by those responsible for its collection. The novel integration of data is one of the principal means of discovering new opportunities in the GeoEconomy. This can only be done in a reliable fashion (i.e. using the right data in the proper way) if there is sufficient knowledge about the characteristics of the data and confidence in its quality. Standards help to ensure this.

Reporting

To coordinate data maintenance as well as later use of the data, a rigorous reporting procedure is used. Metadata, that is data that describes the characteristics of a data set, must be maintained for every data set.

> ***Metadata, that is data that describes the characteristics of a data set, must be maintained for every data set. It is a key component of the rigorous reporting procedure needed to coordinate data maintenance.***

Metadata, as outlined in Section 3.2.6.1, include such information as what the data represents (e.g. a drainage network, a road system), when the data were collected, the geographic area to which it pertains, the responsible agency or individual, the collection standards adhered to, the position or location accuracy, and other particulars. Not only does this ensure the data collection and quality assurance processes are coordinated, the metadata also provides the means for the dataset to be efficiently searched, and then found by users unaware of its existence. The metadata is also of critical importance for the user to evaluate whether it is suitable for the intended use.

The development of rigorous methods to maintain large spatial databases has been a critical component in ensuring the availability of new data.

Volunteered Geospatial Data

Volunteered geospatial information (VGI), sometimes called crowd-sourcing, is data collected by the public and made freely available. It is almost always distributed online.

Wikimapia (www.wikimapia.org) is a site that encourages public contributors to add or update information about locations worldwide. Flickr (www.flickr.com) is a photo sharing site where users can upload photos referenced to location, and OpenStreetMap (www.openstreetmap.org) is an international volunteer effort to provide free digital map data.

VGI represents public data needs that enough people perceive not to be currently met by the traditional authoritative methods employed by government mapping agencies and private industry. Usually the needs being addressed are faster updates of specific information, such as the addition or deletion of roads or consumer services such as restaurants and stores. Sometimes the needs are for information not traditionally provided by other sources, such as a photo sharing site where pictures can be retrieved by location.[134]

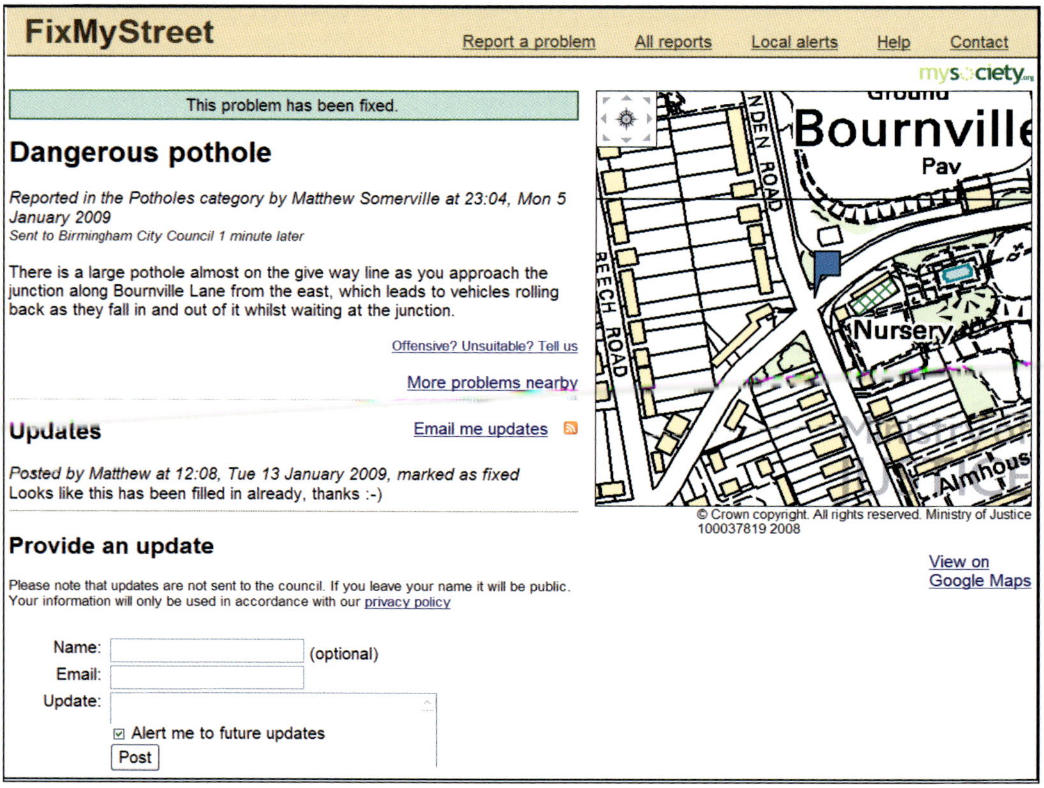

FixMyStreet. The FixMyStreet webservice enables UK citizens to submit the location of potholes and other street damage. As in this example, the submitter can choose to include their name and can also at a later date update their report to note when the repair was completed. Source: FixMyStreet.com courtesy of mySociety.org

VGI can often provide data better suited to a particular need because the uses for which the data are intended are much narrower, and the mandate of the provider is much more limited than for traditional providers such as a government agency. VGI may be provided by a volunteer group or an individual that may or may not assign themselves a formal mandate.

VGI is an important developing source of geospatial data not only among mapping enthusiasts in the public, but as a way to augment existing government and private services. The two types of applications where we see VPI to be particularly relevant are search and updating. If an object or event is easily recognized by the public and is something that large numbers of people will see in the normal course of a day, then VPI can be a much less costly and a more effective way to collect that information than paying people to search for them. For example, the non-profit volunteer organization MySociety in the UK (see www.MySociety.org) developed a webservice called FixMyStreet (www.fixmystreet.com) by which citizens in the UK could report locations where they observed road maintenance was needed. These observations were then forwarded to the responsible municipal maintenance authorities who sent inspectors to verify the observation and then scheduled repairs. The public could also update their observations, such as reporting when a problem they had filed was fixed.

Google, Inc. and other online services provide options for the public to submit map updates such as the location of new streets or abandoned ones. We expect that there will be many more innovative uses of volunteered geographic information. There are many interested people who would like to believe that they are making a contribution, there are many types of geographic information that can be usefully tracked in this way, and there is a lot of money that can be saved and potentially earned by harnessing this activity creatively. We believe that innovative applications of volunteered geospatial information will become commonplace in the evolving GeoEconomy.

3.3.1.4. Data Access

As valuable as it may be to collect and maintain high quality data, it is of limited value unless a wide range of potential users has an effective means to find out what data are available and a suitable means to access the data. Revolutionary developments in the technology of data access, and more open policies on data use, have made new data a driver in the GeoEconomy. Smart search routines use fuzzy logic and other methods to handle inexactly spelled words, search multiple databases, provide explanations to guide the user, and other capabilities that reduce the time and make it easier for less skilled users to find the information they need.

Before the advent of the World Wide Web, data storage and maintenance were highly centralized and access was usually restricted. Remote terminals could access the central database but data transfer rates were slow and the transmission of large files such as images was impractical. Generally when an application required access to a significant amount of data, that data had to reside on the workstation where the application was running.

Early incarnations of the Internet provided connectivity between remote computers, but with the slow data rates, the ability to download datasets was limited. For this reason, remote computing using the type of graphical interface now the standard for Web browsers could not be supported.

The development of Web browsers and high speed Internet access initiated a series of developments that changed the entire landscape of computing and fuelled the rise of the GeoEconomy.

One of the major benefits of using geospatial data comes from its use in integrated analyses.

One of the major benefits of using geospatial data comes from its use in integrated analyses. That is, analyses that use multiple datasets of different factors that pertain to the same geographic area to develop new information. A simple example is the use of data sets on existing roadways, soils, land ownership, vegetation, land use, and topography to plan a new roadway, housing development, or park. The concept is relatively straightforward. Each dataset is selected for its suitability for the project, the different data layers are overlaid, and using a suitable scoring method, those areas with optimal characteristics for the intended use are identified. In practice, collecting all these data sets could be a formidable task. They were likely held by several different agencies. The datasets would have to be found – if they existed. Then arrangements would have to be made to obtain them, and convert them to a standard format for use in the chosen GIS or other analysis program. Even with these many new tools, we have found that data access is still often constrained by outmoded agreements. One government agency will be unable to obtain data from a sister agency, for example, without incurring significant costs. Or another government agency will not be able to easily obtain digital copies of maps – so they will then digitize the maps once again. Lack of access to data can dramatically and needlessly increase costs and reduce the ability to compete or be efficient in the GeoEconomy.

Sharing data, even between departments within the same level of government, can require complex legal agreements to deal with issues such as liability if use of the data causes harm. In developing the Canadian Geospatial Data Infrastructure, this issue was addressed by their legal counsel who developed a set of standard data sharing agreements. These legal documents, developed in consultation with the data providers, covered most situations, making the process of data sharing much easier. In part, this is because organizations have more confidence in using the same agreement used by many similar organizations. As well, over time legal precedents will be developed as the agreements are tested in the courts.

With the development of high speed Internet access and the necessary connectivity services to manage the process, it became possible to make use of datasets housed on remote servers. It also became possible to have the application run remotely as well. With the implementation of geospatial data infrastructures (see Section 3.2.6 above) a centralized online web portal could be developed. It could then be searched to find any geospatial dataset registered with the system, wherever that dataset may reside, and to use the data – either downloading the data required or using an online service to perform an analysis and download only the results.

As previously discussed in the section on Geospatial Data Infrastructures, web portals are now being used to access geospatial information. Examples include the Government of Canada's GeoConnections site GeoConnections Discovery Portal (http://geodiscover.cgdi.ca/) and in the United States the Geospatial One-Stop site (http://gos2.geodata.gov/wps/portal/gos).

3.3.2. What is it used for?

A listing of new geospatial data sets and their uses would itself fill many volumes. While many of them may seem mundane, such as the location of fire hydrants, they are still important. Other data can be quite intriguing, such as interactive 3D models of a city that you can drive through "virtually" using your PC and joy-stick.

We can broadly categorize the new geospatial data by functional areas. Much of the data are what might be called "traditional," such as maps of geographic features traditionally available as paper-based topographic maps. What is new is that they are now in digital form, available online, and often at no charge for the base data commonly produced by government mapping agencies.

With the data in digital form the typical cost to acquire them (at least in progressive jurisdictions) has dramatically changed, as has how quickly they can be obtained. Since the data are now also compatible with standard geospatial analysis tools, the ways that they can be used has also changed. For example, traditionally it would be necessary to find the maps needed for a project, order them, wait for them to arrive, and then use manual drawing methods to add with other information, such as outlines of prospective route locations or new park boundaries.

Today, with the topographic data in digital form, the data can be retrieved online 24/7. There is no need to order and store paper maps.

An added benefit is that the data in the map coverage can be selectively displayed. If all you need is the drainage network and cultural features, you can simply display those data layers. Proposed routes or park boundaries can be drawn as an overlay on top of the topographic data and easily changed as new factors are incorporated into the analysis. If data on soils or ownership is needed, it can be added to the analysis as another digital overlay. Calculations based on the soils data can be incorporated as constraints. The analysis can proceed progressively towards a more and more refined result, adding additional factors as they are deemed important. The entire planning process can be a relatively rapid, collaborative one with the input of different experts incorporated into the analysis.

Fundamentally the data content is not new, it is the fact that they are in digital form, compatible with the analysis tools being used, and can be rapidly tailored to the application, that completely changes the utility of the data. In that sense the data are new, and vastly more useful.

Today geospatial data are used in an ever growing range of endeavours and in ever more sophisticated ways as a result of the new format – digital, new levels of standardization, new access methods, and new technologies available. Following are some of the major categories of data organized by broad subject areas and applications.

Tracking

The widespread adoption of GPS technology has made it practical to track the locations of all manner of individual vehicles (or people, or objects – like cell phones) in real-time or near-real time. For example, a truck fleet operator using GPS equipped vehicles and wireless communication can track, in real-time, the whereabouts of every vehicle in the fleet. In addition, data on the average and maximum speed the vehicle has travelled, how much time and when it has been stationary (to ensure drivers take the mandated number of rest stops), and the route it has travelled can all be recorded and analyzed. This capability allows fleets to be managed much more efficiently. Rather than waiting for a driver to call in and state their whereabouts, the central office can track each vehicle's progress, alter routes to avoid road hazards or bad weather, and gain the flexibility to add new pickups or delete ones no longer required.

Monitoring vehicle speeds can help to optimize fuel consumption (trucks run most efficiently within a certain range of speeds). Overall, the new data on vehicle tracking and monitoring can make any trucking or fleet operation more efficient. This can improve profitability and competitiveness.

Similar tracking technology has been implemented for a wide range of vehicle fleets including ambulance, police, and fire response services, snow removal, trash pick-up, emergency response operations, and courier services. As noted previously, a popular consumer application is the OnStar service offered by General Motors. OnStar equipped vehicles have GPS tracking, detect the status of the vehicle, and provide two way communications. If the vehicle is in an accident, the system detects the incident and can dispatch emergency services with or without driver contact. Other services, such as step by step navigation assistance, can also be provided.

Tracking has of course affected the competitive landscape as well. Those operations that choose not to incorporate the new technology find themselves at a competitive disadvantage, unable to achieve the efficiencies this technology can provide. They risk losing business, reduced profitability, and potential business failure.

Population Characteristics

Governments collect and distribute a wide range of statistical data from population demographics to income levels, cultural origins, education levels, domicile characteristics, family sizes, school selection, immigration and out-migration rates and other data. This information is typically geo-referenced – not to an individual household (to protect privacy) but to relatively small reporting areas. This information is also used to plan future government services, such as health services that will be needed as the population mix changes over time. Much of this information is also valuable to businesses offering products and services. For example, retail outlets targeting the youth segment of the population can assess which areas will have the demographic profile most suited to their enterprise and thus target the most likely areas for expansion.

Private sector firms have developed substantial data collection and preparation services. They supply information on the buying habits of different communities within urban areas with detailed information on the mix of products purchased and their value, as well as information about their employment, income levels, and interests. Businesses use these data for trade area analysis, the planning of store openings and closings, changes to the mix of products and services they offer, the location of distribution facilities, and projections of product sales growth. Such business geographic data has become an important business resource. The data are all geo-referenced and can be easily combined and analyzed using a number of off-the shelf analysis tools specifically designed for these applications, by companies such as Pitney-Bowes that provides the MapInfo product.[135]

Mapping Products

Government mapping agencies have made extensive use of geospatial technologies to fulfill their mandated roles of providing base mapping products for their areas of responsibility. These, of course, are then used by other government agencies, industry, and private citizens. Depending on the jurisdiction, these data may include small scale, country-wide, mapping of physical and cultural features. Geospatial technologies have made these operations more efficient with more rapid update cycles, so that the information can be more current and often more accurate than what was traditionally provided.

Distributing the data online has become a standard service (either at no cost or fee-based). Online distribution offers immediate access and substantial cost savings. It reduces the quantity of physical map products that need to be produced and distributed, and less material is discarded when maps are updated and the old maps are disposed. As well, the online distribution process is largely automated, reducing labour costs.

Online mapping services are commonly provided to allow users to not only download data for their own use, but also to allow them to display and download map products they have customized by selecting only the themes relevant to their application. The National Atlas of Canada (http://atlas.nrcan.gc.ca/sites/english/maps/topo/map) and the United States Government Geodata site (http://gos2.geodata) are two such mapping agencies providing these types of services.

A wide variety of data sets are available from the United States Geological Survey (USGS) including themes such as: administrative boundaries, elevation (height of the land), geology, inland water, oceans, imagery and base-maps, transportation and utilities, agriculture, demographics, etc. Many state and provincial mapping agencies offer similar services.

Status Reporting

Some municipal governments are making use of the new online geospatial data tools to keep citizens informed about the condition of their respective cities. Like many other cities, the City of Minneapolis' website provides information on business services, health and safety, transportation and parking, garbage collection and recycling, and youth services. Some of the information is disseminated in the form of maps. For example, on the City of Minneapolis site maps show crime incidents occurring in the city. Users can construct crime maps of their choice, selecting the city area of interest, the time range, and the types of crimes (see http://www.ci.minneapolis.mn.us/police/crime-statistics/).

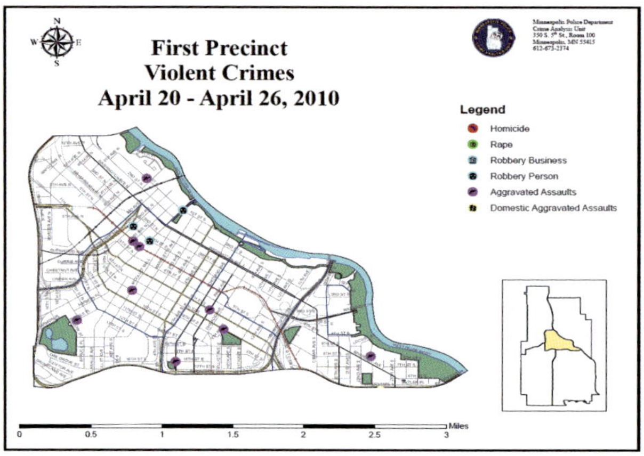

City of Minneapolis Crime Report. Source: City of Minneapolis

Business Support

Many private companies collect extensive geospatial data to support their operations. Power utilities, pipeline companies, forestry, and other resource extraction industries maintain databases of their road networks and other geographically distributed facilities. Data sharing has been another way that non-competitive organizations have leveraged the value of their data investment, to gain access to additional new data sources without having to spend additional resources to acquire them.

As noted previously, there can be legal concerns when data is shared with another organization. The use of standard agreements and establishment of legal precedents is making it easier to develop the mechanisms that allow the sharing of data.

Multiple Subject Areas – Mashups

A mashup is a web application that combines multiple services into a single application (http://www.techterms.com/definition/mashup). Data are obtained from more than one source and the Mashup combines it all to create a new integrated data query application. Mashups can be created relatively quickly and can provide valuable new data and analysis

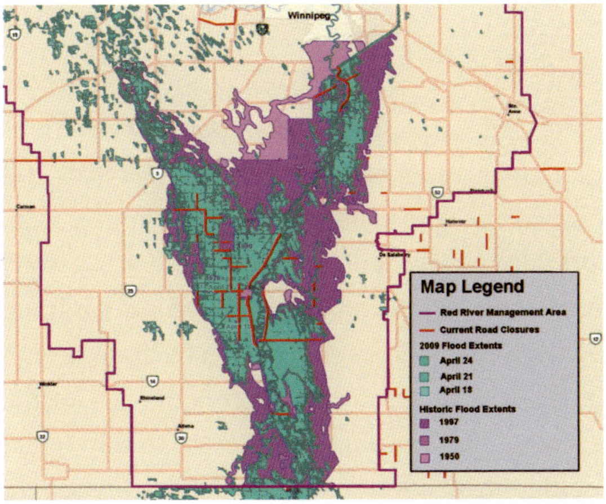

Mashups to Combine Multiple Online Data Sources. This situation awareness application for the Manitoba Office of Disaster Management displays current and historical flood extent derived from recent and historical radar satellite imagery, road closure information from Manitoba Infrastructure and Transportation, and an online base map provided by ArcGIS Online. Updates to any of these data streams are automatically displayed in this mashup application. Source: ESRI Canada, Natural Resources Canada

capabilities. A common form of mashup is to use a mapping tool such as Google Maps combined with a source that lists other data by locations. For example, a mashup to search for houses for sale or rent was created by combining Google Maps with Craigslist (http://www.housingmaps.com/). A user defines criteria for the search (city, price range, number of rooms) which is used by the Craigslist service to query its database and generate a list of houses meeting the criteria, along with their locations. The Google Maps service uses the location data and its map database to create a map displaying the house locations.

Ontario election riding information for the 2007 provincial election was provided as a searchable map-based information retrieval by means of a mashup (see http://www3.thestar.com/static/googlemaps/electionmaps.html?xml=2007_ontario.xml). Map data from Google Maps was combined in a mashup with data from the Toronto Star database to create the online resource.

Mashups are a relatively recent development. However, with the availability of so many web-based data sources, the technique is rapidly developing as a means of creating online maps of specialized geospatial information. Again, caution must be exercised when using information whose pedigree is not well known. The creator must understand the characteristics of the data being combined and the purposes for which the results might be used. The ease with which mashups can be created also makes it easy to produce information which is not sufficiently accurate, is misleading, or is simply wrong. It begs the question: "Who do you trust?" We depend on experts to review all kinds of products we as individuals do not feel sufficiently knowledgeable to judge. We call on house inspectors and read expert reviews on cars when choosing which house or car to buy. So too the public needs to be aware of when they need the opinion of an expert about the geospatial information they are going to use.

Caution must be exercised when using information whose pedigree is not well known.

3.3.3. What are the implications?

3.3.3.1. Raising the Bar

When geographic information system technology was first introduced organizations debated whether adopting a GIS would be cost effective or not. Government services and private sector planning could still be accomplished using existing methods without a GIS. However, once the technology became more prevalent, expectations increased. For example, in submitting applications for resource projects, such as a forest harvest plan or a mining operation, the quantity of information, and the degree of analysis that had to be done to gain environmental approvals by regulatory bodies, made it impractical to compete without the use of geospatial data and (primarily) GIS technology. The minimum standard (the bar) had been raised such that the use of the new technology was no longer optional, it was required. Organizations came to realize that they no longer had the option of whether or not to use geospatial data and associated technology – it was only a question of what would be the best strategy to adopt it. And the bar is continually being raised through introduction of new technologies, friendlier user interfaces, and broader understanding of what is possible by both the public at large and those politicians and public sector employees serving them.

> ***Organizations came to realize that they no longer had the option of whether or not to use geospatial data and related technology – it was only a question of what would be the best strategy to adopt it.***

With the availability of new geospatial data products, there is now the awareness and expectation in government, industry, and the public that these new data sources and the accompanying analysis methods will be used. It is now expected that any reasonably competent job of monitoring, planning, or regulation of activities with a geographic component, whether it be managing a municipality, planning business expansion, or monitoring environmental quality for example, would be carried out using the ever growing geospatial information and technology available.

3.3.3.2. Competitive Pressure

With the easy accessibility of geospatial data delivered online and the wide availability of tools to analyze the data both as commercial products as well as through online services – many of them at no cost – there is less of a barrier for competitors. In the resource sector, competitors can more easily collect information and assess potential new projects to challenge established players.

They can find new opportunities that others had missed by making better use of geospatial information – for example recognizing mineral potential in areas that others had missed, or finding alternative pipeline routes that are less costly or could better satisfy regulations, or discovering a lucrative retailing opportunity by recognizing changing demographics and spending patterns. For the private sector, today's business climate increasingly requires that companies use geospatial information to maintain their competitive position, by improving both the effectiveness of their operations and marketing.

> ***For the private sector, today's business climate increasingly requires that companies use geospatial information to maintain their competitive position, by improving the effectiveness of their operations, marketing and delivery of service.***

3.3.3.3. Accelerating Development of New Ideas

Competitive abilities can be turned to collaborative successes if there is the vision to approach projects in new ways and make use of the geospatial data sources and skills that can now be accessed from anywhere in the world.

> ***Competitive abilities can be turned to collaborative successes if there is the vision to approach projects in new ways and make use of the geospatial data sources and skills that can now be accessed from anywhere in the world.***

One innovative mining exploration company turned this intense competition to advantage. Goldcorp Inc. had made a significant gold discovery at their Red Lake Mine. Their exploration efforts over a three year period determined that the deposit was large enough to be economically mined. Two related questions arose:

1. How big was the deposit? and
2. How long and how costly would it be to find out?

Rob McEwen, Chairman and CEO of Goldcorp Inc., wanted to speed up discovery and improve the odds of success by tapping into the global pool of mining expertise. His approach was unprecedented in the worldwide mining community. Breaking with the mining industry's tradition of confidentiality and secrecy about reserves and exploration data, McEwen made use of the global access to expertise offered by the Internet to attract the best ideas available globally – essentially a global brainstorming session. Goldcorp offered a total of US$500,000 in prizes for the best ideas to determine where they might find the next 6 million ounces of gold at the Red Lake mine – regardless of whether gold was found.[136] The Goldcorp Challenge was issued online and attracted 475,000 hits and 1400 entries from individuals, corporations, universities, and government geological organizations from 50 countries. The exploration proposals identified 110 drilling targets – half of which were new prospects. Eighty percent have identified significant deposits. The mine is now producing 10 times the previous amount for one sixth the effort.

3.3.3.4. Greater Public Scrutiny

New data, such as satellite imagery, is also widely available to the public. Access to this data has made public scrutiny of activities much more transparent.

New data, such as satellite imagery, is also widely available to the public. Access to this data has made public scrutiny of business and government activities much more transparent.

As noted earlier, one of the most dramatic early examples of civilian satellite imagery providing public scrutiny was its use in reporting the explosion at the Chernobyl nuclear reactor in the Soviet Union. The explosion released a radioactive plume that deposited material over areas of Europe and Scandinavia. The Soviet Union refused to release any information about the event. However, imagery from the civilian Landsat and SPOT satellites was widely published in the news media, providing confirmation of the event.

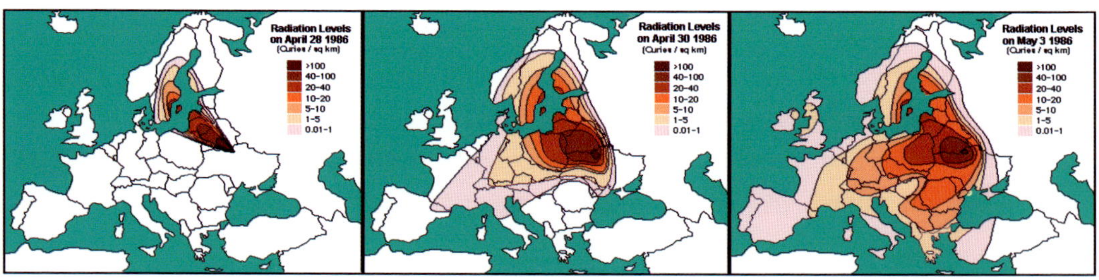

Chernobyl Fallout. Data from diverse sensors tracked the radioactive cloud from the Chernobyl explosion as it spread. The maps for sequential dates powerfully communicate the extent of the problem. Source: David Hulse, University of Oregon

Information about activities that affect land use, for example, was more easily hidden from public view when there weren't public online sources that could be used to scrutinize it. Public interest groups have become skilled at finding this information and combining data sources to build their own analyses and to build cogent arguments supported by quality data, much of it government provided. They have used this power to delay or cancel projects that run counter to their vision of sustainable resource development, loss of rainforests, climate change, and urban redevelopment. In empowering special interest groups, governments and industry have been challenged to provide more carefully considered policy and project justifications. While this may increase costs and cause delays, it can also improve decision-making.

3.4. New Policies

3.4.1. Introduction: Policy is Important

All economies are influenced to some degree or other by government policy – what governments decide to do to try to achieve their objectives. Obviously, governments have a profound impact on their economies – from setting monetary policy, to land tenure policies, and on to environmental regulation and policies dealing with education, health, and food security.

> ***In the GeoEconomy, government policy related to geospatial information has become even more important, as more and more of the information required by governments is tied to location and is, by definition, geospatial.***

Governments also have influence by how and what they invest in infrastructure that will support the economic activities of their people. The importance of governments in geospatial can be seen from historical times when governments did the mapping – and used much of it. That importance is, if anything, greater in today's GeoEconomy. Today, while there are many more players, and while governments do not produce or use all of the data, governments do control who can access what data and they often control what data are produced and how they are used (or not used) to manage the economy.

> ***Those that "get it" are ensuring that their governments can assemble the required information to make better decisions, be more efficient, and be more competitive.***

It is equally obvious that government policy is a complex business requiring vast amounts of information. In the GeoEconomy, government policy related to geospatial information has become even more important, as more and more of the information required by governments is tied to location and is, by definition, geospatial. Those that "get it" are ensuring that their governments are able to assemble the required information to make better decisions, be more efficient, and increase their ability to compete.

But the playing field in the GeoEconomy is not level. While the creation of maps and development of associated geospatial data have long been considered activities for all levels of government, how that data has been created and distributed has varied widely with some intended, and some unintended, results. Government policy associated with geospatial data has become a key factor in determining the level of use of data in almost every jurisdiction in the world. And if geospatial data are not used, participation in the GeoEconomy is limited.

Clearly the decisions about the quantity and quality of resources to invest in developing the geospatial infrastructure depend on the actual and perceived value of the results of its use. Simply put, geospatial capability requires good data, good analysis, good communication, and timely delivery to the right decision makers. But where there is limited understanding of what the data can bring, there tends to be an underinvestment of resources.

There is also an interesting observation we have made about the role of democracy in the development of geospatial information and participation in the GeoEconomy. In general democracies tend to have a better developed sense of the importance of geospatial information than more authoritative regimes. In many discussions with politicians in countries as diverse as the USA, Maldives, India, Canada, and Malaysia we have found that democratically elected politicians tend to have a far better innate understanding of the value of geospatial information than many of the bureaucrats that serve them. Every politician knows where in his or her constituency the people live who vote for them, against them, or who are undecided. To get elected they have to "think geospatially." They also tend to see the value of transparency in decision-making. The more transparent the decision making the less likely the citizens will disagree – and vote them out of office. And of course wide availability of geospatial data leads to more transparency.

The more advanced a society's understanding is of the value of geospatial data, the more successfully they will devote resources where they produce the greatest benefit for the investment. Furthermore, the interplay between geospatial capability and ability to develop effective policies in the relevant areas (e.g. natural resources, sovereignty) is multiplicative. So those that are ahead today will achieve their goals faster and, if they maintain their investment, they will widen their lead on those that are behind. We need go no further than the onslaught of the American retailers in the international market (including the Canadian market) to see this competitive advantage in action. In this sense it is also similar to what we described previously concerning the European age of exploration: those with better maps could explore farther afield, making better maps and gaining more knowledge faster than their competitors, allowing them to project power further and more effectively.

But returning to the realm of geospatial policy, when people involved in the geospatial field begin to speak about policy, particularly government policy, they tend to start the discussion with the price of data – a central element of data policy. Data policy is discussed in the next subsection. But if we only considered data policy we would be ignoring a number of other policies that are at play in the GeoEconomy and which impact our need for and use of geospatial information. There are policies involving research and development, commercialization of that research, foreign aid, international relations, the environment, resource exploration, industrial development, and security to name some of those that are most impacted by geospatial – and which in turn impact geospatial. These other government policy areas are discussed in Section 3.4.3.

3.4.2. Data Policy

3.4.2.1. Introduction

While we believe that data policy is important, most non-specialists (and even some specialists) cannot fathom why many of us in the field spend so much time discussing the subject. We wondered how best to present a section on data policy so it would make compelling reading. In the end we decided to try to use clear language and some real examples – to show that what is often thought a "boring" topic is much more compelling when its importance is more clearly understood.

At the beginning of this chapter we said that geospatial data policy is an area too often ignored by politicians, bureaucrats and even geospatial experts themselves. This clear statement is inflammatory to some – and yet not strong enough for others! The balance of this section will show why geospatial data policy is too important to be left to bureaucrats whose data policy objectives may compromise or even block the most beneficial uses of that costly and valuable asset – the data.[137]

Geospatial data policy is, to us, an exciting and interesting topic because a government's policy on how geospatial data are accessed dictates the degree to which citizens, companies and governments can participate in the GeoEconomy. It can also relate to security and safety issues, as well as to things such as industrial competitiveness and government efficiency which are discussed further in Chapter 4. We have written many hundreds of pages on data policy and its many nuances for clients around the world who wish to ensure that they enter the GeoEconomy without the problems caused by poor policy. What follows is a summary of what we have learned in doing this work.

Geospatial data policy is an area too often ignored by politicians, bureaucrats, and even geospatial experts themselves.

While a general consensus has emerged, there are still those fighting a rear-guard action against what we think has been progress – the opening up of access to geospatial information to the broader public. Here we must say thank goodness for Google Earth and Microsoft's Virtual Earth – they have forced the issue on data policy, by widely distributing geospatial information at no apparent cost to the consumer. Things will never again be the same.

Thank goodness for Google Earth and Microsoft's Virtual Earth – they have forced the issue on data policy by widely distributing geospatial information at no apparent cost to the consumer. Things will never again be the same.

Even all of this (including reference to Google and Microsoft) is likely not enough to kindle much of an interest in geospatial data policy for most people. Government policy – how boring! But these policies may make the difference between your project or your business succeeding or failing, or you being able to outbid your competitor or being outbid and left in the dust. To understand the GeoEconomy, to benefit from it, even to survive in it, it is critical to understand government policies related to geospatial information.

3.4.2.2. What is a Data Policy

Simply stated, a data policy determines what data will be available to whom, when, and what they can do with it.

Simply stated, a data policy determines what data will be available to whom, when, and what they can do with it.

The policy also influences how the resources to capture, process, manage, and distribute data will be allocated. A data policy is a series of statements that guide and/or regulate:

- To what data the policy applies;
- How that data will be obtained and/or controlled;
- Who will have access to the data;
- What they will be allowed to do with the data and information products derived from it and for how long; and
- How the data will be rationed among users.

Some issues that data policy directly controls that highlight the relevance and importance of data policy to everyone

- Some governments' geospatial data policies have resulted in the tax payer paying for the same thing twice.
- In many jurisdictions so-called "cost-recovery" policies for geospatial data actually cost more to operate than the income that they bring in – and they certainly don't cover the costs.
- It has been proven that some government geospatial data policies result in lower economic activity. In one study an inverse relationship was found between the level of cost recovery and growth in the geospatial industry.[138]
- Some bureaucrats have used geospatial data policies to start what are in effect government-subsidized AND government-protected companies – some of which have subsequently been sold for millions of dollars (often to the former bureaucrat's benefit) – along with the monopoly to sell data the tax payer has already paid for.
- Do you think that it is important for a country to have on hand information about its resources and environment that is at least as good as the information possessed by its neighbours or those who would exploit its resources?
- Do you believe (as we do) that basic map information is part of a national/state/city infrastructure?

Data are always rationed in some way because there are limited resources to capture, process, manage, and distribute data. Rationing can be done in many ways. The amount of data and information products derived from analysis of the data that are distributed will depend on (or be rationed by) the price, time, regulation, awareness, technical resources, and skill of those receiving the data. While price is but one factor, it typically receives the most attention.

There are several categories of geospatial data that must be included in a data policy:

- Measurements (e.g. field data such as GPS locations of street lamps and geographic locations of buildings);
- Derived Measurements (e.g. surface water temperature derived from satellite imagery);
- Interpreted Data (e.g. flood hazard map, severity of hurricane damage, and crop condition); and

- Integrated or multi-factor products (e.g. products created from some combination of other data, and observations including data that may simply be "stacked" one on top of the other).

The more processed/analyzed the data, the more judgment is involved in its creation – and the more unique the product. Furthermore, the utility of the data and the value added to the data both tend to increase with the amount of analysis and so these products command a higher price. At the same time the potential liabilities could also increase as there is more analysis required to create information products – and those adding this value could become liable for any errors.

3.4.2.3. Consequences of Data Policy

A geospatial data policy exists within the context of the objectives or goals it is to support. But as one studies the goals and objectives, it is important to realize that data policies have consequences – some subtle, some not so subtle.

A geospatial data policy exists within a context of the objectives or goals it is to support. But as one studies the goals and objectives, it is important to realize that data policies have consequences – some subtle, some not so subtle.

Some consequences are important and some are not. Some of these are intended and many more are unintended. For example, an intended result of providing low price access to geospatial data might be to see increased use. An unintended consequence might be that there is such high demand that the budget for processing is exceeded and/or there are long wait times to obtain data. Of course, the supplier may set higher prices to result in less use and a lower drain on the supplier's resources. The unintended result of high prices might be less use and lower benefits than may otherwise have been anticipated. Higher prices can also be discriminatory to an unacceptable degree if, for example, the data are easily affordable for proponents on one side of a dispute and unaffordable to the other side. Obviously too, with data being made available over the Internet such consequences may take on a different color, as many more people could be using data for many more applications with many more unintended results.

The aim of a data policy is to lead to consequences that support the overall objectives of the jurisdiction (or company) implementing and responsible for the policy. It is also important to recognize that even the best planning won't have perfect results. Ideally, data policies are designed to influence people's behaviour to produce specific desired outcomes. But conditions change over time and so data policies must be monitored and revised both to ensure that they continue to influence people's behaviour appropriately and to ensure that the outcomes are still wanted and still relevant. To identify and correct for negative unintended results, a mechanism must be in place to monitor performance and take action to address what isn't contributing to the overall objectives/goals. Implementing a data policy is implementing an ongoing infrastructure process – it requires continuous monitoring and adjustment. In this sense it can be compared to a road infrastructure where there are continual changes and modifications to improve traffic flow, safety and the like.

In general one can cite the following as areas affected by data policy:

- Quantity of data used;
- Type of applications for which it will be used;
- Rate at which new applications will be developed;
- Distribution of usage among the private sector, government, and academe;
- Products and services that can viably be provided by the private sector;
- Adherence to standards – ease of data integration;
- Perception of fairness and adherence to accepted norms; and
- Social and economic benefits derived from the data and how those benefits will be distributed.

The following Case Studies present real-world examples of how data policy has had both positive and negative impacts.

Data Policy Case Studies

Case Study: Industrial Development: Developed Country

The issue. A multinational company decided that it wanted to locate a major facility that would employ a significant number of local people in either eastern Ontario or western Quebec. To carry out the analysis on optimum location they required digital maps of the regions of interest.

Setting the stage: At the time of this example the Government of Ontario charged a significant amount of money for custom-made digital maps drawn from their data base. The Government of Quebec, on the other hand, would provide the same type of information at little or no cost if it was seen to be beneficial to the Government and people of Quebec.

The Result: When the Province of Ontario was approached the company was told that to get the information would take many weeks and cost many thousands of dollars. On the other hand the Province of Quebec offered to provide custom made maps tailored to the specific needs of the company doing the planning. Guess where the company decided to locate? Quebec! Ontario has since changed its policies.

Case Study: Industrial Development: Developing Country

In Bankura, India the District Magistrate[139] obtained GIS outputs from a UNDP funded program that was designed to develop and use GIS at the more local level. She used this information (on the level of education of residents, availability of land, infrastructure support, etc) to generate information for potential investors in the region's industry. Investment increased. This small and simple application convinced the District Magistrate of the potential importance of the technology in a broader range of activities including land use planning, resource management, etc. This sort of simple but successful demonstration of the value of geospatial technology over a number of years led the Government of India to lessen restrictions on data access.

Case Study: Assessment of Historical Land Use

Aerial photography is available across Canada through the National Air Photo Library (NAPL)[140] and similarly, such imagery is available in many other countries. Ryerson used older imagery from NAPL to evaluate a piece of property he was considering purchasing. The building lot on a former farm property had a dramatic break in slope just behind where the house would have been constructed. A series of older photographs showed the presence of debris at the base of the slope including what appeared at one time to be chemical drums and what might have been paper bags – indicating that the farmer may have used the land as a dump for the chemical waste from his farming operation. An offer to purchase was refused when it was made conditional on cleaning up the site. Historical land use assessments are widely used throughout North America.

Case Study: Copying Maps within the Government of Canada

The Government of Canada's mapping agency, Geomatics Canada, embarked on an aggressive cost recovery policy that led to significant charges for digital topographic map data. As a result of the high cost of the data, the Government of Canada's then Ministry of Agriculture did not have the budget to buy what they needed. Instead they decided to digitize paper maps to create their own version of the topographic maps. These maps were not the same as the "standard" digital maps (with concerns about how they might overlay). That there was such a duplication of effort within the same government seems not to have troubled people at that time, but the fact is that the taxpayer was paying twice for the creation of the same data. Decisions since then have been much more rational and the base data are provided free of additional charges.

Case Study: Use of Remote Sensing Data in Thailand

The Government of Thailand made the decision to subsidize the acquisition of remotely sensed data for government use to ensure that the country could derive the full downstream public-good benefits that are associated with the data's use. This was done through the Geo-Informatics and Space Technology Development Agency (GISTDA). GISTDA has thus developed a special relationship with many data suppliers so that GISTDA could provide data on a COFUR (Cost of Furnishing a User Request) basis. In a study we did,[141] interviews were conducted with officials involved in the management and use of remote sensing data in a number of Thai government agencies responsible for forestry, parks, land use, agricultural economics, fisheries, hydro-electric power generation and distribution, narcotics control, hydrology, defense, mineral resources, and municipal government. As a result the number of applications of satellite remote sensing used in Thailand is truly extraordinary: forest inventory; forest change detection; forest fire; human impacts; crop type mapping (12 major crops); crop area & crop statistics; statistical sampling frame development; land cover mapping; flood zone mapping; landslide mapping; mapping and monitoring fishing activities (freshwater, coastal, marine); aquaculture monitoring; coastal shrimp; vessel detection and monitoring; environmental feasibility studies; security; municipal/local area tax mapping; mineral resource mapping; field work planning; land use mapping, at various scales; disaster and risk management and reduction; soil mapping; route selection and corridor planning and terrain suitability studies; base mapping (topographic); and water management.

One of the most interesting findings was that most users clearly stated that if they did not have access to low cost data they would never be using the information, and hence Thailand would not be deriving the benefits that access to the data could provide. A few million dollars invested in the data has saved many millions of dollars, as this improved access to more data leads to better decision-making. Perhaps the most interesting aspect of this analysis is that more governments do not do this sort of data distribution.

To put the data use within Thailand into a more international context, consider that over 4000 images are provided annually by GISTDA. Given the size of Thailand and the size of Canada, this rate of use would be the same as Canada's governments (federal and provincial) using 77,600 images. Even counting the use of Radarsat imagery for ice monitoring, our information is that Canada comes nowhere near this level of data use. We ascribe this in large part to satellite data's high cost in Canada.

Case Study: Use of Radarsat Satellite Data for Weather Prediction

Using Radarsat data, a small Canadian company developed an approach to quickly map waves, and hence off-shore winds, a major factor in predicting weather off the west coast of Canada. However, the high cost of data was such that the weather agency was unable to afford the data. The data came from a satellite that had largely been paid for by the Canadian tax payer – a number of whom in British Columbia, of course, have a direct interest in improved weather prediction on the west coast of the country. This does not compare very well with the Thai example above.

Case Study: Norwegian Mountain Rescue

In 2004 the Norwegian government gave distribution of electronic maps to a private firm. That firm later decided that it was not profitable to sell digital maps of remote areas and stopped providing them. The Red Cross Mountain Rescue team was using those maps to guide rescue efforts via GPS. As a result the Mountain Rescue Team no longer has updated maps for remote areas and rescue efforts are now said to be impaired.

After these examples, the question must be asked, why do more countries not do what Thailand has done? The simple response is that many do and, through the follow-on to the Earth Observation Summit called in 2003, many countries are involved in trying to share satellite and other related data. However the fact remains that the development of data policy is a complex undertaking. It usually involves multiple stakeholders with varied and often conflicting objectives. These are discussed in Section 3.4.2.4. While we have made the trade-off between data costs and future benefits seem straight-forward, these trade-offs are not so easy if the agency supplying the data has limited budgets – or myopic leadership above them, or an historical approach that has been locked into place by law. In one sense, it comes down to power – who will control who has access to which data and when.

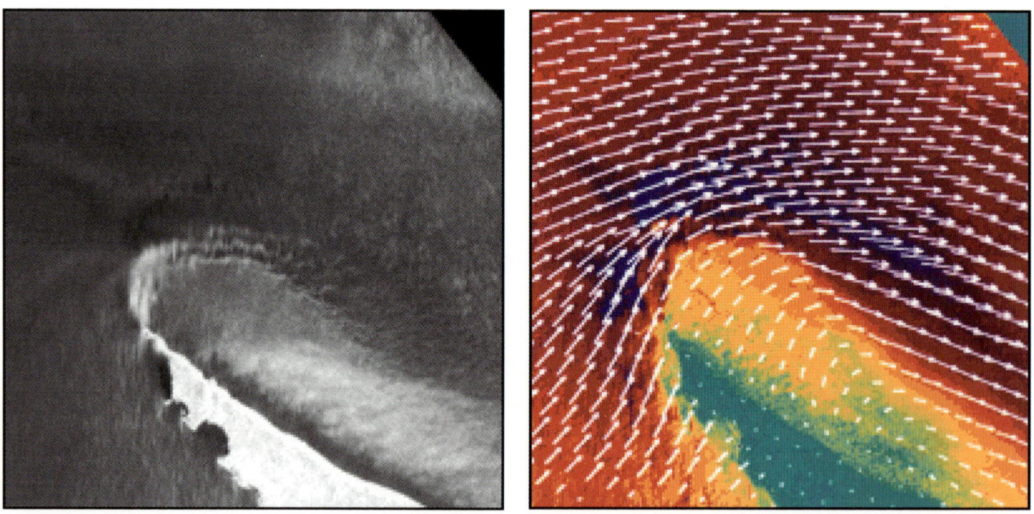

Analysis of Water Currents – Bay of Fundy, Nova Scotia. Image of a wave pattern (left) and a map of water currents (right) produced by interferometric analysis of airborne radar image data. Source: Natural Resources Canada

Another factor adding to the complexity in geospatial data is the concept that was introduced in the mid-late 1980s: cost recovery. The concept is simple. Those requiring and using the data should pay for the costs of creating and supplying the data. The previously cited KPMG study in which we were involved [142] found that cost recovery policies in geospatial data severely limited the use of location data for some players and some applications – especially those related to governments and the "public good." Another interesting finding was that the cost to administer cost recovery is often greater than the funds received. Furthermore, there are significant costs associated with the economic impacts resulting from less use of the data. On average the cost recovery seemed to return just 3% to 30% of the actual data production and distribution costs, depending upon the country. Furthermore, it was also found that cost recovery can inhibit the development of innovative technologies, applications, and industry development in geospatial: the higher the level of cost recovery, the lower the level of development of the industry. On publication of these results that compared the generally lower data prices in the USA to Canada and Australia, both Canada and Australia began to change their cost recovery policies toward lower data prices.

The conclusion on cost recovery is, in our opinion, clear. Given that most information pertaining to human activities has a location component, and given that economic activities can be made more efficient when location data is available, it only makes sense to provide such data at low or no cost. To do otherwise is to deny citizens' access to the GeoEconomy.

3.4.2.4. Data Policy Context

Data policy always exists in a context. Our consulting practice has developed an approach (the Kim Geomatics Data Policy Framework – see the table at the end of this sub-section) to provide an analysis of that context in order to optimize the development of a data policy. The basic context is provided by an analysis of local, regional (in-country), national and international parameters.

Data policy always exists in a context. The basic context is provided by an analysis of local, regional, national and international parameters.

Multiple jurisdictions have a say in data policy. In Canada, for example, there are different government departments at the federal level. Indeed, there are three different ones each with different objectives and different perspectives dealing just with remote sensing, never mind other aspects of geospatial data policy.[143] Adding to the complexity, there are a number of provinces and territories and many municipalities. Each has its own drivers, or important factors, that are considered when data policies are created. As previously mentioned, the need for different levels of detail in different jurisdictions has resulted in a sensible, if sometimes difficult to implement, approach to data collection suggested in Canada by the Interagency Committee on Geomatics:[144] "collect data once, closest to source."

The implication is that governments should work together to collect the most detailed information needed and then generalize that information if it is required at a less detailed level: in other words, municipalities should share their detailed information with higher levels of government, and everyone should share in the cost of acquisition. This approach, while sensible, is hard to implement because of the many different data policy models in use, leaving aside the technical issues surrounding the question of generalizing locally acquired information so that it can be used at smaller scales by higher levels of government.

In addition, there are various agencies involved in land title registration, electricity distribution, etc – each has its own approach designed for how it sees reality. While there has recently been some reconciliation of data policies across certain data sets in Canada, there remains some level of confusion.

As a signatory of various international agreements Canada (like most other countries) sees itself bound by international law as well as customary law – those actions that have been accepted as basic operating principles for so long that they have become customary law.[145]

The acceptability of a data policy may be determined by one or more of the players in data policy in the jurisdiction, the legal framework, regional policy objectives, national policy objectives, international law and, as indicated above, customary international law. The complexity of the data policy issue is underlined by the growing number and range of interests of organizations and others interested in, and offering opinions on, the topic of data policy. If we look just to satellite remote sensing data, the number of players both internationally and domestically is almost overwhelming.

Among those engaged one would expect to find research-oriented scientific and technical organizations such as professional societies involved in remote sensing and mapping – for example the International Society for Photogrammetry and Remote Sensing (ISPRS),[146] and its member organizations such as the American Society for Photogrammetry and Remote Sensing. Both have contributed opinions and information to the on-going policy discussion, as have other similar organizations around the world.

While this civilian discussion has been going on, the military (especially in the United States) has been forced into more public discussions. The line between military and civilian use of remote sensing has increasingly been blurred since the early 1990s, and especially since 9/11. Studies on the potential use of publicly available geospatial information by enemies of the state have been carried out – with somewhat surprising results. Less than one per cent of the data sets studied were deemed to have any value to terrorists – and most of those that did contained data that were likely available elsewhere.[147]

Why is the geospatial data policy situation so complex and why is there this interest by so many different groups world-wide? One can respond with a simple answer: self-interest. Governments around the world are a many-headed hydra with highly variable data policy interests. Some parts of government are concerned about security. Others want access to low-cost data to meet their mandates – be it ice monitoring, weather prediction, geological mapping, base mapping, forest inventory, or some other thematic mapping requirement. Still others in government want to use these data sets to develop industrial capabilities and export sales. Finally, some in government are interested in research – much like their colleagues in academe. In effect, then the different mandates of departments within the same government can result in their actually having conflicting data policy objectives. The one that most often comes to the fore is the desire to minimize access for the perceived enhancement of security versus maximizing access to increase public use of data. There is almost always a need and desire for cooperation, but at the same time there is often intense competition between different sectors and sub-sectors either to use data or restrict its use.

Why is the geospatial data policy situation so complex and why is there this interest by so many different groups world-wide? One can respond with a simple answer: self-interest.

Another less charitable observation is that in some governments there is the inclination of public servants, charged with making "cost recovery" a success, to see themselves as if they were running a business. Such people tend to think that the greater the proportion of costs they recover, the better the job that they are doing. It is almost as if they were playing at being a private sector company – without the risk or personal investment those in the private sector make. These managers miss the fundamental point that if the endeavour was deemed to be profitable in the sense of a private sector firm, then the government shouldn't be doing it in the first place! It is generally the case that governments provide geospatial information as a public good to benefit society at large. There may also be one or more aspects of market failure that make providing the service in the private sector uneconomic - e.g. the cumulative benefits are worthwhile but too widely and thinly distributed to make a pay-for-use model profitable. Cost recovery by government only makes sense in the context of whether or not it supports public policy objectives. Turning a profit is not a valid public policy objective – that is the role of the private sector. User fees do, however, have a place to reduce the drain that may come from unnecessary requests that strain the resources of the provider.

Overlaid on this complex national tableau is an even more complex international situation. Part of the complexity internationally comes from the many different policies adopted by those providing remote sensing data. Another level of complexity is added by demands for transparency. For example, in 2006 the European Parliament adopted the Aarhus Convention on Access to Information.[148] This Convention asserts the right of the public to receive any environmental information held by public authorities within one month of the request and without explaining the reason the data is wanted or how it will be used. This directive applies to information on the state of the environment, policies or measures being taken, or on the state of public health and safety where affected by the environment.

As the potential for remote sensing has developed, many others have weighed in on the topic of data policy. These include a number of international organizations such as the Committee on Earth Observing Satellites (CEOS) which represents those countries and agencies with civilian earth observing space assets or an interest in their application.[149] CEOS has played a key role in organizing science and certain technical issues, and has placed remote sensing in front of major international political fora such as the World Summit on Sustainable Development (WSSD). Several CEOS members have themselves been leaders in the policy discussions. These include the National Aeronautic and Space Administration (NASA), National Oceanic and Atmospheric Administration (NOAA)[150] and the United States Geological Survey (USGS) in the USA, European Space Agency (ESA) in Europe, Japan Aerospace Exploration Agency (JAXA) in Japan, and various agencies of the United Nations (Food and Agricultural Organization – FAO; United Nations Educational, Scientific and Cultural Organization – UNESCO; Office for Outer Space Affairs – OOSA, etc).

However, it has not just been the research community largely housed in academe and agencies of government that has become actively engaged in remote sensing and, by extension, data policy. As noted above, there has been political interest shown at the WSSD, and in two Ministerial conferences held in the ESCAP region of Asia (in Beijing in 1995 and New Delhi in 2000),[151] as well as in the Ministerial level Earth Observation Summit convened by the United States in July of 2003.[152] Political interest continues to grow alongside the range of applications of remote sensing. In Canada the growing political support was shown by the fact that after some years of cut-backs, the first substantial addition to remote sensing funds in some years was approved in the Government of Canada Budget tabled February 23, 2005.

Others who have offered opinions on remote sensing and associated data policies include industry groups such as the Management Association for Private Photogrammetric Surveyors (MAPPS)[153] in the USA, and the Geomatics Industry Association of Canada.[154] As stakes have become higher, and as the potential for commercialization has grown, there has also been an increase in interest by major corporations in the USA (Lockheed-Martin, Boeing, Orbital Sciences, Raytheon, etc), Europe (Dornier, Alcatel, and later Astrium), and Japan (NEC, Itochu, etc). Making life interesting for these have been parastatal companies such as the Israeli entry in the arena of high resolution dual (civilian/military) satellite sensing systems.[155]

We must be reminded that the preceding discussion on the complexity of the context is only for satellite imagery. When one brings in digital map data, national atlases, GPS, and the like, the complexity mushrooms. One need only consider the European, Chinese, Indian and Russian positioning systems competing at some level with the USA's GPS. According to an article in the Moscow Times April 26, 2006, the Russian military has said that "these restrictions will soon be eased." This has come some years after the USA had decided to stop degrading signals so that its industry could be developed. The position in the USA was that by providing more information there would be more users, more commercial services developed, and more economic return. The success of US industry in this field suggests that this was a wise decision – as it was when map data were made freely available in the USA, leading to the success of both geospatial companies and those that use geospatial information.

Kim Geomatics Data Policy Framework Elements Related to Data Use[156]

Cost Recovery and Pricing Policy

- Pricing policy must be stated. Options include: "Free" vs. cost of filling user request (COFUR), vs. cost recovery at some level for government subsidized systems.

Commercialization

- With industrial development a goal of governments, the ideal approach is to balance public and private sector roles based on the level of commercial maturity in the specific jurisdiction.
- The commercial model in remote sensing will change to see suppliers providing information, not just satellite data. This issue tends to be contentious.

Reception, Bulk Purchase and One-Stop Access

- Governments must commit to receive or otherwise obtain satellite remote sensing data.
- Bulk purchasing has been a feature of successful satellite remote sensing programs.
- A central clearing house or single point of access for satellite remote sensing is suggested to reduce confusion, delays, and improve access to data.
- One central government agency may acquire space-borne data from other national or commercial missions for non-commercial government and research and development use.

Data Sharing Among User Categories and Research Missions

- Sharing data between users and researchers is encouraged within the commercial restrictions that may exist with respect to the data and derived proprietary information.

Reproduction and Dissemination for Public Good

- Recognizing that public good is the major driver for national missions, reproduction and access for such applications is encouraged. One option would see data for the general good be freely available (on the Internet), with value-added products made available commercially.

Reproduction and Dissemination for International Aid

- Subsidized access to data will be provided to help developing nations cope with resource and environmental management, disaster response and planning on a project-by-project basis.

Standards

- The country will contribute to the development and setting of international remote sensing data standards and national data will meet international standards for formats, metadata, etc. that are compatible with those used in the geospatial field.

Operational Data Use vs. Developmental Data Use

- Data access policies in support of the development of new applications of remote sensing, both commercial and for the public good, are to be encouraged

---

3.4.3. Other Government Policy Areas that Impact Geospatial and the GeoEconomy

3.4.3.1. Introduction

Almost all areas of government require geospatial data, or are in some way related to such data. Indeed, estimates are that as much as 80% of all government decisions involve geospatial information. However, there are some policy areas that are more directly related to geospatial or that can more directly benefit from the integration of geospatial data into routine activities and thus warrant discussion here. These include research and development and commercialization, environmental policies, international relations and foreign aid, resource exploration, industrial development and safety and security. How geospatial data and technology are related to these fields is outlined in the following subsections.

Almost all areas of government require geospatial data: estimates are that as much as 80% of all government decisions involve geospatial information.

3.4.3.2. Research and Development and Commercialization

A key policy area in the GeoEconomy – as in virtually every other economy – is research and development (R&D). But it is not enough to do the research. In the realm of geospatial, for example, Canada has long been a leader in many research areas including GIS, GPS, and remote sensing. Indeed, most would agree with the statement that GIS was invented and first used in Canada.

But doing the research and developing the tools does little in terms of economic activity unless that activity is commercialized. It is in this area of commercialization of research (geospatial and other research) where Canada and many other countries have fallen short to date, although there are signs that this shortcoming is being addressed.[157] GIS and many other geospatial technologies conceived in Canada or elsewhere have been commercialized in the United States, where investors understand and will more often support the movement from research to commercial.

On the road to making a commercial product or service there are a series of steps. One begins with a theory or idea. This is followed by the R&D, after which a prototype is designed (service or hardware), then built and tested. These steps are then followed by the development of an engineering version, in turn followed by a beta test. The last step is a fully functioning, ready to be sold or used technology or service. All too often the road from research to a commercial system is seen as something that "true academics" don't engage in. In geospatial, the dominance of engineering schools, with their emphasis on building things that work, tends to see less aversion to becoming involved in commercial activities.

Research we have done suggests that there are certain basic factors that point to successful commercialization in geospatial (or other research), no matter what country one is in. The first and most important is the early involvement of industry and users of the product or service, and policies and activities to encourage these are key to success. Ideally industry will be involved as an advisor on key "real" problems to be solved. Industry might also be involved as a user of outputs, and a partner and source of funding – either in-kind or cash supporting the research or supporting contract work. Seeing students hired by industry is another harbinger of potential commercial success. Trying to involve industry and/or potential users and clients after the research has been done, or when a prototype is being developed, is too late.

Policies to encourage early involvement of industry and users in research projects are essential for the successful commercialization of geospatial research and development.

Appropriate research and development and associated commercialization policies are essential to move ideas from the realm of R&D to practical use if a country is to benefit from home-grown R&D in the GeoEconomy. The model developed by GEOIDE in Canada is a useful one to use to ensure a country is better prepared so that research leads to successful participation in the GeoEconomy.[158]

3.4.3.3. Environmental Policies

Environmental policies are by their nature (and virtually by definition) dependent upon geospatial data. Environmental policies – whether related to impact assessments, habitat protection, pollution abatement, or modeling for climate change – all require information on location. Monitoring, for example, depends on the ability to return to the same location to take measurements – and that is possible only by having accurate information on location.

Environmental policies are dependent on geospatial data for impact assessments, habitat protection, pollution abatement, monitoring and modeling for climate change.

In many jurisdictions there is now a requirement for industry managers to exhibit due diligence in the protection of the environment. Geospatial information, including remote sensing, has increasingly become an important part of the demonstration that due diligence has been accorded proper attention. One of the more interesting conversations we have had on this topic occurred some years ago when Ryerson was working as a Vice President of Business Development with a company providing services. A mining company's mine manager stated that that he did not want to buy remote sensing services to monitor his mine site because he did not want to know where there were any problems. If he did know where problems existed, then under the law, he would be obligated to take corrective action. "I would rather not know that there is a problem ... what I don't know won't hurt me." And, we could add, what he didn't know would not cost his company money. While we can hope that such a viewpoint would not be tolerated today, it does give one pause to know that there are people who would prefer not to use the tools available, in case the information yielded would cost them.[159]

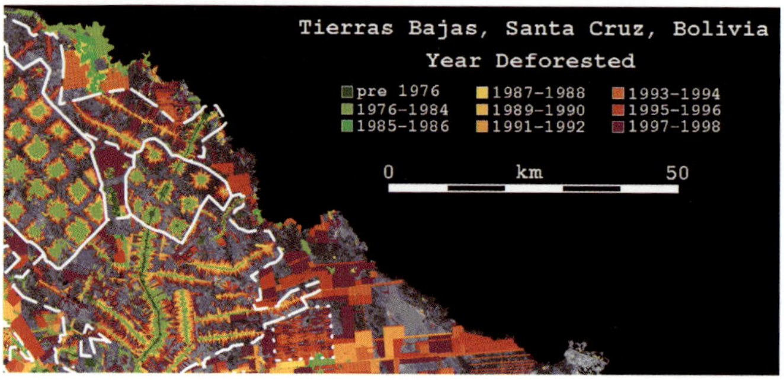

Tracking Deforestation in Bolivia. Satellite imagery from multiple dates was used to produce this map showing the time periods and locations of deforestation. Source: NASA

In closing, an interesting observation we can make on environmental policies related to monitoring and geospatial is that the level of environmental regulation a government can impose is influenced by the cost and technical means by which it monitors compliance. It also affects what can reasonably be demanded of industry to spend. As geospatial technologies make assessment and monitoring less expensive the "bar is raised" on what level of environmental regulation can be imposed. In effect then, geospatial is contributing in a very direct way to improving environmental regulations and their enforcement.

3.4.3.4. International Relations and Foreign Aid

To deal with other countries you must first understand the other country: what makes the other country "tick." One of the keys to unlocking that understanding is a comprehensive overview of the country's geography and how the country's historical and religious background sits on that geography. Obviously playing into the mix is the geography and history of surrounding countries. In the previously cited book *Guns Germs and Steel,* Jared Diamond makes a case that geography not only allows you to understand a country, but also a society's development over time. With globalization tending to make us think that all countries and all societies are somehow closer together than they really are, geographical understanding of the differences that remain becomes even more important.

To understand another country you first need a comprehensive overview of the country's geography and how the country's historical and religious background sits on that geography. Geography not only allows you to understand a country, but also a society's development over time.

With the accompanying rise of the GeoEconomy, geospatial understanding in diplomacy is today even more important than ever. Indeed, some countries have recognized the importance of these geographic differences and their diplomatic implications and have appointed a "Chief Geographer" within their diplomatic service.

International relations and foreign aid policies should encourage the use of geospatial data.

A central part of international relations is foreign aid. At one time foreign aid agencies in most of the developed world were major supporters of the provision of geospatial information and helped train professionals in the data's use in less developed countries. We refer back to Dr. Brandenberger's comment in Section 2.6 that "developed countries do not have good maps because they are developed; they are developed because they have good maps." Almost every developing country was mapped as a wise first step in its development – sometimes by the colonial power associated with the country, and sometimes by a donor. Such work tended to serve the needs of the developing country (for both data and training) while also serving to showcase the donor country's capabilities.

But things have changed. Geospatial understanding in many developed countries' foreign aid agencies has lessened at the very time when it should have increased.

Data Policy Development. Workshops such as this one in Thailand are an effective means of developing data policies that meet the needs of stakeholders with varied interests in both developing and developed countries. Source: R. Ryerson

But things have changed. Geospatial understanding in many developed countries' foreign aid agencies has lessened at the very time when it should have increased. Some countries continue to offer support in geospatial. Other countries, such as Canada, have tended to move away from such support, as interest has increased in new soft issues such as gender equality, health, and the like, where the aid agencies don't seem to see the need for geospatial information. While we agree that these so-called "soft issues" tend to need less geospatial information than the construction of a road or dam, for example, we still must ask: how can one measure requirements for assistance or success of programs that are supported in any of these other new issue areas without geospatial information? Without such data how can one assess the "where"? How can one answer where is health an issue? Where are the most underserved people in need of a hospital? What illnesses are most prevalent in which location? How can we minimize the cost of distributing medical supplies? Where is gender equality a problem? Where are more schools needed first? Where have our programs worked – and where have they not? Yet many international aid agency policies no longer stress the use of geospatial information – neither using such information, nor providing it.

> ***Foreign aid agencies should be a major user and provider of geospatial data and technology – but unfortunately, for a number of reasons, that is often not the case today.***

The lack of interest in geospatial by some donor agencies is something we cannot understand. Perhaps, like the mine manager who doesn't want to know where (and if) his mine is polluting the environment, some aid agencies don't actually want to know where programs are really needed – or where they have failed. Luckily for many developing countries bilateral support continues in geospatial in agencies such as the US Agency for International Development (US-AID), and the aid agencies from Japan, Sweden, and others. We must also quickly say that multilateral support is found in agencies of the United Nations and the development banks. Literally hundreds of projects can be identified in the poorest countries, those closing in on status as developed countries, and everything in between.

We have completed an analysis[160] of over 200 development projects which have been done in the geospatial realm (primarily in remote sensing and GIS) that suggests that, while not all projects are successful, one can predict whether or not they will be successful early on in the process. A pattern has emerged and commonalties have been identified in the more successful projects in which bilateral or multilateral aid was a factor. Successful projects seem to share the following operational characteristics:

- There was some previous experience with satellite or other remote sensing imagery;
- There was a clear focus;
- There was a combination of experienced researchers and bright innovative younger scientists involved in the project;
- There was great attention to detail – technical and financial;
- There was a basic understanding, either explicit or implicit, of the economics of the use of remote sensing for the specific application;
- There were close ties to real users with the interest, dedication, and ability to sustain the work;
- There were close personal ties to scientists from sponsoring countries;
- Training, personal ties, and the experience noted above resulted in the development of confidence and a lack of fear of failure on the part of the in-country project leaders; and
- There was an entrepreneurial "can-do" spirit.

In addition to these operational characteristics, successful projects shared strong institutional arrangements that were suitable for the milieu in which they were located. Another factor leading to insufficient appreciation of the role of geospatial is the experience and background of those deciding how to apportion resources. Those trained and experienced in managing these programs usually do not have much experience or training in doing the job with the help of geospatial tools. There is a lot of information technology and data that these people use routinely, such as computers, calculators, and maps which they wouldn't think to be without. But geospatial data and geospatial databases are not among the familiar tools routinely used to do their jobs. Increased awareness, technology transfer, supportive policies and training are needed – along with appropriate policies. We hope that this book will be a first step in addressing this problem.

By applying this set of guidelines we believe that many more geospatial and remote sensing projects could be successful in less developed countries and aid money could be far better spent by many more aid agencies. Success on the part of developing countries in these geospatial projects will enable them to begin to participate in the GeoEconomy.

3.4.3.5. Resource Exploration

By definition exploration implies finding the location of a resource. A basic element is usually, and in our opinion correctly, thought to be the provision by the government of both base maps and geological maps.

In many countries there are policies that provide special allowances and tax write-offs for exploration activities, including acquisition of data. The reasons for that are clear: the more exploration mapping, that is done, the more resources are found and the more the resulting economic activity. The benefit/cost ratio is over 90:1.

In many countries there are policies that provide special allowances and tax write-offs for exploration activities, including acquisition of data. The reasons for that are clear: the more exploration mapping, the more resources that are found and the more the resulting economic activity. In Australia, studies indicated that for every dollar spent by government on exploration, an additional 3 to 5 dollars are spent by industry. In Australia a total expenditure of about $12 billion (Australian dollars) for exploration mapping (2 from government and 10 from industry) over the decade from 1987 to 1996 led to an increase in economic demonstrated resources (i.e. in ground resources that can be exploited) of approximately $360 billion for metalliferous commodities and $755 billion for black coal. These numbers, yielding a benefit/cost ratio of over 90:1, lend strong support for the creation of the aforementioned map products.[161]

Clearly, where there are policies that support natural resource exploration, including data acquisition, the potential for participation in the GeoEconomy increases.

Clearly, where there are policies that support natural resource exploration, including data acquisition, the potential for participation in the GeoEconomy increases. Where policies do not support exploration, the benefits associated with the resource development often end up by default being ceded to others – or simply remain unrealized. A strong geological survey, and appropriate licensing regimes for resource development, can be the building blocks for a successful entry into the GeoEconomy. Based on the analysis of both geospatial activities and support for resource exploration, it would appear that Australia has been an early entrant into the GeoEconomy.[162]

Without good geospatial information, and the knowledge to apply it, neither good policy nor good decisions can be made.

3.4.3.6. Industrial Development

In many countries of the world – both developing and developed – industrial development is a stated national or state objective, and the "where" question is an important component of that development. In the case studies cited in Section 3.4.2.3, we noted the use and value of geospatial information for industrial development in both Canada and India. Similar stories can be cited in most jurisdictions that are active in the GeoEconomy. Whether trying to attract industry, to develop a cluster in IT, bio-tech, or some other area, or to do an analysis of why certain industrial clusters have developed in a specific location, geospatial information is essential. Indeed, the presence of good geospatial data and the capability to use it is increasingly being seen as one of a number of incentives (such as a good transportation network, skilled workforce or tax incentives) offered to attract industry. Without good geospatial information, and the knowledge to apply it, neither good policy nor good decisions can be made.

The development of centres of excellence in geospatial should be, and in some cases is being, encouraged as a part of government policy.

Geospatial is increasingly being recognized as an enabler in industrial development. As a result, in some cases, geospatial is more closely tied to industrial development than the simple use of geospatial information to attract, analyze, or understand that development. The development of centres of excellence in geospatial should be, and in some cases is being, encouraged as a part of government policy. We did one study for the region of Quebec City that was focused on building a cluster in geospatial – an amalgam of industry, R&D, education, and government activities focusing on geospatial. We have identified some 25 cities or regions world-wide that could make geospatial one of the main foci of the local economy and by so doing help further the locale's success in the GeoEconomy. These cities or areas are listed and discussed in more detail in Section 6.3.9.

We have identified some 25 cities or regions world-wide that could make geospatial one of the main foci of the local economy and by so doing help further the locale's success in the GeoEconomy.

3.4.3.7. Safety and Security

The broad topic of safety and security covers everything from military and police protection to fire fighting, emergency response, and anti-terrorist activities. Each of these has policies associated with it and each of these must have a spatial component. Where is the threat? Where is the fire (and what is where in the building by way of poisonous materials)? Where has the road remained passable after the earthquake (or hurricane, or tsunami, or volcano)? Where is the emergency?

We gloss over military applications by saying that while exciting, many are considered to be secret and thus are not in the public view. However, military applications often do lead to civilian use. As newer technologies are developed for military use, those that were once secret become commercialized – or more likely, knowledge that was embodied in the secret technologies finds its way into the civilian market. Suffice to say that successful military organizations have a mapping component – as they have had for thousands of years. How the links are drawn between the military's mapping and other applications and civilian applications will, to a large extent, determine the success of a nation's entry into the GeoEconomy. One of the most limiting military policies is that which limits wide distribution of geospatial information. Obviously sensitive areas and technologies must be protected, but otherwise a policy that leads to wide distribution and use of geospatial information makes sense.

The broad topic of safety and security covers everything from military and police protection to fire fighting, emergency response, and anti-terrorist activities. Each of these has policies associated with it and each of these must have a spatial component.

Moving on to the civilian use of the technology, one of the more serious issues faced by those involved in safety and security is access to data to prepare for emergencies – be they man or nature induced. Here data policy is important, as is some form of pre-planning. Most agencies do not want to spend money – until the time of a crisis. And then they want data instantaneously. Luckily, through what is called the International Charter on Disasters,[163] rapid and free access to space-borne remote sensing imagery data is possible. Map data is another issue – as is access to the area affected – as was seen in the calamity in Myanmar. There the damage from a tropical storm was so severe that a state of emergency was declared and, after some delay, the government allowed foreign aid workers and relief supplies into the country. However, the ruling military junta's preoccupation with controlling the movement and activities of foreign relief workers, and inadequate access to accurate maps, hampered the delivery of desperately needed emergency medical care and supplies.

The other sort of map data that is often forgotten is the mapping of potential for disaster. For example, what is the potential damage if an earthquake of a certain magnitude hits? On the west coast of the United States and Canada maps have been constructed showing the potential for damage. Building codes vary to take into consideration local conditions and the fact that earthquakes may strike at any time – and they have struck with some severity. The 1989 earthquake that hit San Francisco was of the same magnitude as the one that struck Haiti in early 2010. The San Francisco quake saw 63 people killed throughout northern California, while another 3,757 were injured. In Haiti (with a slightly lower population than Northern California), where poverty and poor building standards go hand in hand with inadequate maps, a similar earthquake has seen a loss of life projected to be in excess of 200,000, a few days after a quake of a similar magnitude occurred.

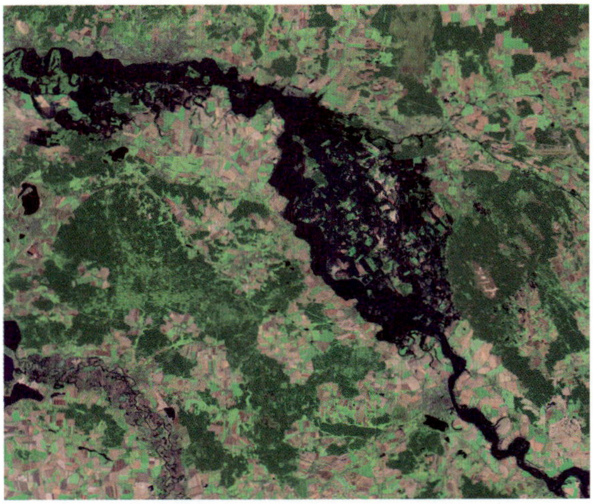

Landsat Satellite Image of Flooding of the Elbe River – August 2002. Heavy flooding killed more than 100 people and caused $20 billion in damage in Austria, the Czech Republic, Germany, Hungary, and Russia. Satellite imagery was valuable in planning emergency response, and recovery, and assessing financial assistance and insurance claims for natural calamities. Source: Image courtesy of J. Allen, NASA Earth Observatory. Data provided by the USGS EROS Data Center Satellite Systems Branch.

There is a saying in the geospatial community that the only thing that recedes faster than the floodwaters is the willingness of government agencies to spend money on geospatial data.

While data access through mechanisms such as the disaster charter are important and commendable, the benefits of policies that support pre-planning for emergencies – from which roads are least likely to flood (to plan evacuation routes or emergency access for example), to where there are likely to be the most serious problems – should be obvious. Those who are truly part of the GeoEconomy can be expected to have such information ready for action.

3.5. New and Expanding Needs for Geo-information

3.5.1. Introduction

The need for geospatial information is growing in at least two ways. First, there is the expansion of what we call existing information needs. This aspect of growth in use is relatively easy to understand – people and organizations simply want more geo-information of the sort that they have been using in the past. We have already highlighted many such uses in the discussion on how geospatial data have been, and are being, used. This expansion along what we might think of as "traditional lines" is based in part on better access (using the Internet to access lower cost data for example), and in part on better understanding of data by traditional users, leading to an increased recognition of the value of the data. To modify a well known phrase – "using is believing" – the more people use the data, the more that they believe that the data are useful. The expansion of existing geo-information needs is discussed in the next section.

The second element of growth is in what we call new needs – the need for geo-information that has not typically been called for in the past. We differentiate "new needs" from old ones not just on the basis that the data have not been called for in the past, but also on the basis of new groups needing (and using) the data, when and where they have not typically used the data in the past. For example, today individuals are expecting easy and fast access to data that once were only available to governments (such as digital maps and imagery), and small companies now routinely seek out information that was once only accessible to big business (such as detailed consumer information). Vans roam up and down our streets providing street views for prospective homebuyers, door-to-door salespeople – and perhaps for sophisticated burglars. We consider these to be new needs, for while the actual data may not be new, its use and user certainly are. These "new needs" present the most opportunities as well as the most challenges – both for those supplying the information and those using it. The expansion of new uses is discussed in the second sub-section below. The final sub-section reviews the impact of this anticipated continuing growth in the GeoEconomy.

3.5.2. Expansion of Existing Needs

3.5.2.1. Introduction

We have suggested that it should be easier to understand and hence easier to predict what the future needs will be for existing users of geospatial or geo-information, than it would be to predict new uses for new users. One might assume that existing uses tend to follow more traditional lines, and by definition are not "outside the box." However, those "inside the box" are as difficult to project as those "outside."

Previous studies of the expansion of traditional markets typically assume that there is a straight-line increase from the past to the present and on to the future. While straight line projections may be convenient, even in this relatively simple case we do not think that they will yield assessments that are really accurate. There are complexities caused by the technologies we have been discussing, how they interact with each other (see Section 4.2. for further elaboration on convergence), the changing model of data delivery, the experience of the users, and (not surprisingly) by where the users are located. Together these factors lead us to conclusions on where we will see expansion in traditional uses of geo-information and what the impact will be on the GeoEconomy.

3.5.2.2. Technologies Affecting Traditional Use of Geo-information

Introduction

Usually when new technologies are first adopted they are used to replace some aspect of an existing process. In the second phase or level, the new technology is used to change the process, not just make some aspect of it quicker. This sort of change has been happening in various aspects of mapping for several decades. The third phase or level of adoption sees the capabilities of the new technology used to devise an entirely different means of doing the job. Those that develop this third phase or level of innovative uses of geospatial information have a disproportionate advantage, because they have essentially re-engineered how a problem is addressed to take maximum advantage of the best tools available. While levels 1 and 2 are less risky, they carry more of the baggage of legacy systems and "old" ways of thinking. Over the past few years a number of third level uses of geospatial have been developed – and often by those outside of the traditional geospatial world. Google's use of imagery and map information in advertising is a prime example of the third phase of adoption.

> ***The technologies that impact traditional use of geo-information data are headlined by the Internet. But they also include an increasingly growing and wide range of digital distribution and storage media, mass market distributor applications (such as Google Earth), lower cost GPS receivers, and print-on-demand.***

One reason for the recent rapid rise in new approaches is the convergence of the factors listed below, while another is that the risk and cost of developing new applications has been dramatically lowered. The reduction of cost and risk has been done through intelligent data policy (lower cost more easily accessible data), access to inexpensive software, and business incubators.

The technologies that impact traditional use of geo-information data are headlined by the Internet. But they also include an increasingly growing and wide range of digital distribution and storage media, mass market distributors (such as Google Earth), lower cost GPS receivers, and print-on-demand. Each of these has played a role in increasing traditional use of geo-information. Together they have resulted in a dramatically changed geospatial market that can be fairly described as having been turned on its head over the past few years.

The Internet

The Internet alone has been important, but what has really changed the geospatial world has been the use of broadband technologies. Broadband allows rapid access to the huge files that one typically finds with geospatial data. Simply stated, broadband rapid access to the Internet has been the key that has unlocked access to more data for more people – especially traditional users who used to rely on paper maps or digital files sent by courier or regular post. File sizes that would have led an Internet provider to "pull the plug" on access to e-mail just a few years ago, are now sent back and forth without a thought. Even more amazing is the fact that one can acquire data, analyze it on-line, and bring the result into a desk-top system without ever downloading the data, using technology built by companies such as CubeWerx. With current Internet capabilities, limitation of storage capacity which was once a major issue is beginning to fade in importance.

Distribution and Storage Media

Media used to both distribute and store data can now hold more, do so more reliably, and there are more options in terms of types of media. At conferences, when we see a map or image that catches our fancy, we ask if we can have a copy – and then out comes our 16 GB USB "stick" for which we paid less than $50 in October 2009. Ten short years ago we were carrying discs that held less than 2 MB – not enough to hold a single high quality image from today's typical digital camera…never mind imagery from a digital aerial camera. We can expect that improvements will continue and will likely keep pace with the volume of data being used. Indeed, the advances will likely contribute to pushing the volumes of data we want ever higher.

Mass Market Distributors and the Media

While the appetite for imagery from satellites and high-flying aircraft began with weather forecasts, it was truly whetted by imagery that was shown by the media during the first Iraq war. More recently both Google and Microsoft have had a profound impact on the visibility of geospatial data – both map data and imagery. In effect they have given a degree of legitimacy to the use of imagery and maps in traditional ways – but an even greater impetus to the much broader use of such data for everything from real-estate sales to advertising local services. Added to this is a growing interest in the environment and climate change. Geospatial information in the form of satellite imagery has been essential in demonstrating that there have been changes, and in both explaining and measuring them.

Lower Cost GPS Receivers Plus Cell Phones

When the cost of GPS receivers was high, GPS use was the purview of professionals in the geomatics field or those to whom knowing their location was of high value – such as an exploration geologist. Now units can be had for $100 or less. What is more, there are now GPS chips in some cell phones, so the position of individual cell phones can be accurately tracked. The location of anyone carrying a GPS equipped cell phone can now be continuously monitored A boon for emergency dispatch, municipal work crews, and delivery companies to improve response times and efficiency, and for parents to keep tabs on their teen-aged children – and their friends!

Media Use of Satellite Imagery. High resolution commercial satellite imagery was widely used to illustrate stories about the 2002 Iraq War in the print and broadcast media. Source: GeoEye satellite image.

Print-on-Demand

As noted elsewhere, maps can be printed on demand from a desk-top computer in our offices or in our homes. The speed with which maps can be produced means that many more maps or other representations of data can be produced for the traditional and non-traditional user alike. A psychological barrier has been crossed. Maps used to be considered a product that was created, used and archived. Today a map is a consumable – just as likely to be created, used, and disposed of in the recycle bin. The mystery, or special aura, associated with the creation of a map has disappeared. Maps have gone main-stream.

3.5.2.3. Where Expansion in Traditional Uses Will Occur

We will see a significant increase in the use of geospatial data and technology across all traditional application areas. While not an exhaustive list, these include areas such as urban and regional planning, development (of land, of resources, and infrastructure of all types), land and resource management, environmental reporting, and fleet management. It should be noted that these increases in use may not necessarily result in a concomitant increase in the purchase of data. This is because of the very different models of data distribution involving, as they do, a significant amount of data sharing and so-called "free" data.

We will see a significant increase in data use across all traditional application areas. But these increases may not result in concomitant significant increases in the purchase of data.

The trend towards providing more data at no or low cost can be expected to continue. There are many who advocate the provision of comprehensive data sets that would be made available for all to use.[164] Furthermore, as data have become more digital, and as the technology to use digital data becomes ubiquitous, we also believe that much of that increased use will not involve traditional map products, although data will still be used in what are recognizable so-called "traditional" applications.

There will be a proportionately greater increase in the use of data at more and lower levels of government for applications that are easily understood and better documented.

There will be a proportionately greater increase in the use of geospatial data at more and lower levels of government. Small and medium sized municipalities will have access to data and technologies in ways and at a cost that will, for the first time, make the data's use cost-effective for them. Data will be used by these smaller municipalities and states in ways that are easily understood and generally (but not always) well documented, following on, as they do, from applications in larger municipalities and larger states. On the lack of documentation, we have found, in some cases, that those using geospatial data in an operational fashion are often not interested in taking the time to write about the application that they have developed. Two reasons are often identified for the reticence to share their work. The first is because private sector groups often see their work as giving them a competitive advantage. The second is that government employees lack the time and resources to "write up" their work. "We are too busy using the data to write about it." Is how one public servant in the Province of Prince Edward Island once explained it to us. With the coming of more casual forms of reporting such as blogs and web sites, this reticence to document applications is rapidly disappearing.

One thing that history teaches us is that wide exposure to the data can be expected to lead to more and different uses. Section 6.4 documents several examples of the sort of questions that companies can answer with geospatial information and technologies.

There will also be wider use of data by a far greater range of players within the private sector. No longer will its use be restricted to banks, insurance companies, large retailers, large transportation and engineering companies, multinationals, and political parties. The democratization of geospatial in the GeoEconomy will see small companies, from corner stores to local consulting and service firms, applying the technology to offer more targeted services in their local markets. Industry use will, however, remain largely invisible to most of the public. The only result will be in slightly different forms of unaddressed mail arriving at our homes or mailboxes.

While the increased use of simple GIS systems by the private sector and the public has not been as dramatic as some had thought that it would, we believe that the recent financial troubles will lead to much greater use – confirming an observation Ryerson made some years ago during a recession. It seemed then that bad economic times led to an increased use of geospatial data for two reasons. First, those familiar with the technology and its application try to improve their competitive position through increased use. This leads others to try to follow. Second, barriers to adopting new technologies diminish as companies become willing to adopt new technology as they try to survive. As desperate people in the water either drown or learn to swim, in desperate economic times, people either begin to use new technologies and approaches such as those associated with geospatial – or they die. Indeed, Ryerson (2009) gave a paper titled "Economic Panic is a Good Thing – for the Geospatial Industry." This growth in geomatics in tough times is world-wide. In a recent conference held in China, Vice-Minister Xu DeMing stated that in China the geomatics industry has increased its sales during the current downturn. Some 10,000 companies (many of them spun off from government agencies) share an annual income in the vicinity of $10 billion.[165]

Bad economic times lead to an increased use of geospatial data as those familiar with the technology and its application try to improve their competitive position. Those that don't, die.

Non governmental agencies (NGOs) and various advocacy groups have long used geospatial data. For example, at one time Ducks Unlimited was North America's largest civilian purchaser of satellite imagery. They used the imagery to map wetlands and their changes. That sort of use of geospatial information can be expected to continue, with a projected significant increase in use by environmental groups. The world has begun to understand (largely through Google and Microsoft) that a map or image is a useful tool to simplify the understanding and depiction of the environment that surrounds us. One of the interesting aspects of the use of data by such groups is that if the advocacy groups on one side, or the government or developers on the other, use geospatial data, it forces the opposing side to use the same or similar data to make their case. One of the strategies employed by some imagery sales people, is to sell to one or other of the sides in any major development or environmental dispute. That almost automatically results in the other side also buying the same or similar data. Such "dual use" will continue to expand.

Geospatial data used by the public will increase in one way – but decrease in another. The increase comes from the ease with which routes can be planned – and the instantaneous production of specialized maps which are usually only used once. As noted above, maps are now being regarded as consumables. However, with the increased desire to reduce fuel use, coupled with a downturn in the economy, we can expect to see less travel and fewer new roads between towns and cities in certain parts of the world (that ratty old map will suffice). As a result we can expect to see a decrease in maps produced in bulk (i.e. in large print-runs) – such as road maps.

3.5.3. The Growth in New Needs for Geo-information

The expansion of new needs for geospatial data is both more exciting and more difficult to understand and predict than the expansion of existing needs. To understand this non-traditional expansion we must understand the technologies we have reviewed, and the policy environment – what governments will allow us to do.

The expansion of new needs for geospatial data is both more exciting and more difficult to understand and predict than the expansion of existing needs.

We also need to understand what the fundamental role that geospatial information can (and might) play in the lives of people, and in the activities (including management) of companies, municipalities, state and national governments, as well as international organizations. We must then take all of this information and project it into the future.

Our somewhat murky crystal ball tells us that:

- Tracking evidence of global change will go main-stream as more people become interested in what has, to date, been an area of interest only to scientists. Data on global change will be widely sought after;
- There will be a resurgence of interest in the role of geospatial in international development, spurred on by the successes demonstrated with zoonotic diseases (diseases that go from insects or animals to humans) and the pressures in this area coming with global climate change and natural disasters;

- Monitoring carbon offsets and their performance (such as the success of tree planting) will become a major application – borrowing from work on monitoring forests;
- Agriculture will become an ever bigger user of the technology – from GPS to imaging. This will expand into less developed countries as pressures on the food supply mounts. The UN (especially the FAO) will continue to play a major role;
- Hyperspectral remote sensing will become important for monitoring of high value crops, forest damage, and environmental impacts;
- Commercial banks and others involved in investing will pay more attention to geospatial information when making investments – whether in real estate or financing resource processing factories.[166] The sub-prime mortgage fiasco will contribute to that interest;
- Mobile imaging vans will create streetscape imagery and information for broad use in planning and maintenance of urban infrastructure, for tax assessment, for private sector marketing (including real-estate), policing, emergency response, mortgage lending, etc;
- Calls for transparency in government will lead to the use of more geospatial data across the board related to planning and environmental assessment;
- GIS and image analysis tools will increasingly be used by the general public – as add-ons to mass-market tools, provided by either the mass market players or by those who create add-on products to work with the data they provide;
- Geospatial capabilities will become part of society's knowledge landscape;
- As natural disasters become more prevalent and cause more damage we will see more work done to plan for response before events take place and these will depend in large measure on geospatial information;
- More data will be used as scientists attempt to model disasters to lead to prediction; and
- In some cases privacy issues will be trumped by society's desire to protect its people and environment.

This list can be added to and if history teaches us anything, some of the major new uses will be in areas that we, the so-called experts, have not yet thought of.

3.5.4. Conclusion: Impact on the GeoEconomy

If we are, as we contend in this book, already in the early stages of the GeoEconomy, then the increase in the use of geospatial information, whether traditional uses or new ones, will lead to a broadening of the benefits of geospatial to a much wider segment of society.

The primary beneficiaries of an increase in traditional uses will be those who have the greatest stake in traditional applications – governments, NGOs, advocacy groups, and some segments of industry. In one sense then, the greatest beneficiaries will be the general population served by governments, NGOs and advocacy groups. These will benefit because they will have access to more and better information upon which to base decisions.

New uses will tend more to benefit selected groups within society – farmers, forest owners, etc. There is, however, one major new application that will benefit society at large and that is in the provision of health services and the study of the spread of disease. Recent work has seen some interesting demonstrations of the technology to better manage the spread of contamination inside hospitals, as well as the spread of illness across a city, state, country, or region of the world.

You can have the best information in the world, but if decision makers are unable to properly assess, digest, and use the information presented to them, the quality of the information won't matter.

As noted above, because it tends to be done quietly, industry use will have little impact on the use of the data and technology by the public. However, the use of the data by advocacy groups and government, as well as use in the media, leads to more exposure. But none of that exposure can compare to the impact of Google and Microsoft. Because of these, more people than ever before are looking at and using geospatial information.

With increased exposure, however, come some problems. New users, without the appropriate background, may not understand the implications of accuracy (accuracy of both the geometry of the data and of the information that might be extracted from an image or map product). There will be increasing demands placed on the limited number of people in the geomatics field. Indeed, the importance of the field and concerns about the number of qualified people has been noted in a number of studies and statements by a variety of policy groups, think tanks and others. And while some might see the benefit of the data and be willing to pay more for data acquired especially for them, others may not be aware of the costs of obtaining such data – or even how to specify what data are required. In addition, the pace of change in geospatial is daunting. We believe that in time these issues will be overcome as more intelligence is applied to the data and a better understanding of technology emerges.

While the preceding paragraph suggests that solutions can be found for most issues, there is one major concern yet to be adequately addressed. Standing between today and the promised bright future of tomorrow, where geospatial data are widely and easily available and used, is one major problem. We are getting better at using various geospatial tools to acquire and develop information, and the younger decision makers and users of tomorrow will be familiar with services such as Google Earth and geospatial information in general. But we have not yet developed the tools to help decision makers use geospatial information to improve their decision making. You can have the best information in the world, but if decision makers are unable to properly assess, digest, and use the information presented to them, the quality of the information won't matter. We believe that more research on geospatial decision making is needed so this decision-making process can rest on as solid a foundation as other well-developed ones like financial decision making.

CONVERGENCE AND SOCIETAL IMPACTS OF GEOGRAPHIC INFORMATION IN THE GEOECONOMY: TODAY

4.1. Introduction

To this point we have reviewed the history of, and background to, the GeoEconomy – with emphasis on geospatial technology's development and use, as well as on the resulting data. We have also discussed the implications of geospatial or geographic data for governments, industry and individuals. Along the way we have given a number of examples of the application of geospatial tools and data across government, and indeed across the economy, as well as where we think uses of geospatial data will increase – but our view has been from the perspective of the technology, its history, and its application.

Now we step back to look at geospatial technology and data and the GeoEconomy from a broader perspective, to assess why things have moved so quickly as of late, and what the impacts have been on society.

In this section we deal with two important elements associated with the GeoEconomy:

1. The convergence in technology, convergence within and between various disciplines, and convergence in the realm of policy that have affected the GeoEconomy. Together this convergence along a number of disparate axes has led to a situation where geospatial data HAD to become more important than it had hitherto been. This is discussed in the next sub-section; and
2. The result – the societal impacts "writ large" of geospatial is covered in the second sub-section below.

This section closes with some philosophical musings on the impact of different generations of geospatial data and how they have been developed and used. One might see convergence here too.

4.2. Rapid Convergence and Implications

4.2.1. Introduction to Convergence

Success in the GeoEconomy is characterized by a more integrated and usually spatially-based approach to issues. This in turn drives the growing demand for more diverse and higher quality data that can be analyzed in a more integrated way. Simply stated, the location component serves as a means to integrate diverse data sets. For example, the associations between health, income, and environmental quality can be assessed using datasets for each of those factors, but only if they all include location.

> ***Success in the GeoEconomy is characterized by a more integrated and usually spatially-based approach to issues. This in turn drives the growing demand for more diverse and higher quality data that can be analyzed in a more integrated way.***

While this need to integrate information has been gaining momentum in its own right, it has been given further impetus by convergence along a number of disparate axes that we have identified – technologies, disciplines, and policies. With the need for integrated analyses and the growing convergence along these axes, we have reached a position where we have both the need to be, and the ability to be, geo-aware.

4.2.2. Technology Convergence

There has been rapid convergence of technology across a wide spectrum driven by a number of factors, only the most important of which are listed here:

- The development of more powerful and lower cost computer systems;
- Increased data storage capacity;
- Improved data bases (including ones with the ability to handle geospatial data);
- Broad-band communications;
- The Internet;
- Interoperability (within and across technologies);
- Mobile communications devices;
- Precise positioning capabilities;
- Improved sensor technology (LiDARs, radar systems, digital cameras, higher resolution satellite data);
- Better data security;
- Vastly improved gaming software (a source of high quality graphics and 3-D animation); and
- Enterprise resource planning software.

Each and every one of these has had an impact on, and in some cases has driven, developments leading to the emergence of the GeoEconomy. It is also useful to keep in mind that the greater the differences between the technologies being combined, then the greater the risks and the more difficult the challenge to produce something useful.

There has been rapid convergence of technology across a wide spectrum that has driven interest in and need for geospatial information and technology.

Adding GPS to cell phones is the convergence of very different technologies, as is bringing together video and telephone. Some decades ago video and telephones were combined – and never caught on. Yet today, because of increased communications capabilities and the Internet, we now have video-conferencing and video tools that can link us to our associates and families across the city or around the world. There is a greater business risk in adding GPS to cell phones and cameras because there are more difficult technical issues to solve, but where it has been successful, there is tremendous opportunity and financial rewards if the new applications are successful. Such new applications typically produce a wide range of consequences both intended and unintended. For example, there are growing privacy concerns associated with tracking people by their cell phones. Google's Streetview images show detailed digital photographs that were taken from a van driving down the street. When this service was first released, people could be identified in locations where they did not want to be seen. There was even a report of a fist-fight being recorded on the streets of a major US City! Since then faces and license plates have been obscured on Streetview. Of course, the effect of the 9/11 terrorist attack has tended to make the public much more willing to tolerate compromises in personal privacy. This has enabled a much more invasive collection of personal information. While some of this data is captured to meet legitimate security needs, much of it could be used to serve private commercial interests or those of "big brother governments" seeking to uncover information about individuals that they were not previously able to obtain.

Convergence – A Driving Force in the GeoEconomy. Source: S. Aronoff

The process of incorporating location data into our societal information infrastructure is proceeding rapidly in some areas, lagging in others. Where there is the potential for significant financial gain the new data and technologies can be expected to be adopted rapidly. One need look no further than how consumer products, GPS, Facebook, and Twitter have been adopted. The components of our societal infrastructure designed to manage these services in a socially acceptable way – such as privacy policies – tend to be left behind, trying to catch up as services like Facebook and Streetview are introduced.

As the value of using and integrating these diverse and improving technologies together, or using components of one or another of them in geospatial data collection or analysis, is recognized they are soon made to function more easily and effectively together. In many cases they are even combined into a single device. Several highly visible examples can be cited. For example Nokia, a major cell phone manufacturer, has introduced a cell phone with built-in GPS and map display of the unit's location. Garmin, a major GPS receiver manufacturer is developing a GPS unit that is also a cell phone. Handheld personal digital assistants (PDA) initially were designed to provide basic information handling and storage of contact information and appointment schedules. With the addition of GPS capability and road information databases they became Personal Navigation Devices (PND) as well. With wireless capability and more computing power the PDA evolved into a handheld computer that is also a cell phone that provides email access and Internet browsing.

One of the most widely known examples of rapid incorporation of spatial data into a completely different field was Google's introduction of free online access to maps and both satellite and airborne imagery with coverage of most of the Earth.

We have seen the dramatic rise of Google in the geomatics field. According to *Wired Magazine*[167] and Google's own web site,[168] Google Earth was downloaded an average of 400,000 times per day over a period of about 600 days. The total in one week – 2.8 million – is likely more than the number of geomatics specialists in the world today. The Google example also illustrates how the convergence of very different technologies and services can result in a dramatic re-assessment and re-pricing of assets, because suddenly in their new role these assets serve a different purpose in an entirely different business model. Instead of the satellite imagery being sold in low volumes at high prices it was given away in massive volumes in order to generate even greater profits by attracting viewers to its advertising.

The importance of geospatial data and analysis for geo-enabled devices has led to the purchase of data providers by hardware and software manufacturers and service providers from fields once considered unrelated to geospatial.

The importance of geospatial data and analysis for geo-enabled devices has led to the purchase of data providers by hardware and software manufacturers and service providers in fields once considered unrelated to geospatial. This has led to new players with new approaches to business, new ideas, and more money than has been available to the more traditional geospatial industry. For example, as noted elsewhere in this book, the hardware and software manufacturer Nokia purchased NAVTEQ, a company that creates digital maps and map content for navigation and location based services. In May 2008 the European Union approved the GPS provider TomTom's takeover of Tele Atlas for 2.9 billion Euros. TomTom is a manufacturer of in-vehicle navigation systems – like the GPS in your car. Its principal rationale for the takeover was to obtain an asset of tremendous value – TeleAtlas' very large high quality geospatial database of street networks and associated information.[169] While TomTom eventually wrote down the value of TeleAtlas by 1 billion Euros, they still considered it to have been a good deal. Its chief executive said that "the rationale behind the acquisition still stands."[170]

Perhaps one of the most widely known examples of the rapid incorporation of spatial data into a completely different field was Google's introduction of free online access to maps and satellite imagery. For the first time, easily accessible satellite imagery covering of most of the Earth was made available for free to anyone with an Internet connection. Image quality for most urban areas was high enough for people to see the house where they lived and cars parked on the street. Areas of less interest to the general public tend to be at a lower resolution. The source of much of the geospatial data for this application was Keyhole Inc., a company Google purchased in 2004 for its massive database of map information and aerial and satellite imagery. While Keyhole was finding it difficult to grow its business selling high quality products at premium prices, Google saw the opportunity to spur tremendous growth in its advertising income by giving the data away!

For some time now big and successful global industrial players have not only been readying themselves for the GeoEconomy, they have been actively investing in it.

The growing importance of imagery and image analysis has recently been underscored in several other ways including outright purchase of, or special teaming arrangements with, companies for the purpose of gaining access to their geospatial technology and expertise. Imagery is being incorporated into many areas where only a few years ago it would never have been considered. For example, Microsoft bought Vexcel to obtain their digital camera systems. They also purchased GeoTango, which had developed technology to display 3-D geographic information on the web. We can also consider the fact that almost all of the formerly independent producers of image analysis software have been bought by major players from outside the geospatial world. With its interest in measurement and the growing importance of imagery, to access image analysis software Hexagon, a major European engineering firm, bought ERDAS and ER-Mapper through their Leica subsidiary, while ENVI was bought by ITT. Of the larger suppliers of image analysis software fifteen to twenty years ago, only PCI Geomatics remains independent at the time of writing – and PCI has bought several companies. Another indication is the way in which software producers have incorporated image-handling functionality in systems ranging from those dealing with engineering (Autodesk, for example) to GIS (ESRI), by either buying systems or software, or through licensing agreements with producers of such tools.

Perhaps the biggest contribution of convergence is that it has led to the opening of a large window on the geospatial world for investors and entrepreneurs who have been able to see the value of geospatial in reshaping the world into the GeoEconomy. Paradoxically, the geospatial specialists who created the basic geospatial tools and capabilities that sit at the base of the GeoEconomy have generally not benefited to the same extent as the more recent arrivals in the industry. These new entrants into geospatial are developing new applications and more profitable business models than most of the original creators of geospatial tools had ever imagined.

Paradoxically, the geospatial specialists who created the basic geospatial tools and capabilities that sit at the base of the GeoEconomy have generally not benefited to the same extent as the more recent arrivals in the industry.

4.2.3. Discipline Convergence

There has been what we call discipline convergence operating at two levels. The first is within the many disciplines that have come to rely on geospatial data in one form or other, while the second is within the geospatial world itself.

Governments the world over are now making decisions and following processes that in effect drive many disciplines together and as a result drive everyone into using geospatial information and tools. This discipline convergence is yet another factor underpinning the development of the GeoEconomy.

Land use planning, and road engineering and construction, are two examples of disciplines that have long relied on geospatial information, drawing on technical information (where can the road be built, where will the gravel come from) as well as economic analysis (which route will minimize the cost). Many years ago, technical and economic issues were the only ones considered. For example, in 1939 a new highway (now called the Queen Elizabeth Way or QEW) was opened in the province of Ontario between Toronto, Hamilton, St. Catharines, and Niagara Falls, ending at Fort Erie across the border from Buffalo, New York. The highway was built through unique fruit growing land and over time it became wider, busier, and has contributed to the development of a solid belt of suburbia along almost half or more of its length – destroying much of the unique fruit growing land. It is doubtful that the highway would be approved today.

Today geospatial technologies provide the means to assess issues from a broader perspective. Some years ago, as much as we may have wanted to look at these many environmental and land use competition aspects of a road engineering project, there wasn't the data and there wasn't the means to do what we can today. Today the road would likely have been built on the Niagara Escarpment, where it may have been more expensive to build but where its environmental impacts would have been less. Today, road engineering and construction must also consider environmental impact statements, endangered species, environmental damage during construction and after, noise pollution, public consultation, ground water hydrology impacts, and even archaeology. Why? Because there have been changes in social values. Fruit growing land is now seen as a scarce and more valuable resource – in part because widely available geospatial information, such as land use maps and statistics about the quantity and productivity of agricultural land, quantified how scarce and valuable it was. As a result, governments introduced land use regulations to control the loss of this agricultural land. Much more sophisticated analysis of projects that would reduce this resource (like highway construction) are now required. The sophisticated planning analysis needed to meet these regulatory needs can be required in part because geospatial data allows these many factors to be examined with relative ease. So the availability of geospatial data has contributed to the impetus to preserve scarce resources, the identification of what to preserve and where it is, as well as providing the means to control its loss and evaluate competing uses.

If we look at land use planning in the broader context there are even more factors – and disciplines – to consider. To work in this new model requires the integration of many disciplines and a wide variety of information that must all be relatable one to the other. As noted elsewhere in this book, the best way (and often only way) to do so is by bringing all of the information together on a geographic basis. Governments the world over are now making decisions and following processes that in effect drive many disciplines together and, as a result, drive everyone into using geospatial information and tools. This discipline convergence is yet another factor underpinning the development of the GeoEconomy.

The second part of discipline convergence is within the geospatial field itself. Integrating geospatial information in ever more inventive ways has had a profound impact on the disciplines that directly support the GeoEconomy. The terms geomatics (developed in Canada and in use in much of the English and French speaking worlds outside of the USA) or geospatial (used in the USA and almost everywhere else too) have been coined to refer to the group of disciplines and technologies that include cartography, elements of geography, geographic information systems (GIS), geospatial data infrastructures, global positioning systems (GPS), photogrammetry, remote sensing, surveying, and geodesy. (Geodesy is primarily concerned with precise measurement of the Earth on a global scale. Among its uses is the measurement of small changes in the earth's surface – such as the movement of continents, the rise of mountain ranges, or the rise in land surfaces that were covered by glaciers during the last ice age (termed rebound) in such areas as Canada's north). A formal definition of geomatics is "the science and technology of gathering, analyzing, interpreting, distributing and using geographic information."[171]

The sub-fields of geospatial or geomatics used to operate in silos of splendid isolation. It used to be that the remote sensing specialist rarely talked to the cartographer – but occasionally spoke with the GIS specialist. The land surveyor would occasionally speak with the geodesist, but largely ignored the rest. This is no longer the case. GPS and geospatial data infrastructures simply did not exist – although GPS is usually considered to be within geodesy. In recent years the overlap between these once independent disciplines has increased rapidly, driven by the benefits of using them together and supported by increased computing power and the development of the processing methods to implement them.

The sub-fields of geospatial or geomatics used to operate in splendid isolation in silos. This is no longer the case.

The importance of convergence can be seen from several examples. The first example involves geographic information systems or GISs. Until relatively recently GISs had little or no capability to handle remotely sensed data such as aerial photography and satellite imagery. Today, the image handling capability of geographic information systems is not only standard, it is expected that basic image enhancement and analysis functions are provided. Secondly, when the first image analysis systems were developed they did not include any provision to produce maps to photogrammetric standards. Today, these systems typically have at least basic map generation capabilities as well as photogrammetric display and analysis features, such as the display of stereo imagery and the measurement of elevation. They also can use data from geographic information systems for display and analysis, and output data in GIS-readable form.

While the teaching of these disciplines has expanded to include an awareness of the contribution of the others, increasingly one finds that GIS and remote sensing courses are taught in the user disciplines such as forestry, planning, and the like. At the same time, enrolments in the geomatics field seem to be decreasing in North America. This is a troubling trend as the rest of the world is entering the GeoEconomy and increasing the number of graduates in this field. While geomatics experts are needed, perhaps this apparent decline is a natural outgrowth of technology adoption across a wide range of applications areas. Today few people go to conferences or focus their studies on calculators and spreadsheets, but their use is ubiquitous. Similarly we expect that the application of geomatics to various disciplines will become ubiquitous – in forestry, agriculture, and engineering for example. Geospatial concepts and applications will increasingly be included in the courses of study in these other disciplines, and not just within the traditional geospatial technology fields.

4.2.4. Policy Convergence

Introduction

The complexity of societies has increased, in both developed and developing countries. However, the same policy issues mentioned in speeches outlining a government's priority seem to arise in both developed and many developing nations. These issues about which policies must be developed include security and privacy, responsible use of natural resources, sustainable development, protection of sovereignty, environmental quality, climate change mitigation, disaster mitigation and response, education and health care, and economic security of individuals. Democracies also usually express the desire to foster transparency in government and everyone seems to want to have efficiency in provision of government services. In effect, policies worldwide are converging on the same key issues.

> ***The same policy issues seem to arise in both developed and many developing nations: security and privacy, use of natural resources, sustainable development, protection of sovereignty, environmental quality, climate change mitigation, disaster mitigation and response, education and health care, and economic security of individuals. Democracies also usually express the desire to foster transparency in government and everyone seems to want to have efficiency in provision of government services.***

In order to successfully address these issues geospatial information and the capability to use it effectively will be essential. Geospatial information or "location" is the unifying tool that allows these policies to be viewed and discussed together. As well, as these other policies mature, they in turn will impact the further development of geospatial policy. The following discussion considers what we see as these key issues and illustrates the contribution of geospatial information. What is especially important here is that the key issues faced by society are almost all tied together by the need for geospatial information, as the following discussion will illustrate.

Security and Privacy

In democracies, balancing the need for security with demands to safeguard privacy rights, along with the call for transparency in government decision-making, is complex with many tradeoffs, and many of these tradeoffs affect geospatial data and associated technologies. Geospatial data can be used to view security, privacy, and transparency issues from a common perspective and thus resulting policies can be effectively developed and/or integrated.

At the same time, the convergence we have been discussing is speeding up the delivery of capabilities we aren't always prepared to handle as a society. Consider social networking for example. We have a need to develop privacy policies suited to the capabilities of new technology at the same time that our society's values are changing. As people adopt social networking technologies, those who do so are changing their attitudes about what personal information should be private. And as attitudes and technologies change, the newly developed privacy policies and associated regulations continue to affect the geospatial data policies that are being created.

In democracies, balancing the need for security with demands to safeguard privacy rights, along with the call for transparency in government decision-making, is complex with many tradeoffs, and many of these tradeoffs affect geospatial data and associated technologies.

For example, privacy and security policies, each supported by geospatial information, are both converging and conflicting. An individual's right to privacy often depends on where they are. People expect that what they do and say in public is somewhat constrained, but that their own home is private. But what if in the privacy of their home they are plotting a terrorist attack? Obviously too, different levels of security are needed for different locations. While both may be subject to terrorist attacks, a nuclear power plant is different from an amusement park. It is important to know where important facilities are located, as well as information that could affect the vulnerability of that location. But that information in itself can be of potential use to terrorists.

As well, the question of security and geospatial information, and the policies that have been developed for each, have been the subject of a number of studies in the USA and elsewhere with some fascinating results. As noted previously, a US study of over 600 government databases found that fewer than 6% would be of any use to an attacker and less than 1% contained data that was both of value and not obtainable elsewhere. It is more likely that limiting access to these databases would impede public services, such as emergency first responders, while doing little to improve security.[172] The net conclusion was that only a handful of the 600 data bases that were examined should have access to them restricted and be removed from the public domain – most were actually judged to be of more use to the responders to an attack than to the attackers.[173]

Responsible Use of Natural Resources

Most countries now consider it in their best interest to regulate natural resource use to control unwanted impacts on the environment and the people who depend on it. Policies and regulations differ among jurisdictions, but in setting limits on the way natural resources can be extracted and processed they define what the society considers "responsible use".

Responsible natural resource development makes extensive use of geospatial at every stage – from exploration (in terms of oil, gas and minerals) and feasibility studies, to submission of project applications, regulatory review and monitoring, to project completion and reclamation, geospatial information and analysis is integral to all these activities.

Clearly, sustainable development, however it is defined, depends on geospatial data and technology.

For example, to monitor natural resource use and its environmental impacts – be it ore from an open pit mine, fish from the ocean, or timber from a forest – requires information tied to location. Thus, natural resource policy development, must by definition, heavily rely on and be tied to the use of geospatial data. Samples collected in the field or observations made during inspections are only useful when tied to location. New geospatial technologies such as remote sensing are used to "see" what is going on. This might include measuring the volume of material extracted from an open pit mine using LiDAR data or the size and location of a forest clear-cut using aerial photography or satellite imagery. Remote sensing techniques can collect data on water temperature, sediment loads, erosion, and other factors that may indicate problems with resource development operations.

As well, as the world's population increases and populous nations like China and India strive to raise their standard of living, there is growing demand for natural resources of all kinds. The easy-to-develop non-renewable resources like minerals, oil, and gas were taken long ago. Now ever more sophisticated methods – of which geospatial technology is a part – are needed to find and develop new sources. Renewable resources, from forest products to wind power depend on the astute use of geospatial information to manage them effectively. Those with the geospatial know how will have an advantage over those without it.

In conclusion, without geospatial information and technology responsible use of natural resources is not possible. Thus policies governing natural resource development are increasingly converging with, and dependent on, geospatial policies, governing the capture of information and use of GIS technologies.

Sustainable Development

Sustainable development is a mandated objective (i.e. required by law) in many jurisdictions. Assessing the sustainable development effects of policies and projects necessarily requires the consideration of diverse environmental factors as well as economic, legal, and cultural ones. Most, if not all, of this data is tied to location, allowing the use of powerful geospatial analysis methods to analyse this diverse information together – rapidly and effectively. Assessing the justification for sustainable development projects and their planning and execution, has and will continue to depend on geospatial data and technology. Thus, following the discussion on resource management, policy development in the sustainable development area relies to a great extent on the use of available geospatial information.

Protection of Sovereignty

One measure of sovereignty is the degree to which a country has mapped the territory it has claimed. Typically policy relating to sovereignty requires some form of map information to illustrate the country's territorial claim. When territorial claims are made, such claims are increasingly being backed up through the use of a GIS to display the many different data layers supporting the claim. Canada pioneered the use of GIS in territorial claims in a case concerning an off-shore dispute in the George's Bank with the USA that Canada won before the international courts.

Having claimed borders and asserted their location is of little value unless this claim of sovereignty can be enforced. The protection of sovereignty can be as simple as patrolling coastal waters with gunboats, or patrolling inland borders with ground or air-based military forces and surveillance systems. In some countries more complex technology solutions are employed, such as coastal radars that identify the presence of incoming ships, or all-weather radar satellite surveillance. In all cases, geospatial information in the form of a map or other more complex technology is used to both make and enforce sovereignty claims.

> ***Traditionally, military and civilian geospatial data collection has been done separately – although this too is changing in part because of convergence.***

Traditionally, policies in some countries asserted that military and civilian geospatial data collection be done separately because, while their information needs overlapped, they were also different – and the military wanted to maintain security of sensitive information – although this too is changing in part because of convergence. Many countries had policies that restricted public access to large scale maps and high resolution aerial photography for reasons of security and some still do. Countries with their own reconnaissance satellites typically kept information about their technology and the imagery generated secret. As the number of countries with earth resource satellites has increased and the technology has improved, competition has made high resolution satellite imagery widely available in the civilian domain, albeit sometimes with limits on which countries can access the data. We see convergence of sovereignty policy and geospatial policy, affecting the level of things that can be mapped, and to whom they may be distributed.

For example, in the United States, a Presidential Directive mandated that the military purchase imagery from civilian sources whenever it could suit the application. Several earth observation satellites have been designed as dual use platforms, providing lower resolution imagery for civilian use and higher resolutions for military intelligence. In other cases the military has simply used the civilian data. Cost saving has been a major factor in these decisions, as has the desire to provide imagery to the soldier on the ground – but not imagery that is regarded as secret in case it falls into the wrong hands.[174] There has also been a convergence of needs between the civilian and military communities, which is supported by geospatial information. Both require, and can now afford, high resolution imagery where ground detail is optimized for mapping applications. The convergence in needs is now being reflected in the increased sharing of data and some technology – a trend that is likely to continue both to save costs and also because the collaboration has proved mutually beneficial.

Environmental Quality

Remote sensing is widely used to monitor environmental quality from global to local scales. From monitoring the thinning of the ozone layer over Antarctica and detection of oil spills at sea to the monitoring of resource extraction, society has come to depend on the information produced by this technology. Effective use of this and other environmental data depends on a wide range of geospatial technology from image analysis systems to extract information from imagery to geographic information systems that integrate and analyze the diverse environmental data sets needed to assess environmental quality and develop action plans such as mitigation measures.

In 2009 there was considerable discussion within the United States about the carbon output of the oil sands of Alberta, Canada. A map was subsequently produced by the National Atlas of Canada that showed far greater carbon emissions from coal-fired electricity generating plants in a number of US states. Just one map (shown below) removed an irritant from bilateral discussions and clearly demonstrates the power of geographic information.

Measurement of environmental factors, such as UV levels or contaminants causing pollution, generally requires some form of direct measurement in the field. An essential element in measuring any indicator of the quality of the environment is the measurement of "where." The "where" factor becomes even more important if one wishes to return to exactly the same location to measure the same factor again. As an example, remote sensing is used to verify the condition of secondary oil containment to enforce, for example, the USA's 1973 Oil Pollution Prevention regulation for the Secondary Containment of Oil. This regulation is found under the Oil Spill Prevention, Control, and the Countermeasures Program that addresses the oil spill prevention provisions contained in the Clean Water Act of 1972.[175]

To some extent the monitoring and regulation of environmental quality has become institutionalized. Governments now have major departments or even separate organizations such as the US Environmental Protection Agency (EPA), responsible for developing environmental policies and associated regulations, and ensuring compliance with them. These organizations could not function unless they were geospatially-enabled and, to at least some extent, the use of both geospatial technology and data has been built into most environmental policies.

Measuring and Mitigating Climate Change

Although the causes of climate change may be controversial, governments around the world have recognized the need to address the climate changes that are occurring. There has been a dramatic reduction in the size of massive glaciers, measurable rise in sea level, changes in permafrost and the extent of Arctic ice, and shifts in weather patterns.

The climate change issue involves global processes and has effects that range from the global to the local scale. A rise in sea level affects all coastal cities, some more severely than others depending on local characteristics. Changes in glaciers and snow pack will affect the availability of water in what are often otherwise near deserts in Western Canada, Chile,

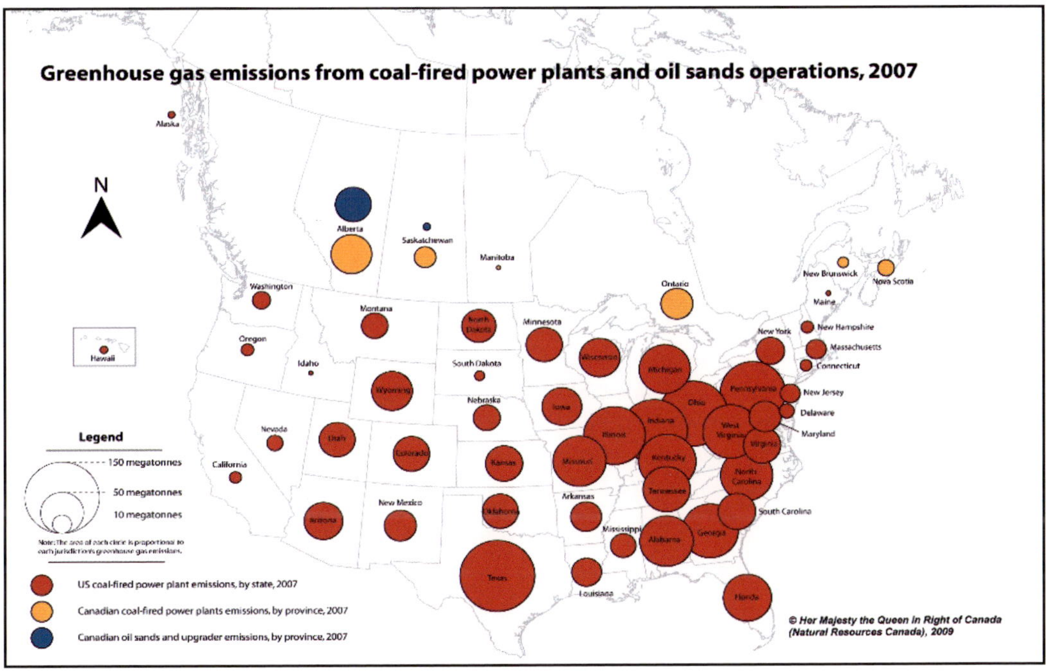

Greenhouse gas emissions from coal-fired power plants and oil sands operations 2007. Source: Map created by National Atlas of Canada, used with permission, Natural Resources Canada, Ottawa.

and other locations in Europe and Asia. The demand for global action has dramatically increased the use of existing global data sets and the demand to produce additional ones has encouraged cooperation in data collection and dissemination. The low cost of online data dissemination, public demand for digital geospatial data, and the notion of public benefit have led many governments to make much of their data available free to the public and in doing so provide access to the world's researchers. Mitigating climate change is even more difficult than monitoring it, involving more complex research, advocacy, and negotiation. Geospatial information plays an important role in them all and its use has become an integral part of governmental climate change and adaptation policies.

Disaster Mitigation and Response

Natural disasters occur in every country and some industrial accidents attain disaster proportions as well. The 1986 explosion at the Chernobyl nuclear reactor in what was then the Soviet Union dispersed airborne radioactive material far beyond that nation's borders.

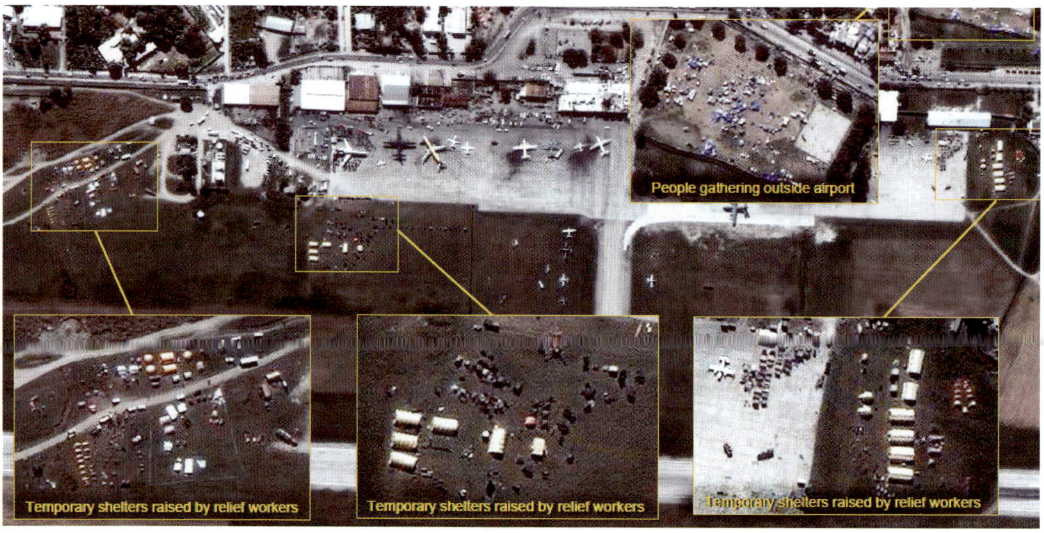

Satellite Imagery in Support of the Haitian Relief Effort 2010. The devastating 2010 earthquake in Haiti crippled the country's infrastructure. Commercial and government satellite image providers tasked their systems and donated imagery that provided emergency responders with multiple images each day that helped identify locations of congregated survivors and areas in greatest need. Imagery was also used in both delivery and monitoring of relief efforts. Source: DigitalGlobe

Events like the 2004 Asian Tsunami that struck 13 countries with devastating effect, killing a quarter of a million people and injuring half a million more,[176] as well as floods, forest fires, earthquakes and landslides cause severe losses of life and property. Many countries have response policies and plans to address such events and the use of geospatial information significantly improves the speed and effectiveness of emergency response, the planning of relief efforts, and damage assessment – especially in the world's coastal zones. Whether it is GPS positioning to coordinate delivery of medical aid and supplies, or the use of imagery and GIS to assess damage and direct relief, the use of geospatial data and tools enables relief efforts to be more quickly delivered and more effective – saving lives and realizing greater benefits from limited resources. Indeed, the amount of geospatial data now available is itself becoming a problem. Following the recent earthquake in Haiti one of the major requirements was to bring in a consultant who could coordinate the avalanche of geospatial data that was being made available. Use of geospatial information and technology is an integral part of the disaster response policies of all countries.

Education and Health Care

All countries provide some level of health care and education for their citizens. Most countries give high priority to improving these services, although their specific needs may be very different. Developing nations tend to have rapidly growing populations of predominantly younger people, straining what are commonly limited health and educational facilities. Developed nations tend to have populations with an older average age, increasing more slowly, but requiring more health services as they age. With the use of geospatial information, planning and delivery of both education and health care in developed and developing countries improves the effectiveness and appropriateness of the services delivered. Where should the school (or hospital) go? What diseases are most prevalent and where? Which areas have the greatest need? How far are they from distribution centres, and how long would it take to deliver emergency care? Where is the young population concentrated? Are there enough schools close enough for them to reach? Where should new schools be built? Where should schools be converted to other uses? These are all questions geospatial information is being used to address. Not surprisingly, geospatial information is now routinely being used in many states and countries in the development of both health care and education policies.

Economic Security of Individuals

As the GeoEconomy has become more widely recognized so too has the importance of geospatial information in maintaining the economic security of individuals – be they located in Thailand, Bermuda, the USA, or elsewhere. Professionals in geomatics have long argued that location-based or "geospatial" information (such as digital maps and satellite images) and related technologies play an important role in enhancing public and private sector decision-making and in the management of policy, social, environmental and economic development issues. In Canada, workshops on this topic sponsored by the Federal Government and various provincial agencies have been given by the Geomatics Industry Association of Canada. The investments by industry in this field suggest that industry is listening and is reaping the benefits of investing in geospatial data and technologies. But are governments listening as intently as they should? We suggest that there should be more attention paid to linking economic policy with geospatial capabilities.

We have performed several analyses on this topic – including the development of economic clusters based on geospatial data and technologies. (Economic clusters are groupings of business, university, and government organizations that possess expertise and resources that could be shared for mutual benefit. By encouraging such organizations to locate near each other and develop collaborative relationships they can be more successful, increase economic activity, and attract new investment.)

Those governments that don't understand the importance of geospatial information are putting the economic and physical security of their people at risk.

We have found that many governments, while somewhat familiar with geospatial information (you can't run a government without it), do not fully grasp how their planning and decision making processes would be changed if they incorporated the use of more sophisticated spatial analysis tools into their ongoing operations. Because competition becomes ever more global and more intense each year, choices governments make have far reaching economic impacts on their citizens. Take for example a jurisdiction's decisions about royalties or taxes. It not only has financial effects on the citizens and companies operating there, it could affect whether resident companies choose to expand or downsize, whether foreign investment is attracted, or whether neighbouring jurisdictions will appear more attractive and lure away investment and the jobs investment brings. A more closely developed linkage between geospatial and economic policies would allow more sophisticated use of geospatial tools that can broaden the decision making perspective to take more of these issues into account. There is always the option of not changing what has worked in the past, or changing slowly. However, in a globally competitive world these options risk being placed at an economic disadvantage. We see geospatial capabilities as a key factor for governments to recognize and assess the economic impacts of the choices they make.

Simply stated, those governments that do not understand the importance of geospatial information are putting the economic and physical security of their people at risk.

Transparency

Geospatial technologies have changed the degree to which governments, businesses, and individuals can control the information that can be publicly known about them and their activities. And organizations have to take this lack of control of information into consideration when developing policies. Not long ago, authoritarian governments could effectively suppress information about activities within their borders. The build-up of military forces and weapons systems, radioactive contamination of large areas, release of toxic wastes, genocide, forced relocation of populations, and mass starvation are but some of the many secrets governments have traditionally been able to hide. There has always been some "leakage" of information by word of mouth and organized surveillance. Spies have probably existed as long as there has been human conflict. However modern electronic communication, radio, and today's geospatial technologies, such as aerial photography and satellite imagery, have given the public access to information on what governments and other organizations are actually doing. These technologies have increased transparency. Today, policies surrounding access to information are continually being shaped by advances in geospatial capabilities.

Perhaps the first major advance in transparency provided by geospatial technology was the launch of civilian earth resource satellites that produced imagery of almost the entire planet for public purchase. We have already referred to the use of SPOT data for monitoring Chernobyl. The emergency was eventually acknowledged and countries in the path of the cloud of airborne radioactive material scrambled to advise their citizens on ways to minimize their exposure. Policies preventing distribution of information were no longer realistic.

Today services such as those provided by Google and Microsoft provide historical imagery (generally 1 to 3 years old) of most of the planet online for free. Civilian satellite image providers such as GeoEye and Digitalglobe, in addition to their commercial sale of imagery, post news worthy imagery on their websites viewable at no charge and rapidly provide imagery for emergency relief at no charge. Many geospatial data policies now encourage the ready dissemination of imagery for public use.

With camera equipped cell phones now common place, some GPS-equipped, and a plethora of sympathetic supporters with Internet access and expertise, it has become virtually impossible to prevent global dissemination of public events – a fact affecting all policy development. Short of violent suppression or the complete stoppage of cell phone use (both of which do occur) there is now almost no way to suppress information that the public knows is available. In 2009, after a contentious election in Iran, the government forbid media within the country from reporting on the public demonstrations. Yet the authorities could not stem the flow of pictures, video, and verbal reports of these events from citizens on the scene. Assessing these eye witness images and tying them to their locations provided the world with a flood of geospatial information with which to analyse the scope of the unrest.

Access to geospatial data is also a key component of transparency, enabling public awareness and analysis of controversial issues. The demand for transparency has affected data policy decisions in some jurisdictions. For example, in 2006 the European Parliament adopted the Aarhus Convention on Access to Information. As noted previously, the Convention asserts the right of the public to receive any environmental information held by public authorities within one month of the request, without explaining the reason the data is wanted or how it will be used. This directive applies to information on the state of the environment, policies or measures being taken, or on the state of public health and safety where affected by the environment. Much of this information is referenced to location and hence can be used in geospatial analyses.

4.2.5. The Active Use of Converging Geospatial Information

There are entire books devoted to the use of geospatial in many of the subject areas mentioned in the preceding subsections. However, it is interesting that there have been few broad-brush studies on what geospatial data have actually been used for at the national level. This may, of course, be one of the reasons why geospatial agencies have had such a difficult time convincing their political masters that investments in the technology are important – and will pay off.

One study we did which was national in scope saw interviews conducted across the Government of Thailand, to assess the extent to which remote sensing data were integrated into the daily management of Thai resources. The results may be surprising to those unaware of the sophistication of both Thai science and technology and the Thai Government. In fact, based on almost any measure one can come up with, it would appear that more satellite data are used by Government agencies in Thailand than in Canada. In some agencies as many as 70 to 100 people were users of the data, while in others it was fewer than twenty. In virtually all of the agencies interviewed, operational needs demanded continuing access to satellite remote sensing data. This demand led to the launch of Thailand's own satellite, THEOS, which is now providing imagery around the world.

Operational Satellite Remote Sensing Applications in Thailand[1]	
• Forest inventory • Forest change detection • Forest fire • Human impacts • Mapping and monitoring fishing activities (freshwater, coastal, marine) • Aquaculture monitoring • Coastal shrimp • Vessel detection and monitoring • Environmental feasibility studies • Security • Municipal/local area tax mapping • Mineral resource mapping • Field work planning	• Land use mapping at various scales • Crop type mapping (12 major crops) • Crop area & crop statistics • Statistical sampling frame development • Land cover mapping • Flood zone mapping • Landslide mapping • Disaster and risk management and reduction • Soil mapping • Route selection and corridor planning and terrain suitability studies • Base mapping (topographic) • Water management
[1]Source: Ryerson, RA. And C. Peanvijarnpong (2007)	

Another study on the value of geospatial data done in Australia and cited previously[177] clearly demonstrated the importance of geospatial information to the economy. Careful reading of that study shows the importance of the convergence of a number of technologies in the generation of significant economic benefits.

4.2.6. Conclusion

This section has shown how convergence of many types has occurred (policy, technology, discipline), and resulted in a growing and seemingly insatiable demand for integrated geospatial information.

> ***The GeoEconomy has been joined – the GeoEconomy train has left the station. Those who use geospatial data inefficiently (or not at all) will be left behind.***

The bottom line here is simple: geospatial data are now being used routinely around the world to address major policy issues of governments and industry. We have shown a direct link between geospatial information and major policy issues faced by almost every government in the world. Together these dependencies and the demand for integrated geospatial information are harbingers of the GeoEconomy. The GeoEconomy has been joined – the GeoEconomy train has left the station. Those who use geospatial data inefficiently (or not at all) will be left behind.

4.3. The Result: Societal Impacts

4.3.1. Introduction

Geospatial technologies have transformed our society and its infrastructure. They have changed the way people think, what they perceive to be possible, and what actions they choose to take. People now think differently. Many people no longer look at a map. They look at a Blackberry or smart phone and expect to get an accurate answer. As a result what is being sold (and what can be sold) by the mapping industry is forever changed.

> ***Geospatial technologies have transformed our society and its infrastructure. They have changed the way people think, what they perceive to be possible, and what actions they choose to take. They have also changed the balance of power, proffering advantage to countries, enterprises, or individuals who discover, develop and implement applications that give them a competitive advantage.***

Geospatial technologies have also changed the balance of power, proffering advantage to countries, enterprises, or individuals who discover, develop and implement applications that give them a competitive advantage. Simply stated, having better geospatial capabilities than your competitor (be it a company or a country), not only puts you ahead, it makes the rate of development faster so the gap becomes greater. The gap between the have and have not countries could widen. Some developing countries are skipping ahead – which is good for them. The converse is that those that don't skip ahead will fall farther behind economically and the gap will widen over time.

In general, more developed nations tend to have a better developed geospatial infrastructure. In part, this is related to the cost of the technology, availability of trained personnel, and the foresight of government leaders. To some degree, it may also be that societies with a better developed geospatial infrastructure have an advantage that contributes to their continued success.

While geospatial technologies are not as widely implemented in the developing world, the incremental advantages of adopting them can be proportionately greater than for developed countries – when you begin with very little mapping, for example, producing state of the art maps for urban areas can be of inestimable value.

For this reason, less developed countries have, in many cases, passed over some of the traditional approaches to mapping – approaches that simply did not work for them in the past. In some respects this has given less developed countries an opportunity to catch up, seeing as they did not have investments in the older technologies to which more developed countries are still often tied. In many cases ties in so-called developed countries, to both less efficient and less accurate geospatial technologies, lead to a certain degree of inertia. Some of the developing countries moving from 1960s technologies to technologies of the new millennium are the same ones that skipped over land lines for telephones and went directly to cell phones.

Following is a general overview of the societal impacts of geospatial technologies in terms of raised expectations, competitive advantage, and a more holistic approach to decision making. Societal impacts on the major issues affecting people and governments today are also addressed.

4.3.2. Raised Expectations

The increased service levels that geospatial technologies provide have raised expectations about the level of service that should be delivered and the capabilities of the information infrastructure provided by governments. For example, in urban centres throughout Canada and the United States, the public expects that a 911 phone call will connect them to an emergency response coordinator with access to fire, police and ambulance services. Adoption of geospatial technologies enables emergency responders to immediately have the address of the caller from the phone number location of the incoming call (if it's a landline – the tracking of cell phone locations is just beginning to be implemented). It is geospatial technology that can determine which responder can reach the scene most quickly, determine the optimal route to use, and, if necessary, provide step by step voice command driving instructions. While the public may not be aware of all the technology supporting their emergency response services, their expectations have been shaped by what they experience.

> ***The increased service levels that geospatial technologies provide have raised expectations about the level of service that should be delivered and the capabilities of the information infrastructure provided by governments.***

Similarly, it is common knowledge that there are services constantly providing weather information for locations around the globe, information that is available from the radio, television, and on the Internet. So, when concerns about climate change prompted research and calls for action, it was accepted that the technology existed or could be developed to measure and predict the Earth's climate. Without geospatial technologies these activities would not be possible. Yet with little knowledge of the technologies supporting these efforts, the public expects them to be "doable." If society expects these capabilities to be available the debate then switches from whether or not to use geospatial technology to how best to provide the technology.[178]

4.3.3. Competitive Advantage

It is often said that knowledge is power. Today power is shifting towards those with the knowledge and skills to effectively use geospatial information. Some understand the fundamentals of geospatial technology better than others and have the creativity to imagine novel applications. A now familiar example is that of Google Inc. providing free online access to maps, images, and navigation aids as a way of attracting viewers to the advertising that Google sells, tied to the locations for which imagery or maps are requested. Others seek to duplicate their success, but the multi-year lead of the first to market has given Google a strong competitive advantage that Microsoft and others are striving to match.

> ***It is often said that knowledge is power. Today power is shifting towards those with the knowledge and skills to effectively use geospatial information.***

There are innumerable small scale examples of clever applications that deliver valuable results and provide a competitive advantage. For example, in the 1970s and early 1980s, the uses of imagery from earth resource satellites such as Landsat were not widely known by the public. A small western Canadian company had developed a business harvesting wild edible mushrooms. The business depended on looking for mushrooms in areas where they are most likely to grow. These areas were in certain types of forests the year after a forest fire. By chance they saw an image in a magazine that showed forests and forest fires. They went to the author, who was at that time involved in advising people on the use of the imagery, to ask if they might obtain and use such imagery. In effect they developed a simple map of where the forests were that would produce "good mushrooms" after a fire and then mapped where there had been forest fires over the past year. The application was simple and it worked. Using satellite imagery they could rapidly survey an area to find promising sites and determine the best way to access them. They did not understand satellites. They were not experts in image interpretation. They did not care about geospatial technologies. But they did understand mushrooms and how to use the technology to improve their odds of finding them, giving them a competitive advantage.

Image of Magong City, Taiwan Acquired by Thailand's, THEOS Satellite. Source: © 2009 GISTDA

More recently the government of Canada decided that it wanted to have better results when the winter Olympics were held in Vancouver in February 2010, than it had when the Olympics were held in Montreal in 1976 and Calgary in 1988. Canada had never seen one of its athletes win a gold medal on home soil. To end this drought a significant pool of money was created, some of which went to fund research. In one of those research projects Gerard Lachapelle, one of Canada's leading experts on GPS, was involved in a research project using an advanced GPS carried by ski-racers. It continuously measures a skier's location on the ski hill to within a few centimetres. The information that the system collected on training runs was used to select the fastest equipment and the best way to attack gates and turns. Perhaps in some respects the results were too good – a number of the best skiers were hurt in accidents, skiing faster than expected, before the games even began! On the curling ice, heat sensing and other equipment and technologies were used to improve both the delivery of the curling stone and sweeping (which helps control how far a curling rock's path curves as it moves down the ice and how far it travels).

The fact that Canada won 14 gold medals, more gold medals than any country has taken home in the history of the winter Olympics, attests to Canadians' skill and determination – to which home-grown geospatial know-how contributed![179]

The geo-advantage can be measured in several ways, most of which come down to a monetary advantage.

The geo-advantage can be measured in several ways, most of which come down to a monetary advantage. It can be measured in time – the ability to act sooner, or with greater efficiency. In other cases it may be measured in capability – the ability to do more with less. The geo-advantage can provide improved quality of product or service for the same price, and it can improve our standard of living. Sometimes data are made publicly available as a public good, such as free access to the national map data that many jurisdictions provide. In other cases data can be treated as a trade secret allowing an entrepreneur to secure mineral rights to valuable property by knowing prime locations ahead of the competition. This explains the interest in remote sensing by exploration geologists (and their willingness to pay for it).

Governments can (and often do) gain advantage by better planning of military actions to counter insurgency, or by advising farmers to use specific fertilizer applications to increase crop yield and reduce costs thereby using land more efficiently. We discuss in more detail in Chapter 6 how farmers in Bangladesh are advised to plant certain crops after flooding to maximize food production.

4.3.4. A More Holistic View

As geospatial technologies have been adopted by government, business, and individuals they have had a profound effect not only on the quality of information that is produced but also on the decision processes that are used.

As geospatial technologies have been adopted by government, business, and individuals they have had a profound effect not only on the quality of information that is produced but also on the decision processes that are used.

Geospatial technologies promote a geographic approach which is by definition a holistic view that helps us to understand our surroundings and the interrelationships that exist. For example, understanding the relationship between climate, soils, and topography is used to predict agricultural productivity.

Geospatial technologies (such as GIS) allow more factors to be taken into account in making trade-offs among competing objectives. For example, in choosing a route for a new roadway a geographic approach promotes consideration of physical and human factors such as the environment, existing land use, terrain characteristics, social impacts, engineering constraints, and cost. The ability to rapidly analyze and re-analyze different designs in a GIS allows a large number of scenarios to be tried and modified, working towards an optimal solution. Such an iterative approach would be too costly to develop using manual techniques.

Using a broader perspective increases the range of possibilities that are considered and increases the chance of finding novel and workable solutions to the problems we face.

4.3.5. Environmental Assessment and Monitoring

Geospatial technologies have made possible the collection and analysis of environmental data not possible using other methods. For example, satellite imagery is used to generate global sea surface temperature maps on a daily basis. These data are used in analyses ranging from weather forecasting to marine ecosystem studies. Before this remote sensing technology was available there was simply no other way to map global sea surface temperature – a critical factor in understanding our environment.

Geospatial technologies have made possible the collection and analysis of environmental data not possible using other methods.

Much of the environmental data now routinely used is generated from remotely sensed imagery. A constellation of earth observation and meteorological satellites continuously collects image data and other sensor measurements, providing regular coverage of most, if not all, of the Earth's surface with spatial resolutions ranging from a few kilometres to better than half a metre. Airborne sensors are used to collect data with higher spatial resolutions. A wide range of environmental data is derived from this imagery including land use and land cover, vegetation health, terrain characteristics, water quality, sea surface parameters, ocean currents, and sea ice extent and characteristics – to name but a few data types. Data sets with global coverage have been of particular value in climate change and other studies of global-scale processes.

Geospatial technologies have also revolutionized the collection of field data. First, remote sensing data allowed one to select the best sample sites. Then GPS allowed the field worker to return to exactly the same location. Instead of hand written notes and sample site locations estimated from viewing maps and air-photos, data collection can be recorded digitally into handheld or portable computers, with locations automatically determined using GPS satellite positioning. The data can be examined immediately, together with other data sets using GIS software and, if necessary, the data can be used to update a remote database in real-time when a wireless connection is available.

The procedures used for environmental assessment and monitoring (which are fundamentally a series of repeated assessments) are dependent on geospatial technologies both to collect the data and perform the analyses. The technology has improved the detection and quantification of environmental degradation and the prediction and mitigation of environmental problems before they occur. Satellite imagery has been particularly valuable to assess and monitor environmental conditions in remote areas and areas where access is denied. For example, remotely sensed data and GIS technology are commonly used to monitor and quantify the loss of rainforest in the Amazon Basin, the changes in the Aral Sea, impact of the development of the oil sands in Alberta, and the advance of the Sahara Desert.

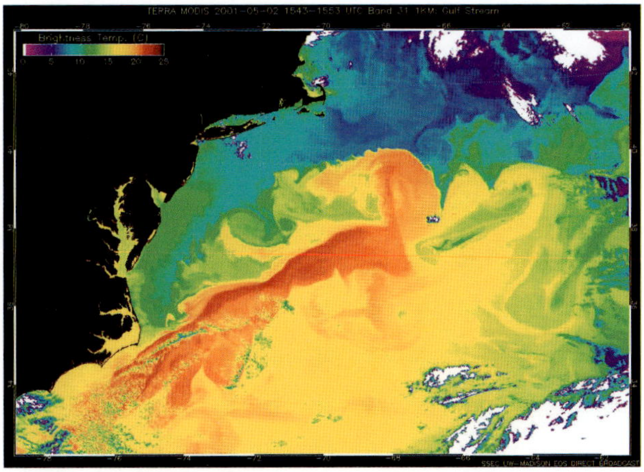

Satellite Image of The Gulf Stream. The Gulf Stream, a warm water current off the Atlantic Coast of North America is clearly visible as an orange colored feature in this thermal infrared image from NASA's MODIS sensor. Source: University of Wisconsin-Madison Space Science and Engineering Center

4.3.6. Government Efficiency and Transparency

Geospatial technologies have enabled governments to improve the dissemination of useful spatially-based information. Web-based mapping and data access allow data to be shared among organizations and offer public access to information. For example, in Canada with web-based access to the National Atlas, the public can retrieve a wide range of information and statistics as well as produce customized maps. Government sponsored information portals such as GeoConnections in Canada and the GeoData site in the United States offer powerful geospatial search capabilities with which to find geospatial data residing in diverse databases both nationally and internationally.

Geospatial technologies have enabled governments to improve the dissemination of useful spatially-based information, leading to transparency and increased efficiency.

Geospatial technologies are also used extensively in the management of urban infrastructure. It is standard practice for GIS functionality to be incorporated into municipal information infrastructures. This has improved the efficiency of operations, the delivery of services to the public, and planning. For example, maintenance tasks that require excavation of road and sidewalk areas need to take into account the multiple facilities at the location. Water mains, sewer lines, electrical conduits, gas lines, and other infrastructure may all be present and potentially damaged. As well, by optimizing work order planning, multiple tasks to be done in the same location can be scheduled together.

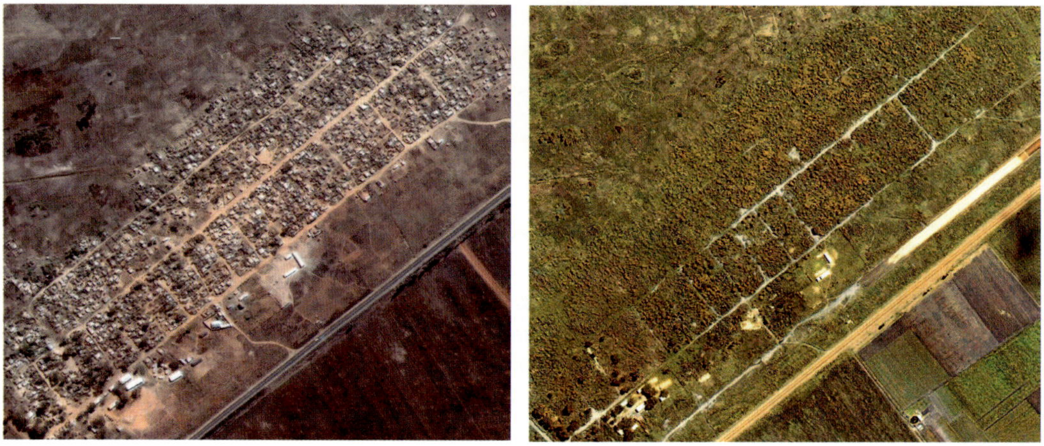

Satellite Images of Porta Farm, Zimbabwe. Amnesty International used satellite imagery from June 2002 (left) and April 2006 (right) and expert analysis by the American Association for the Advancement of Science to corroborate and document reports that in June 2005 the government had forcibly evicted people and destroyed their homes[181] Porta Farms was one of the locations that Amnesty International documented using satellite imagery. Source: Digitalglobe

Public advocacy has been empowered with the availability of online geospatial data sources and analysis tools. For example, a group of organizations working together as the Conservation Commons developed the Web-based Conservation GeoPortal. This free Web site, operated with the support of the National Geographic and ESRI, provides services to find, evaluate, and share data and maps for conservation. The portal does not store or maintain data. It provides access to data sets and resources held by contributing organizations. In addition to powerful search capabilities, the site provides a map viewer to display, combine, and share live map services. In other words, it provides direct access to the remote databases where the information is being stored and maintained. The site provides information about protected areas, conservation priorities, species distribution, habitats, socioeconomic data, satellite imagery, base-map layers and threats to biodiversity. Particularly valuable is the capability of the Map Viewer to overlay data layers obtained from different sources, manipulate the data, and output the results. In effect, conservation advocates the world over are provided with access to datasets hosted by contributing organizations worldwide. They then have basic display and output tools to generate information tailored to their needs. Providing such broad access to information facilitates incorporation of the best and most current available information into conservation decisions and communications, improving the benefits of conservation efforts.[180]

Public advocacy has been empowered with the availability of online geospatial data sources and analysis tools.

Geospatial technology and information has been a critical factor in the progressive increase in transparency of human activities to public scrutiny by the world at large. Despite security concerns heightened by terrorist attacks and other threats to society, increased transparency is not only a powerful check on the irresponsible use of power. In making deception more difficult it contributes more to enhancing global security and world peace than any malevolent uses it might have. It is useful to recall that only because their spy satellites could verify compliance were the United States and the Soviet Union in 1972 able to sign the first Strategic Arms Limitation Treaty.

Watch dog organizations like Amnesty International and many environmental advocacy groups make extensive use of satellite imagery to confirm activities that authorities would like to hide. The extensive historical record of satellite imagery dating back to the 1970s is a valuable public repository readily drawn upon to document where and when events occurred.

4.3.7. Health and Safety

Geospatial technologies have improved the delivery of health services and the monitoring and prediction of disease spread. Emergency case information, together with street network data, can be used to optimize the allocation of emergency equipment and services. Similarly data on crime events are used to optimize police patrol areas and coverage. Geography has long been used in the study of disease occurrence and spread. The link between cholera infections and the use of water from certain public water pumps in London was discovered by John Snow, said to be the founder of epidemiology, by analyzing a map showing the locations where cholera patients lived and the water pumps they used.[182] With today's geospatial technologies and the availability of digital records of disease incidence, these studies are more easily done. For example, West Nile virus has spread gradually northward in North America. Using data on vegetation cover, climate, disease incidence and other information, predictions of the likely spread of the disease could be made. As mentioned elsewhere in this book, the FAO has looked at avian influenza in much the same way. Such applications enable public health services to better anticipate service demands and plan mitigation measures.

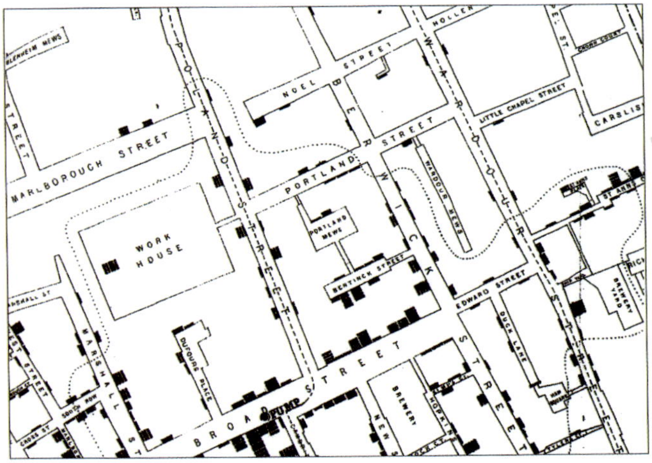

Snow's Map of Cholera Deaths Near the Broad Street Pump in London published in 1855. Source: Adapted from 'Dr. Snow's Report (1855)' at 'The John Snow Archive and Research Companion' online at http://johnsnow.matrix.msu.edu. Used with permission

Emergency response by police, fire, and ambulance services, without the aid of geospatial technologies, depended on the drivers choosing the best route to the site based on their experience and perhaps anecdotal traffic reports. Geospatial technologies can now provide in-vehicle routing, with real-time updating of traffic conditions to determine an optimal route. The location of the vehicle can be tracked and displayed on a map showing the route and progress of the vehicle, and voice commands can give explicit instructions to navigate to the emergency scene. As a result, an emergency vehicle can be on site significantly faster than was the case in the past. For many emergencies, minutes count. For example, the chance of surviving cardiac arrest is decreased by 7-10 percent for each minute that passes after collapse.[183]

Geospatially enabling emergency response dispatchers has dramatically improved emergency response times. Incident address locations are generated from the telephone numbers of incoming calls and used to optimize the routing of responders. While easily implemented for land lines, determining the location of mobile phones requires service providers to implement enhancements to their networks. Recognizing the importance of geographic positioning for emergency response, the United States' 1999 Wireless Communications and Public Safety Act mandated that cell phone service providers implement the capability to determine the geographic position of a cell phone within 300 metres.

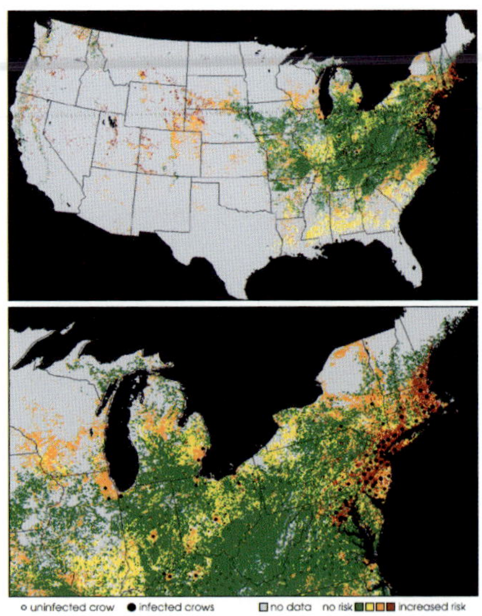

Modern Disease Mapping – West Nile Virus Incidence in the United States. Satellite imagery and data on infection rates from national and state health departments were used to generate this incidence map for West Nile Virus. Source: NASA

Another aspect of emergency response is management of the recovery process. The use of geospatial data improves the effectiveness of relief operations by optimizing the allocation of the limited resources and their delivery, and providing information to help prioritize relief activities.

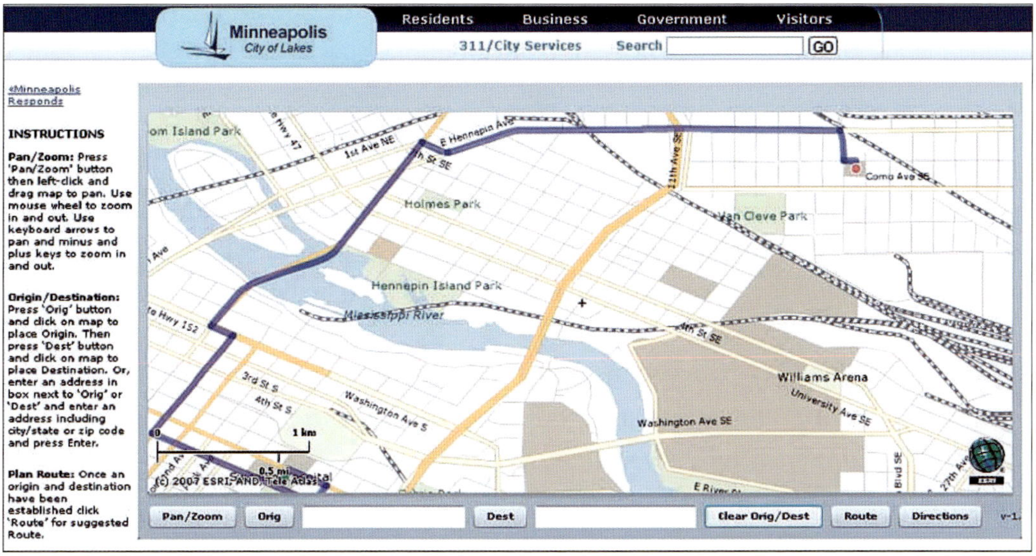

Minneapolis Online Alternate Routing Service. After a downtown bridge collapse, effective re-routing of commuter traffic was essential to maintain the economic productivity of city center enterprises. An online service was developed in less than a week that allowed commuters to obtain a map and driving instructions to and from their home and work place which was updated in real-time as traffic was re-routed during repair work. Source: City of Minneapolis

A novel example of the use of geospatial capabilities to improve municipal operations and public safety was a rerouting application developed for the city of Minneapolis after the collapse of a major bridge in 2007. The bridge was a major transportation artery in the city. The disaster recovery operations required street closures that changed over the course of a day, creating traffic tie ups and long delays. To keep businesses open and get people to work, information on current street closures and alternate routing needed to be disseminated. Since street blockages changed frequently, it was difficult to provide up-to-the-minute information. To address this need, web-based applications were quickly developed that tracked street closures and provided routing information. Citizens could access the web site, see the most recent barrier updates, and create personalized travel routes by entering addresses or by selecting start and end points on a map display. A route was then instantly calculated and drawn that could be printed out and taken on the road.[184]

4.3.8. Industrial Efficiency

Powerful applications of geospatial data in business (termed business geographics), have become standard practice to improve operational efficiency, reduce business risk, and improve competitiveness. Integrated analyses of business data coded to location such as demographics, consumer spending, and business locations, together with geography and infrastructure datasets such as street networks, postal codes, political boundaries, municipal zoning and imagery can be performed in a GIS environment. Typical business applications include market and competition analysis, finding optimal sites for new business locations, optimizing inventory storage and delivery routing, evaluating the success and profitability of existing facilities, and analyzing customer data to improve operations, identify new markets, define sales territories, and target customer development and advertising. For example, by combining information such as sales data, demographics, and competitor locations with geographic data such as census boundaries, sales territories, and store locations, businesses can better understand their market, their customers, and their competition.

> ***With increasingly broad and easy access to geospatial tools, the "big brother" watching us need no longer be big business or big government – it can be your local corner store, the dry cleaners on the corner, or your friendly hardware store or pharmacy.***

With increasingly broad and easy access to geospatial tools, the "big brother" watching us need no longer be big business or big government – it can be (and likely is) your local corner store, the dry cleaners on the corner, or your friendly hardware store or pharmacy.

4.3.9. Agriculture

Food is a critical resource. Monitoring its production is a critical activity almost everywhere. National and international organizations use remotely sensed imagery together with ground sampling and other data to monitor the growth of important food crops during the year. For example, in Canada the Crop Condition Assessment Program provides cropland and pasture monitoring information for agricultural regions in Canada and a portion of the northern United States. Current conditions are compared with the 21 year average to identify areas under stress and those showing more advanced development than average. Updates are done weekly and provided free as a web application. As well, in July a spring wheat yield forecast is prepared for the Canadian

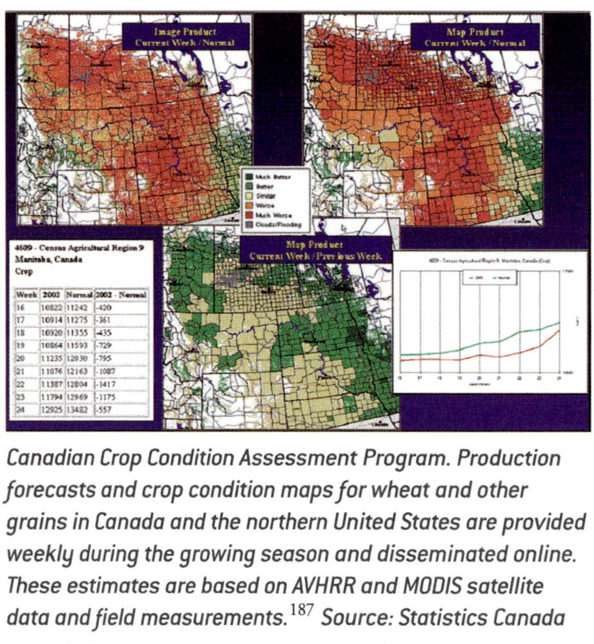

Canadian Crop Condition Assessment Program. Production forecasts and crop condition maps for wheat and other grains in Canada and the northern United States are provided weekly during the growing season and disseminated online. These estimates are based on AVHRR and MODIS satellite data and field measurements.[187] *Source: Statistics Canada*

Prairies using historical yield data and the current crop condition assessment to forecast crop yields. The data are used by governments, grain marketing agencies, crop insurance companies, farm producers and the public at large.[185] The Global Information and Early Warning System (GIEWS) is a similar program that has been used by the United Nations Food and Agriculture Organization to monitor crops in the Sahel region of Africa.[186] By predicting food shortages months in advance, the agency is given a head start in finding food donations and preparing relief efforts when a famine risk is predicted.

A more localized application is precision farming – a farming technique that makes use of geospatial technologies to modify treatments within a field to compensate for natural variations. Remote sensing and GIS technology are used to detect and map these variations. Further discussion on this application is given in Section 6.4.

4.3.10. Military Operations

Military planning, logistics, and war-fighting have been revolutionized by geospatial technologies. Military operations depend on the integration of multiple information sources such as topography, building infrastructure, the location of military installations and their characteristics, the distribution of friendly and enemy forces and equipment, and sites of recent actions. Remotely sensed data are used extensively, including high resolution satellite imagery, airborne imagery, and real-time images collected from aircraft and Unmanned Airborne Vehicles (UAVs). These data must be integrated with field reports and terrain and infrastructure information. Specialized situational awareness systems are used to analyze and integrate these data to identify targets, enemy location and strength, and other information. Satellite positioning enables field commanders to have accurate information on their position, facilitating their own operations and the coordination of support.

4.3.11. Natural Resource Management

It is a requirement in the management of natural resources that multiple ecosystem factors and processes are taken into account. Forestry, land use planning, agriculture and associated irrigation, mining, oil and gas exploration development and control of soil erosion, all depend on geospatial technologies.

For example, in planning forest harvest operations, new roads and staging areas for collecting, preparing, and loading timber onto trucks may have to be built. These facilities can in turn increase water runoff, soil erosion and consequent sedimentation of local streams and rivers, impacting fish and other aquatic life.

Government regulations of forestry operations set standards of practice such as the width of buffer zones around water courses. In order to take these factors into account, satisfy regulatory guidelines, and optimize forestry operations, geographic information systems, remotely sensed data, and satellite positioning have become standard tools for the planning and management of forestry operations. Geospatial data are also essential in the study of societal impacts of natural resource management.

4.3.12. Security

Geospatial data are used in the design and assessment of security measures. Providing protection and controlling access is inherently a spatial analysis problem. Such factors as line-of-sight, positioning of emergency and policing resources, and response times, require geospatial analyses. As well, the impact of specific events such as explosions and the release of contaminants are usually modeled geospatially.

4.3.13. Conclusion

Geospatial technologies have become an integral part of our society's information infrastructure. They directly impact the cost, efficiency, and level of service with which the fundamentals of society are provided, including food, shelter, safety and security, health, and economic prosperity. As a result, people's daily life experience is predicated on the geospatial capabilities of their society, their government, their industry, and at the individual level, the amenities to which they have become accustomed.

Geospatial technologies have become an integral part of our society's information infrastructure. They directly impact the cost, efficiency, and level of service with which the fundamentals of society are provided.

As geospatial technologies have become more powerful and less expensive and data has become more easily available, they have had profound impacts on the organizations that have chosen to use them – and will have equally profound impacts on those who chose not to use them. With these better tools and data we can now make our own geo-luck – or maximize the potential of the location or situation in which we find ourselves.

4.4. Changing Social Values

The idea of using multiple information sources to analyze a problem is not new. The first hunters and farmers tried to use as many sources of information as they could to make decisions about where best to hunt or farm. Throughout history people have tried to avail themselves of as much information as they could get their hands on.

What has changed in the GeoEconomy is that we now have technologies that can be used to collect, store, and analyze more information more quickly than was ever possible before.

What has changed in the GeoEconomy is that we now have technologies that can be used to collect, store, and analyze (and thus understand) more information more quickly than was ever possible before. The first Earth Day was held in 1970. It challenged societal trade-offs that valued environmental quality less than development and resource extraction. It was also a call to examine value trade-offs more broadly, taking into account as many consequences as could be predicted. This approach was expressed in planning, for example, by expanding the criteria for choosing a road location to include the environmental impacts of alternate routes such as levels of erosion, water quality, and the ability of some ecosystems to recover from disturbance more quickly than others. It was this change in societal values that led to considering more factors in making value trade-offs. One of the objectives of the developers of geographic information systems was to use their new tool to address the broad consequences of planning policies and activities. The generation that grew up during this period developed the initial GIS technology and lobbied for policy changes that have since become institutionalized in public policy. The world owes these early GIS pioneers a great deal.

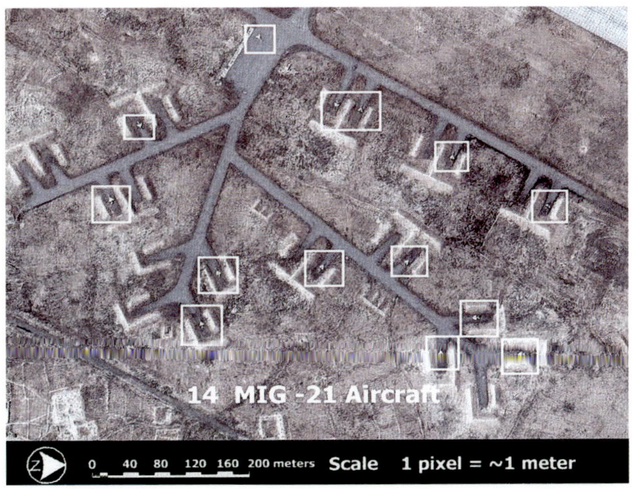

Commercial Satellite Imagery for Military Intelligence. MIG-21 fighter aircraft were identified by military analysts on this 1m resolution IKONOS image acquired August 2001 of the Bagram Airbase, Afghanistan, then under Russian control. Source: GeoEye

While the technology and associated technical disciplines have developed rapidly, often the senior bureaucratic or political levels limit the use of geospatial technology for decision making.

The requirement for environmental assessments, and the objective of sustainable development, arose from the beliefs of the generation that grew up in the 1970s. The subsequent generation developed more advanced geospatial processing tools, connectivity technology that enabled high speed data transfer over the Internet, and the spatial data infrastructure concept. This has made geospatial data more widely available and easier to use by adopting standards for reporting and storing data. While the technology and associated technical disciplines have developed rapidly, it is often the senior bureaucratic or political levels who limit the use of geospatial technology for decision making. This is done by either not supporting the creation of sufficient base-line information, or in some cases, by "not wanting to know" the information the technology would uncover.[188]

There have been important political land marks in the recognition of the national importance of geospatial data. It was US President Clinton who in 1994 issued Executive Order 12906 that "established executive branch leadership for development of the coordinated National Spatial Data Infrastructure, and called for development of a National Geospatial Data Clearinghouse, spatial data standards, a National Digital Geospatial Data Framework and partnerships for data acquisition."[189] As noted elsewhere, in 1999 the United Nations adopted a motion asserting that geospatial data is part of a nation's infrastructure.

Yet at the grass roots political level many politicians don't understand the far reaching benefits of incorporating geospatial data into their decision making – although most have some direct experience of its utility. While most politicians have detailed information about where constituents who vote for and against them are concentrated, many seem not to have translated this concept into the broader operations of government.

4.5. Conclusion

The demand for more integrated use of data is a characteristic of the GeoEconomy, as is the convergence of technologies, of technical disciplines, and of policies. The results for societies or parts of societies that have embraced the GeoEconomy are clear and measurable – they have benefited. While there has clearly been an increase in the availability of geospatial technology and data and increasing investment in the technologies for commercial gain, investments by the public sector seem in many instances to have lagged behind.

Failure to invest in geospatial information and technologies will lead to failure in the GeoEconomy.

There seems to have been a "disconnect" with the political level in many jurisdictions. Many of the politicians whose support is needed to truly reap the benefits of the GeoEconomy seem to be ignoring or are unaware of the importance of geospatial information – or the bureaucrats who serve them are unable to articulate it. No matter the reason, failure to invest in geospatial information and technologies will lead to failure in the GeoEconomy.

FUTURE SOCIETAL AND ENVIRONMENTAL TRENDS DRIVING THE NEED FOR GEO-INFORMATION USE AND TECHNOLOGIES

5.1. Introduction

In the previous chapter we considered the convergence of policy and technologies and modes of thought that have led to the current situation in the GeoEconomy. We now look ahead with an analysis of some of the key trends that seem to be providing guidance for the future in the GeoEconomy.

Throughout this book we have said that geospatial knowledge is today, and will increasingly in the future, offer advantages to those who know how to harness it. At the beginning of this book we made the case that, throughout history, societies that made use of geospatial technologies and knowledge gained competitive advantages. These included greater access to resources, the ability to generate larger food surpluses to support higher population densities, the ability to ensure their physical shelter and security, and the capacity to project power.

Glacier Retreat. Muir Inlet Alaska 1941 (left) and 2004 (right) in Glacier Bay National Park. During the 63 year interval Muir Glacier retreated 7 km to the northwest out of the field of view in the 2004 image. Riggs Glacier is still visible to the Northeast. Source: USGS.

While in 2009 the general economy retreated into recession, many companies in the geospatial field grew. We have already noted that this paradox is recognition of the value of geospatial information and technology. Simply put, in difficult economic times companies and governments seek to improve their efficiency and decision making capability.

We believe that to face tomorrow's challenges geo-knowledge will be of even greater importance – both at the macro level in dealing with the major global changes that seem to be upon us, as well as at the local level where individual enterprises compete to survive. To illustrate, we consider some of the major challenges now emerging for our environment, our society, and for us as individuals.

In all of the trends discussed in the seven following sections of this chapter, the power of geospatial information plays a major role today and one that we believe will increase in the future. We believe that given the trends that seem to be prevalent throughout our world, geospatial information and the analytical skills to use such information offer a "knowledge-advantage." That "knowledge-advantage" will make those who know how to make use of it more powerful, influential, and likely wealthier.

5.2. The Need to Monitor Environmental Processes

There has been a growing public awareness that changes in environmental conditions, especially climate change, appear to be having significant social and financial impacts on the global population. For example, no longer is the monitoring of global climate seen to be primarily of "just" academic interest. Climate change is becoming accepted as having potentially severe impacts on food production, an increased incidence of natural disasters, and progressive changes in plant and animal distributions globally. Some small island nations like the Maldives may even disappear, while the potential for dislocation of many millions of people in low-lying countries like Bangladesh are staggering. Obviously the effects will have broad impacts on a number of aspects of society, including national security and public health.

Paul Epstein of the Harvard Medical School[190] has predicted that changes to the climate that result in global warming will have impacts on nature and humanity that will swamp all other environmental factors. Climate change will lead to mass migrations, epidemics will increase as temperate zones become warmer and tropical diseases like malaria spread, storms will overwhelm sewer systems, and heat waves will raise ozone levels. Indeed, many of these changes are already upon us.

Paul Epstein of the Harvard Medical School has predicted that changes to the climate that result in global warming will have impacts on nature and humanity that will swamp all other environmental factors.

The World Bank estimates that the economic burden on society caused by bad environmental health is between 2% – 5% of Gross Domestic Product (GDP). A team of researchers from Yale University, in their conclusions delivered at the World Economic Forum in Davos, Switzerland, noted that although a country's standing in their Annual Suitability Index is strongly related to per capita income, there are many exceptions. Their analysis suggests that as poor countries become richer they invest more heavily in environmental improvements such as cleaner water supplies and better sanitation. But their research concluded that money alone was not sufficient. Good governance was an important determining factor. Accurate data was one of the factors that showed a clear statistical relationship with environmental performance. Other significant factors were transparent administration, lack of corruption, and checks and balances within governmental decision making.[191]

It has also been reported that the rapid industrialization of countries like India and China has had broad environmental ramifications. For example, in China, rapid industrialization has led to the depletion of local resources and increase in imports of oil, gas, and even coal. But it has also led to the rapid depletion of resources China cannot import, such as clean air and water. The former was certainly given a great deal of media attention during the 2008 Beijing Olympics. Environmental degradation exacerbates the problem. Emissions can result in acid rain that reduces agricultural yields, and can contribute to a variety of human health complaints.[192]

These impacts are more easily identified in certain populations that receive a lot of public scrutiny, such as the decline of polar bear populations in the Arctic due to a reduction in sea ice. However, these more obvious species reductions are indicative of much larger disruptions to species composition and distribution. The public is concerned and is willing to vote their conscience, as the ever increasing stringency of environmental regulations and calls to curb greenhouse gas emissions attest. However, the United States is reported to be the largest contributor of green house gases, roughly a fifth of the world's emissions.[193]

Climate Change. Ice break up is occurring earlier as Arctic temperatures have been warmer in recent years. This polar bear is sprawled out to cross an area of thin ice. Polar bears depend on access to ice flows thick enough to support their weight in order to hunt seals, their principle food. Source: USGS

Geospatial technologies have been instrumental in identifying and documenting climate change. Remote sensing has been used to document important climate change indicators such as changes in the size of glaciers, ice extent in polar waters, the changing size of major lakes, and changes in vegetation patterns and condition. Geographic information technology has also provided the means to store and analyze the large data sets needed to study and model climate change phenomena. But its biggest future role will be in developing predictive models that anticipate environmental changes, assess management alternatives, and plan mitigation measures.

5.3. Management of Strategic Commodities

5.3.1. Food

Food production and food shortages are issues that can become the source of serious social unrest and even armed conflict in the future – as they have in the past. To meet world population food requirements it will become increasingly important to monitor climate change and environmental deterioration of food producing areas. If food supply tightening becomes a recurring problem it will have international ramifications. The need for geospatial data will be critical to monitor and manage the food resource at the national and international level.

Food production and food shortages are issues that can become the source of serious social unrest and even armed conflict in the future – as they have in the past.

The usefulness of geospatial information has long been recognized by the Food and Agricultural Organization (FAO) of the United Nations. For some time now the FAO has had a highly specialized group of geospatial specialists in their Rome headquarters. In 2003 this group published a monograph on the application of geo-information to agricultural development.[194] The importance accorded the technology in the FAO, as well as its visibility, can be seen from the fact that a former leader of the GIS/remote sensing group in Rome was promoted to serve as Deputy Director General of the FAO.[195]

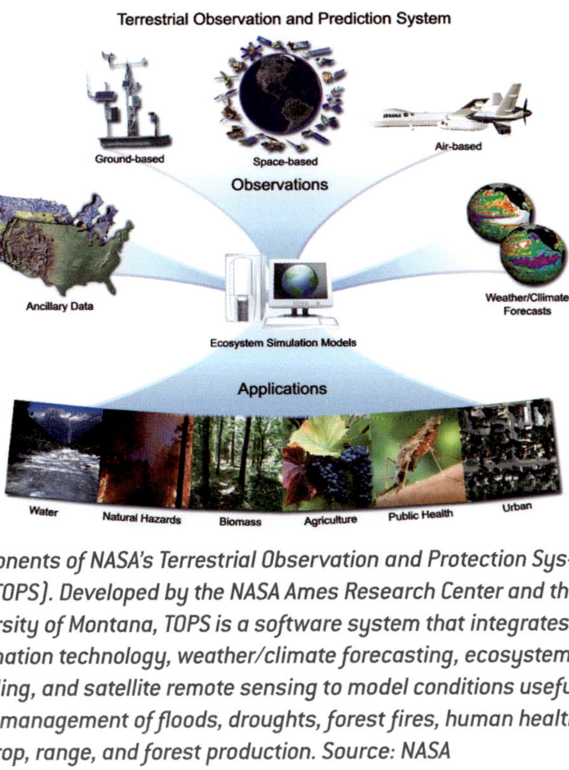

Components of NASA's Terrestrial Observation and Protection System (TOPS). Developed by the NASA Ames Research Center and the University of Montana, TOPS is a software system that integrates information technology, weather/climate forecasting, ecosystem modeling, and satellite remote sensing to model conditions useful in the management of floods, droughts, forest fires, human health, and crop, range, and forest production. Source: NASA

As countries become more industrialized and per capita incomes increase, they tend to consume more resources. China, for example, has about a fifth of the world's population. As the Chinese have become more prosperous, they have changed their diet by adding more meat. China now consumes half of the world's pork. Increased meat consumption requires increased production of feed grain. Since it takes more grain to produce meat than to feed people directly on grain and derived products, such as bread, increased consumption of meat will require greater yields, more land devoted to production, or increased imports – or some combination of the three. Simply stated, with current resources and practices China will not be able to grow enough grain to support the meat production that the changing diet in China now demands.[196] China is not alone, the same situation is developing in India, although in that country the impact is somewhat cushioned by the number of vegetarians found there.

Further contributing to the problem is the prediction that Northern China will become more arid. Between 1990 and 2000 there were 36,000 sq km of agricultural land lost to desertification. It is predicted that rice yields will drop by 10% for every degree rise in temperature. Rising sea levels and the associated intrusion of sea water are expected to further reduce the amount of arable land in China, at a time when its population is demanding more and better food.[197] Again, similar situations can be expected to develop in India. While the increased use of genetically modified (GM) crops may be expected to lessen the impact of these demands on the world's food supply,[198] the continued sustainability of agricultural systems everywhere, with or without GM crops, will require monitoring and management that can only be done with geospatial data.

5.3.2. Other Strategic Commodities and Political Influence

Other strategic commodities such as oil, minerals and base metals will see increasing demand as major population areas such as China and India seek to improve their standard of living and use political influence to gain access to these commodities.

> ***Other strategic commodities such as oil, minerals and base metals will see increasing demand as major population areas such as China and India seek to improve their standard of living and use political influence to gain access to these commodities.***

In 2008, during a massive build-up of infrastructure, China was identified as having consumed half the world's cement, a third of its steel, and over a quarter of its aluminum. Compared with 1999, in 2008 China spent 35 times more on imports of soybeans and crude oil, and 23 times as much on copper.[199]

With growing prosperity, countries like China have a growing middle class that is buying inexpensive cars targeted to that market. Between 2000 and 2006 the number of new cars sold in China grew by 37% a year, making China the second largest market for automobiles. In 1990 China consumed 2.4 million barrels of oil per day (m b/d) leaving 400,000 b/d of oil for export. In 2008, China consumed 7 m b/d, about half of which is imported and it is predicted that by 2030 it will consume 16.5 m b/d of which about 13 m b/d will be imported.[200]

The increased consumption of consumer goods and services by a burgeoning middle class will strain the conventional exploration and development yields of non-renewable resources in countries like China and India. Resource exploration and development methods will continue to be improved by ever more sophisticated analysis of geospatial data. Skilful use of these analytical methods will offer significant competitive advantage – especially as most of the easy-to-find deposits have already been found.

The demand for resources can also be expected to have significant political implications. Resource hungry countries like China have been investing heavily in countries that supply the raw materials they need. In poorer countries, the economic benefits can lead to significant political influence as well. In 2007, the Chinese agreed to finance $6.5 billion for improvements to the Congo's infrastructure and another $2 billion for construction and refurbishment of mines, with the Congo's mineral reserves serving as collateral. Similarly China has been active in Myanmar, the Sudan, and Angola. In Western nations political pressure by their citizens has restricted companies from doing business in countries suspected of human rights abuses.

5.3.3. Water

Simply put, we need water to survive. But the sufficiency of water is not always "the" issue with water. Another concern is too much water – flooding. Both can lead to conflict and can be man-induced or nature's response to some of man's more intrusive activities. The impact of desertification on agriculture has been mentioned previously.

Simply put, we need water to survive.

Niagara Falls. At peak flow, 168,000 cubic m (6 million cubic ft) of water flows over the falls carrying water from the four upstream Great Lakes to the fifth, Lake Ontario. The five Great Lakes comprise the world's largest surface freshwater system, with about 18% of the world's supply. Formed by glaciation and filled by melt-water from ice sheets 2-3 kilometres thick, less than one percent of the water is renewable on an annual basis (precipitation and groundwater). Climate change is expected to reduce the annual water recharge into the lakes resulting in reduced flow over the falls.[201] Source: S. Aronoff

Disappearing aquifers and reduction in the Aral Sea, reduction in wetlands, the amount of water consumed to produce a barrel of oil in Canada's oil sands, and changes in mangrove forests in tropical coastal zones, all presage issues in the future concerning both the quality and quantity of water. These needs have resulted in advanced water management practices used in areas with limited water supplies. These practices used in countries such as Israel are based on geospatial information.

The trend is clear: water will be a much scarcer and more valuable resource in the future. As water's value increases, so too will the value of the geospatial information related to water's availability and management. The increased value of geospatial information is yet another harbinger of the GeoEconomy.

5.4. Transparency of Government and Industry Activities

In the future, the actions of government and industry will encounter ever more stringent public scrutiny with increased availability of geospatial data, such as satellite imagery, topographic and resource mapping, and location-specific government statistical data. It can be expected that public use of these data by special interest groups will become increasingly sophisticated. In some cases the results will be politically de-stabilizing, while in others the results will lead to a consensus and agreement on how to solve the problems that have been identified.

In the future, the actions of government and industry will encounter ever more stringent public scrutiny in part due to increased availability of geospatial data.

Consider for instance the severe and widespread damage to Myanmar in May 2008 caused by a cyclone, as was noted previously. Large segments of the population were in dire need of basic assistance, yet the military government sought to control the flow of outside emergency response teams and the delivery of humanitarian aid. With active international citizens groups monitoring developments, it was difficult for the military to operate with anonymity. The ability to monitor the situation caused external political pressure on the regime. One can expect that this ability to render many decisions transparent – whether a government wishes to do so or not – will, in the future, lead to better government.

The use of geospatial information will also promote smarter development – in the sense that it becomes far less costly to anticipate the many concerns of stakeholders and, to the extent possible, mitigate or eliminate them before they become an issue in the national and international media. This openness is an approach that has been advocated by democratic activists for decades. While improved geospatial data and analytical tools will never eliminate disagreements among stakeholders, the availability of higher quality data and collaborative tools will be helpful in the future, as it has been in the past, to reach decisions that are at least understood, if not universally agreed upon.[202]

5.5. Global Competition

Global competition will intensify. This will not only be among businesses, but among governments, and government economic interests represented by such vehicles as well-endowed sovereign funds. Businesses and countries who know less about their resources than those who might wish to buy or exploit them will be at a distinct disadvantage. Sophisticated use of geospatial data and analysis will give purchasers far more detailed information about resource and development potential than many owners themselves can appreciate. With the wide dissemination of information to a global marketplace, the talent that can be brought to bear is almost limitless – especially for wealthy foreign companies purchasing the resources of a less developed country.

Global competition will intensify. Businesses and countries who know less about their resources than those who might wish to buy or exploit them will be at a distinct disadvantage. The requisite knowledge will come from geospatial analyses.

Resource owners who have limited geo-awareness will not be able to accurately assess the true value of their holdings – even with the explosion of information. The way things are progressing now, it will also become increasingly difficult in the future to find skilled personnel to conduct and manage effective geospatial information gathering/intelligence services to meet strategic and tactical objectives. This will be especially problematic in less developed countries. The demand for skilled personnel in the applied geospatial fields continues to greatly outstrip supply, even with the rapid increase in relevant university and technical school programs.

5.6. Commoditization of Geospatial Data

The quantity of geospatial data available is increasing at an exponential rate. In the first fifteen years of earth resource satellites (1972 to 1987) only six were launched that provided imagery to the civilian community.[203] A recent projection for new satellites in civil and commercial remote sensing is staggering in comparison. In a 2006 study, *"The Market for Civil & Commercial Remote Sensing Satellites,* Forecast International[204] projected deliveries of approximately 139 imaging satellites worth $16.3 billion over the next 10 years. The first half of the period will be more active than the second, with 97 spacecraft slated for production within the next five years." Of interest is that they project few commercial satellites, as most of these will be put up by government agencies. However, the imagery will be used and, even if used in government, this will require additional capabilities to those that already exist in government. This will create more opportunities in the future for those providing the services and technology to use the imagery.

It will be interesting to see which governments will be ready for this informational onslaught – and which will not.

Having many earth resource satellites in orbit at the same time provides needed redundancy of coverage and characteristics.There are currently several medium and low resolution satellites that generate near complete global data acquisitions every 1 to 3 days depending on location. (Coverage tends to be less at the equator and greater at the poles.) There are also several civilian radar satellites systematically collecting earth imagery. Similarly, monitoring of ocean temperature and currents, ozone levels, and other parameters are monitored by multiple sensors.

The number of earth observing satellites launched for primarily civilian use rose from 6 between 1972 to 1987, to a projected 139 between 2006 and 2016.

While geospatial information may become increasingly available and affordable, the skilled application of that data will remain highly valued. With the flood of data the importance of trusted "arbiters of quality" will become ever more important. Those with the expertise to distinguish misleading trends or common place conclusions from the truly insightful and competitively valuable will be highly influential and richly rewarded.

5.7. Sovereignty and National Interests

With multiple geospatial data sources there can be competition for data sales on the basis of rapid delivery times, ease of scheduling, reliability of acquisition of the data, and other factors. However, it is not an entirely free-market operation. National interests commonly take precedence over purely financial decisions. This is, of course, entirely understandable. If my country is paying for a satellite, I do not want my country's enemies or my competitors gaining access as easily as I do to the data it produces, especially if my tax dollars have paid for it.

Many countries have realized that although there may be financial gains to be won by selling operating satellites, or the companies that build and control them, national security considerations, including issues of sovereignty may take precedence. This occurred in Canada when the government of Canada blocked the sale of MacDonald Dettwiler's space assets to a US firm. Part of the concern stemmed from the US legal structure that might allow the US to control Canada's access to a satellite paid for, in large part, by the people of Canada. US laws give the US government preferential access and tasking of any satellite operated by a US-licensed commercial provider. Further details are given in the end notes.[205]

A fundamental conclusion from this case is that in the GeoEconomy geospatial technologies and expertise are not benign. The information they generate can have significant economic, political, and military implications. As such, there are "public good" considerations which will often take precedence over purely financial costs and benefits. Thus, while the data may become increasingly commodity-like in their abundance, and to some extent are interchangeable, there are other important factors which influence governments to assure their control of the technology and the analytical capabilities to use that technology.

> ***A conclusion that must be reached, especially as we consider the GeoEconomy, is that geospatial technologies and expertise are not benign. The information they generate can have significant economic, political, and military implications.***

Governments will seek exclusive and rapid access to state of the art geospatial information and to tightly control its distribution whenever it is in the nation's best interest. Their objectives may be to defend against unwanted foreign action and to project power – be it political, economic, or military. But exclusive access will become ever more difficult to secure because there are a growing number of sources for most geospatial data. It will become increasingly difficult to gain advantage simply by having access to data others can't obtain. When competitors, be they businesses or national governments, have the same data, the advantage will go to those who can better analyze and use it. It is for this reason that research in the use and application of geospatial data is so important. And, as has been amply demonstrated here, the important applications are not just military ones.

There will be shifts in the balance of power among nations as populous nations become more developed industrially and technologically, and as their need for resources rapidly expands in a world with limited ability to increase supply at the attractive prices that prevailed in the past.

With global warming, along with the other demands on agriculture, there will be increased demand for dwindling commodities and rising food prices will create shortages – and there may well be an increase in military conflicts, as has been suggested by Paul Epstein of the Harvard Medical School.[206]

5.8. Infrastructure Renewal

Rapidly developing nations like China with its large trade surpluses in manufactured goods can fund an unprecedented expansion of its infrastructure. Between 2001 through 2005, China spent more on roads, railways, and other fixed assets than in the preceding 50 years. Between 2006 and 2010, $200 billion is expected to be invested in railways alone. Over the same period, China is expecting to build 300,000 km of rural roads and expand its expressway system to 70,000 km from the approximately 53,000 km existing in 2007. Similar rapid growth in aviation facilities is also planned. Passenger traffic has expanded from 7 million passengers in 1985 to over 185 million in 2007. To handle this increase, 97 airports are to be built by the year 2020 – adding to the 142 that existed in 2006.[207]

The Forbidden City from Tiananmen Square, Beijing – Then and Now. In 1984 (left) bicycle transportation dominated, with bicycle lanes as wide as those for motorized traffic. Twenty-five years later in 2009 (right) cars and buses dominate and the bicycle lanes have almost disappeared. Source: R. Ryerson

Industrialized nations face a different and yet still difficult infrastructure challenge. Roads, bridges, canals, and levees are aging to the point that some are failing before they can be repaired. The American Society of Civil Engineer's 2005 Report Card for America's Infrastructure estimated that $1.6 trillion was needed over a five-year period to bring the nation's infrastructure into good condition. The U.S. Bureau of Transportation Statistics estimated that 53% of Rhode Island's bridges were structurally deficient or functionally obsolete.[208] The situation may be somewhat better in Western Europe, but it is certainly worse in most of the developing world and within the former Soviet Union. While not directly part of the geospatial world, in situ sensors are now being embedded in new bridges to provide information on their condition. Such information, gathered from over a large geographic area, can provide insights into changes that might be required to building codes across a jurisdiction as the climate changes.

The number of engineering works that are anticipated certainly explains the purchase of geospatial companies by major engineering companies. Simply stated, to successfully complete the expected future work, they want geospatial expertise accessible, and today that often means in-house.

5.9. Conclusion

It is clear that based on societal and environmental trends there will continue to be strong and increasing demand for, and use of, geospatial information for the foreseeable future. While many of these demands will be economically driven and come from industry, local government, and individuals, even greater demands will come from nations seeking the resources and expertise to compete for increasingly scarce fuel, food and water. Scarcity, and the "where" of that scarcity, is another descriptor that will come into play in the GeoEconomy.

We repeat what has been said previously: what has been surprising to many observers of industry in general, and the geospatial industry in particular, is that during the recent economic turmoil the geospatial industry enjoyed a period of growth that in some cases saw companies double their sales. Why has there been this growth in the geospatial sector when in general the economy was collapsing?

Some years ago, one of the authors made a startling observation related to sales statistics for satellite data – the worse the economy got, the more satellite data and geospatial information was sold! The rationale was clear – with an economic downturn, industry and government sought ways to become more efficient and they became more likely to embrace change than compared to when things were going well. If satellite data allowed them to save some money, they would use it. In one sense the GeoEconomy is a reflection of that premise written on a larger canvas. During economic downturns, such as the recent recession, times are not as good. Competition is harder. We need to save money. We must be more efficient. The fact that at these times we use more geospatial technology and information serves as a clear signal that we have entered the GeoEconomy.

The accelerated move into the GeoEconomy during bad times is a response to the need to be more efficient and use geo-information more effectively when times are not as good, competition is harder, and we must save money and be more efficient. We must use geospatial technology and information.
This is a clear signal that we have entered the GeoEconomy.

GAINING THE GEO-ADVANTAGE

6.1. Introduction

This chapter looks at how individuals, communities, companies, states and countries can use geospatial data to gain a "geo-advantage." As previously noted, the GeoEconomy has come about, in part, as a relatively rapid evolution of our societal information infrastructure. In many respects this evolution is akin to the introduction of computers in the 1960s and 1970s. During that period governments, financial services, and some businesses rapidly adopted the technology – and some didn't. Some implemented the technology in effective and creative ways that catapulted them ahead of their competition. Some started off poorly and then reinvented themselves. Some didn't understand that computerization was not an option but an imperative, and when they failed to adopt the technology quickly enough they failed and disappeared. The history of the introduction of computers also tells us that among the suppliers of the technology there were some hugely successful long-term survivors and myriad small players (and a few large ones) that disappeared. Those who benefited the most were the ones who discovered how to use it before others caught on. The GeoEconomy is no different.

As in all other significant shifts in economies, some of the major advances and some of the major success stories will come from unexpected sources. That, of course, is what makes a true paradigm shift so fascinating – such shifts are at the same time fraught with danger for some and present wonderful opportunities for others who "catch on" sooner. Those who first figure out (or who have already figured out) how to create and use geo-information will do better in the GeoEconomy than those who don't.

Who will "catch on" sooner? What can people or organizations do to be in that special group? One of the key messages we have is that we cannot be prescriptive about geospatial: there is no formula to implement perfect applications. Each application, like each individual, community, company or country is unique and each may well call for a different approach. For that reason, what we present here is not a recipe that we promise will lead to success. It is rather a set of suggestions for the reader to consider as an approach that offers options to be evaluated and expanded to suit needs and perspectives that are always to some extent unique.

In all cases it starts with one simple question: where?

We have segmented the potential applications into groupings of the individual, community, company and country. In each case we have attempted to identify some of the key issues and then explain how geospatial information is being used or can be used to address them. Again, to achieve the full benefits of beginning to think spatially, the reader must incorporate his or her own situation and values on the process. In effect we are encouraging the reader to think and act strategically in using geospatial information.[209]

What this chapter tries to accomplish is to present a broad sampling of applications of the technology to help the reader begin to think spatially at home and at work. It is based on our observations, experience and the understanding that we have gained through the hundreds of projects we have led or evaluated, and many more that we have read about during our combined 70 years of experience in the field. For those many years we have been asking the "where" questions. We expect that what we present in this Chapter, and in this book, will lead people and organizations to be more efficient, make better decisions, keep better records and, we hope, succeed in the GeoEconomy.

In all cases it starts with one simple question: where?

6.2. Individuals

6.2.1. Introduction

Geospatial data can be used by individuals for everything from the mundane "where is 1212 Mulberry Lane, the location of the party on Saturday night," to planning holidays, deciding if the area where that hotel is located is a safe part of the city to visit and stay in (see Section 6.3.8) or making far more important decisions: for example, where should I send my children to school, in which neighbourhood should I live, which neighbourhood should I avoid, which house should we buy, where should I buy a retirement home, or where should I buy an investment or recreation property?

The most important and least often faced questions for individuals involving geospatial decision making are outlined in the following table.

Questions Individuals May Answer with Geospatial Information and Technology
• Where should I send my children to school? • Where should I work? • Where should I live (or expressed another way, should I buy a house and if so, where)? • Where should I retire? • Where should I buy a recreation property? and • Where should I invest my money?

These decisions require an assessment of a complex set of factors, often under time pressures, and with the potential for severe and negative ramifications if the wrong decision is made. We need look no further than the recent financial meltdown in the USA – too many people made the wrong "where" decision related to mortgages. The fact that such decisions are only faced occasionally makes the problems all the more challenging. We can suggest several ways to improve such decisions – especially important in light of the recent sub-prime mortgage fiasco in the United States and the variable quality of schools.

There are many more decisions and questions concerning location that are faced by individuals that involve less severe penalties if a bad choice is made. For example:

- Where should I buy a car (or other major purchase)?
- Where should we holiday?
- How can I best get to place x, y, or z?

Some of these decisions can be changed – unlike a house, a school, recreation property or retirement home, a car or piece of furniture is not tied to a specific location – they can be moved to a better location. But holidays and the daily decisions we make that involve location, and the geography through which we move, can have significant cumulative impacts in terms of time, stress, and cost. One need only think of that last trip to a vacation spot – in the wrong season when it rained the entire week – or taking the wrong road home at rush hour.

6.2.2. Choosing a School

Where should I live and where should I send my children to school are often linked – where you live usually dictates where children would go to school – unless they are in a private school – and even then you do not want children facing too long a daily commute. Until my children went off to university I (Ryerson) have always chosen my neighbourhood by the quality of school my children would attend and access to school buses.

In most jurisdictions, schools are ranked by quality through some means or other (often by an independent review system) and most jurisdictions prepare a map of school locations and the areas from which each school draws its students. These school boundaries are important as they identify areas where you may want to localize your search. But boundaries may change – and schools may close – even good ones.[210]

Proximity to Quality Education. The location of a child's home affects the school they are likely to attend and the quality of education they will receive. Choosing a home thus has long term implications for the success of our children. Source: R. Ryerson

Obviously, it seems to make sense to buy (or build) a home close to the chosen school and to make sure that the school is sustainable – that it has sufficient numbers of children within its catchment area to ensure that it won't close. To do this one needs some information on ages of children in the area. Again, depending on the area, such maps of the demography (or population characteristics) may be available from city hall, the school board, or the most recent census. One can also find out what the average class sizes are in a school, what the recent trends have been, and what the criteria are for sustainability. Lastly, one can assess what types of neighbourhoods are drawn into a specific school, if the character of a neighbourhood is a concern. One quick way of doing the local check of the school district in many cities is through the mobile van imagery provided by mass market geomatics services like Google

Earth[211] or by a simple drive through the areas to determine where there may be undesirable elements. From my computer in Ottawa I can scan areas of San Francisco – or almost any other urban area in the USA, as well as Australia and parts of Europe. As we write, streetscapes of Canada and many other countries are coming soon to a web site near you.

Armed with this information you can begin to identify areas on a map (or on an image from one of the mass market suppliers) for your house search to ensure that you are in the right school district. One can simply draw lines on a map or image, or get quite fancy – using the easy-to-use GIS and mapping tools that are readily available on the web.

There are other concerns one might also have in selecting a specific school. For example, there are, literally, thousands of sites in the United States and Canada that are contaminated by toxic chemicals. It may be from an old garbage dump, or something worse – an area contaminated by coal tar, chemicals, leaking storage tanks, and in some cases, wilful discharge of noxious chemicals. Such sites are known to be dangerous to people's health: many types of toxins and many types of impacts can be cited. Unfortunately, some schools have been located on contaminated sites – or close to factories that pollute the atmosphere, or too close to other illness-causing problems. In many cases old aerial photography is available and whether or not there are potential contaminants can be easily discerned – a former gas station, old drums, evidence of garbage disposal, old factories, and filled-in valleys, among other things, give an indication that all might not be well with a specific site. In some areas, lots that have long been vacant (often used as parking lots) are often an indication of problems.

6.2.3. Choosing a Work Location

Choosing a work location is obviously related to choosing a school and a place to live. For those with the luxury of choosing a place to work, some of the same location factors are important as when choosing a place to live or buy a house – safe, clean environment, etc. In addition, it is likely a good idea to select somewhere where there are multiple opportunities for the same type of work, assuming that one is an employee, given the fact that few people today start and end their careers with the same employer. When a spouse or life partner is involved in the decision, and both are in specialized careers, then there must be opportunities for both, and typically larger cities are selected that have a number of companies in the same business area. This sort of decision making is also followed by companies seeking a pool of employees, and together the decisions by the employees and companies tend to lead to geographic areas of specialization. For this reason high technology clusters are found in a number of locations in the USA, Europe, Canada, and elsewhere. If one is a geospatial specialist, for example, the cities noted in Section 6.3.9 would be a good place to start to find employment!

6.2.4. Choosing a Place to Live

The same sort of tools one uses to look at a school site can also be used to look at a house site. When you are buying a house you want a property that is at the same time safe **and** a wise investment. At the worst, contamination can cause illness and even death. At the very least it can make the property difficult to sell and cause the buyer to be potentially responsible for an expensive clean-up and/or legal costs. How can such errors be avoided? Judicious use of readily available geospatial data offers valuable assistance. Here are some suggestions:

Aerial Photography: In a very real example, Ryerson was looking at a new house site south of the city of Ottawa, Canada. He was looking at a 2 acre lot in the countryside with the road and front of the house on a higher elevation than the rear. The back of the lot dropped down about fifteen to twenty feet to a small stream and wooded area. "Perfect!" he thought. He obtained old aerial photographs and found traces of what looked like old oil drums and bright-appearing debris (on the black and white imagery) that might have been paper bags of the sort used to package farm chemicals. When he went to the site and dug around he found exactly that: debris from farm chemicals. "Our beautiful site was in fact a farmer's dump for his chemical debris." The offer to buy the property was conditional on the debris being removed and the site cleaned up. The seller would not agree, and Ryerson did not buy the land.

Ownership Records: The history of the property as documented in ownership records can be valuable. Past uses that could have left hidden, and potentially detrimental, conditions behind might be revealed. Perhaps the land was previously an industrial site. Depending on your location (of course!), ownership records can be obtained from land registry offices that are usually run by municipal, state or provincial governments.

Municipal Plans: Another important factor is (especially in new areas) what the plans are for the surrounding area. Will it become a big box shopping mall, a tennis court (with lights and noise all night!), low-income housing, a school, a fire station,[212] or a major highway? Zoning maps at the local municipal offices (and sometimes even on-line) will disclose what the plans are...and whether this is the ideal house for you. Of course another concern might be if the municipality or jurisdiction in question tended not to respect zoning. In that case it might be wise to buy a property in a developed area, or one that is well along in the process of development to ensure that nothing undesirable will happen in the future to the idyllic location you have selected for your "dream home." Close to Ryerson's own home, efforts by a large developer (Minto) to develop land within a small village (Manotick) created serious divisions and controversy. For the details do an online search for "Manotick Minto Dispute."

Topographic Maps: Natural topography can be usefully studied. To the practiced eye it would seem strange that, as you follow streams into (or out of) an older neighbourhood, they

House Site and Orientation. Geographic location affects the type of site and orientation most desirable for a home. In northern climates, a sheltered site oriented to maximize heat gain from sun exposure during winter months and minimize wind patterns that cause snow accumulation on walkways and driveways is a benefit. Source: R. Ryerson

seem to disappear – or were not accompanied by valleys as were neighbouring stream courses through the same sort of landscape in the same area. In a number of cities the explanation is simple – former valleys or stream courses were used as garbage dumps – and completely filled in. Needless to say, the water keeps coming down the valleys, and as the garbage compacted the land shifted – and houses tended to sink or otherwise move.

Geology Maps: Geology maps can be used to identify regions with increased geological risk that could affect the stability of the land or the security of its resources. For example, in limestone areas with subterranean water large underground caverns can form. If these caverns are near enough to the surface that they collapse, they form what are known as *sinkholes*. Florida is one area where these limestone conditions occur and occasionally one of these caverns collapses forming a sinkhole into which everything above it falls – including houses. In fact the problem is so common that there are lawyers in that state who specialize in sinkhole damage.[213]

For homes that depend on a well for their water supply the amount of available water may decrease or the well could run dry in certain seasons if the underground aquifer that supplies the well is not being replenished as fast as it is being depleted. The geology affects the characteristics of the aquifer as do other factors, such as increased withdrawal of water by other wells or additional wells resulting from new development. That beautiful home may need to have water trucked in.

And then there are geologic structures that pose an increased risk, such as fault lines where the land is prone to gradual movement over time or sudden shifts during earthquakes. Surprisingly, fault lines are often completely invisible on the ground, but reveal themselves as linear features on satellite imagery. Of course you have to be certain that those straight lines you see on the imagery are not simply roads or property boundaries!

Insurability and Flood Risk Maps: It seems that both floods and wildfires are striking more often. In the USA and many other countries there are flood risk maps (put out by FEMA or the equivalent). That peaceful stream may become a raging torrent, that uninterrupted view of the beach may be an uninterrupted path for high waters from a hurricane or tsunami. The risks of any of these events occurring have usually been mapped – often by governments or by insurance companies. If an insurance company will not insure you against a specific peril at the location of your house, do not buy the property! Similarly, wildfires could make that wonderful forest the tinder box that, when set off, will see your house burn down. Being geo-aware will protect your investment and perhaps even your life.

Health Maps: If the study of these various sources hasn't scared you off that home or home site you might also want to investigate the closest hospitals and ambulance service. Look at health maps of the area. From these maps you can see where cases are more numerous for a specific disease. Is there an elevated risk of certain diseases in the area you are considering? Why? In some cases it may be related to air or ground water pollution. One need look no further than the true case in California that was popularized by the Hollywood movie *Erin Brockovich* starring Julia Roberts as Erin Brockovich.[214] Between 1952 and 1966 toxic industrial waste water had been discharged to an unlined pond and percolated into the ground water, affecting an area about 3 km long and almost 1.5 km wide. The resulting court case was finally settled in 1996 for $333 million, reportedly the largest settlement ever paid in a direct action lawsuit in US history. The incident is a dramatic illustration of the potential impact of chemicals in the environment and the use of geospatial information. In the movie there is clear reference to the use of maps and aerial photography in scene 27, in exactly the manner we suggest here – the use of historical imagery to assess the situation over time. Even Hollywood has moved into the GeoEconomy!

Crime Incidence Maps: Earlier we mentioned the use of police records to assess the incidence of crime. Many cities have posted local crime statistics on the web.[215] In one area where Ryerson lived, houses of the same style as his, built along a park on the opposite side of the road from his home, were somewhat more expensive because the parkland was considered a bonus. But over the years the beauty (and value) of "backing on the park" dimmed considerably. Every one of the homes on the park-side of the street was burgled (every one!), but never one on the other side of the street. Why? The paths through the park provided a direct access to and from a low-income subsidized housing development a short walk away. Eventually footprints in the snow led police to the home of a burglar who lived in the area.

6.2.5. Choosing a Place to Retire

While the same factors noted above should be examined when looking at a retirement property, there are even more ways that geospatial information can be used for this sort of application by the individual. For a retirement area it would seem wise to know how far you are from a hospital or health treatment centre that is accessible to you – and how fast an ambulance will get you to that hospital. It has been proven that slow responses leads to increased mortality rates and slower recovery. In the region where we live, a number of small towns have seen manufacturing companies or mines shutting down. Those with hospitals are re-inventing themselves as retirement communities. The best known of these in Ontario is likely Elliot Lake – which runs full page advertisements in national media emphasizing how senior-friendly the community is.[216]

6.2.6. Buying a Recreation Property

There are jokes about people buying swampland in Florida. But if you were the person who bought such land, it would not be much of a joke. Buying the proverbial swampland is not all that rare, but it is preventable. All of the preceding advice is applicable to buying recreation properties, and to that we can add even more useful information.

While the link to geospatial may seem tenuous for some of the following questions, all of them have location implications – all of the answers can be mapped, overlaid on one map, and the resulting map can then be used to help make a decision. Some questions to ask about that wonderful building lot on which you will build your family get-away:

- If the lot is vacant, what are the rules about building – can you build a home of the size you want where you want?
- If the lot isn't vacant, can you tear down what is there, or do you have to maintain the existing building footprint and location? (We have seen a case where if you tear down a building you cannot rebuild anything!)
- How far must you build from the water?
- Is the land affected by a nearby nature reserve in some fashion – are there minimum distances from that nature reserve that must be maintained for a building?
- If you are a boater, what is the boat access like, how deep is the water?
- What variability can be expected in water depth?
- Are you legally allowed to build a dock – can you get the necessary approvals at this location?
- Is the water deep enough and is the access sufficient for you to build a dock?
- Has the property got any illegal improvements on it you must remove or repair?
- What shoreline improvements are allowed?
- Can you build a septic system?
- The water may be clear of algae and weeds in the spring, but check it in late summer after there has been a growing season for it to accumulate. Is the water still clear of unwanted vegetation?

Many of these questions can be answered by the study of maps and/or aerial photos at the municipal offices, at government departments of the environment, or agencies responsible for navigable waterways, septic systems, etc. Again, aerial photography as well as navigation maps and environmental protected area maps will prove useful. The results can then be plotted on a map of the property to see if what you want to do can be done where you want to do it. Mind you, none of this information will end up being useful if the rules are changed. For example, Ryerson bought a piece of lakefront property that was perfect for a summer home or cottage. It was easily accessible, had a nice beach which the cottage would face, and bordered on a nature reserve that would never be built upon. But before construction began the laws on cottage location in Ontario were changed such that there was no location on the property sufficiently far from water to allow building! It took several years and many meetings with government officials to finally secure approval to build – and then the cottage did not face the beach and was, as a consequence, much less desirable. It is therefore also important to determine where proposed zoning changes or environmental designations are under discussion. Once again, "where" is the keyword, and "where" is geospatial.

Vacation Property. Source: S. Aronoff

6.2.7. Where to Invest

For those depending upon investments for retirement income a critical question is where should I invest my money? By where, we don't mean just the stock market. We mean where have the companies you are thinking of investing in been carrying on their business, and what are the implications of that location? Those who chose to invest in housing in the USA in 2007 and 2008 made a mistake. Those who bought low-lying waterfront property on the west coast of Phuket before the famous tsunami, or those who have recently invested in the stock market in certain regions of the world, lost and lost heavily.[217]

While these are obvious examples, some of the guidelines for industry, discussed below, would also be relevant and point the way to other considerations for this application. Less obvious among these are important factors that vary by country or region. While the stock price of a company will usually reflect the strengths and weaknesses of the specific company, how a company performs is often tied to fundamentals than can be explained geographically. These in turn may simply be a reflection of the relative strength of financial oversight in certain countries, the amount of corruption, or some other factor that varies from one place to another. But vary they do.

Two pieces of investment advice given to us by our brokers is useful to pass along. First, if an opportunity sounds too good to be true, it usually is. Second, don't invest in an area unless you, or someone you know and trust with your life savings, is thoroughly familiar with the potential downsides of an investment in that location. We use the plural of downside here for good reason – there are many location-related downsides from potential floods, to natural disasters, crime, and health concerns.

6.2.8. Buying Large Ticket Items: Cars and Furniture

What does geospatial have to do with buying a car or a sofa? Potentially a great deal – and it sheds some interesting light on a phenomenon well established in many parts of the world and growing in North America. The concept of specialized areas in cities and towns where one goes to buy certain goods or services is not unique to automobiles – or home furnishing malls. In many regions of the world one finds certain types of services located in the same general area (or areas) within a city. In India, for example, we have seen several blocks of cities taken up by trucking companies. In Bangkok there is a huge centralized agricultural wholesale fruit and vegetable market through which almost all produce seems to flow. In the same city there are areas where one finds machine shops, garden supplies, or other products that makes comparison shopping easy.

Clustered Retailing - Durian Fruit Stand in a Wholesale Fruit and Vegetable Market, Bangkok. When retailers selling similar products are located close to each other, comparison shopping is easier for the consumer. Competition may depress prices but sellers can also benefit from the increased number of buyers attracted by the clustering of shops. Source: R. Ryerson

In North America in particular, the concept of "automalls" has arisen. Car dealers tend to concentrate in one area. The idea is that the consumer can comparison shop from dealer to dealer seeing different cars (or furniture in a furniture mall) more easily and in less time than having to go to locations all over the city. Less travel even saves you money on fuel and reduces your carbon footprint. In principle this sounds like a good idea – although one result may be that having all of the dealers together tends to re-enforce the idea of buying a car – which not all of those looking at cars are really intending to do. In this case the more serious car buyer can comparison shop over the Internet and, using services such as Car Cost Canada, the buyer can offer the dealer a price based on wholesale invoice price reports. For furniture the Internet isn't quite there yet since we usually want to actually sit on a couch before we buy it!

And once you have bought that new car, boat, or recreational vehicle, you do not want it stolen. You can now subscribe to a service that will track your possession whenever it moves. If the movement has not been authorized, the vehicle can be disabled, and the police called to the location where the property has been moved.

6.2.9. Getting to Where You Want to Go

Case Study:

In April 2010 Ryerson visited his son Chris, daughter-in-law Alana and Grandson Ben in San Francisco. The plan for that sunny weekend day was simple. We would drop Chris at his noon-time soccer game across the city, and we would have a leisurely lunch. According to a well known on-line mapping system several restaurants were within easy walking distance. So off we went, pushing Ben in his stroller with Alana navigating with her smart phone. The first thing that we learned was that the routing instructions didn't take into consideration the changes in elevation: the shortest route (which we were advised to take) went up over the top of a fair sized hill. It turned out that a much easier, and only fractionally longer, route was all on flat land. The next problem was that there were no restaurants. The system took us to a recycling depot in the midst of an industrial area. In fact, there were no restaurants within a few kilometres of the location to which we were directed! An always well-prepared Alana had some food for two year old Ben, and we managed to make it back to the car (this time not over the hill) and drove around until we found an accessible and child-friendly restaurant. The moral of the story is clear, be careful how much you rely on geospatial systems in unfamiliar territory – even if you are travelling with a geospatial expert!

Geospatial Search Errors. No database is without errors. This family arrived for lunch at a location their online search identified as a restaurant only to find it was actually a recycling depot. Source: E.A. Blair

---

Online route planning services like MapQuest or Google Maps are easily learned and widely used, whether it is a short drive to that store to buy that neat item you saw advertised or to plan a long holiday. But remember that all of these mapping services carry some fine print saying, in effect, that their information may be in error. Indeed, we were travelling to a location in Frederick Maryland and found that a road that we were told to use had been renamed. Needless to say, since a geographer was driving we became completely lost!

In another case, a meeting on the GeoEconomy at which we were to speak was held in a building that our map showed on one side of a major thoroughfare – but in fact it was on the **other side** of the road and we had no easy means of crossing. So it took us a little extra time. When using maps it is a good idea to build in some contingencies for map inaccuracy – like allowing that little extra travel time just in case. No data sets are perfect – be they maps or tables of data. If you use a lot of data you'll inevitably encounter errors or inaccuracies at least occasionally – just like sometimes your car won't start.

When driving in some regions or weather conditions extra caution is warranted and a little extra preparation, including better information, can save a lot of grief and in some cases may even save your life. Examples are winter driving in rural areas or driving in the rainy season in regions subject to flooding. Before we venture out into the rural North Country in winter we check a weather web site that reports which roads are snow-covered or ice-covered, and we often consult the highway department to check on road closures. In the USA one might want to consult http://www.theweathernetwork.com/index.php?product=ushwyfx&pagecontent=usahwy – a site that describes the short term weather forecast along the entire Interstate Highway System.

Are you a hiker, motorcyclist or cyclist? Imagery and elevation maps are readily available for most regions of most developed countries. These, along with road maps, are useful for planning bicycle, motor bike outings, or hiking. Again, for specific routes it may be wise to verify that the expected road surface is in fact what you will get. In less developed countries maps are also useful to determine where it may be safe to travel – and where not.

And if you are meeting a good friend in congested or very large areas like the Ginza area of Tokyo, the Toronto Zoo, or in New York's Central Park, you might make yourselves easier to find if you both have GPS-enabled cell phones set up to track each other!

6.2.10. Conclusion

This range of activities to which an individual can apply geospatial knowledge may seem limited and perhaps even self evident, but the applications are important and, for most people, will determine their financial security and that of their children long into the future. While choosing the right schools, work location and neighbourhoods together are not a guarantee of future success, the right choices can reduce the probability of either financial loss or failure. Those of us who have selected our neighbourhoods and schools well, know the benefits that good geo-decisions can bring. And as we move further into the GeoEconomy we are equally certain that many more applications for individuals will be identified, providing fodder for future authors.

6.3. Communities

6.3.1 Introduction

Communities of all sizes are facing increasing pressures on their budgets, as well as new demands on their managers and decision makers. Tax payers want their municipalities to offer better services for less cost, while at the same time operating in a more sustainable and environmentally aware fashion. This all at a time when infrastructure is decaying, security concerns are growing, and an increasingly unpredictable climate is constantly changing the parameters. Well-run municipalities, from the biggest to the smallest, are turning to the intelligent application of geospatial information to accomplish these seemingly disparate goals – and much more.

For communities, the most important "where" questions involving geospatial decision making can be grouped under a few major headings, such as understanding the community, infrastructure, land use planning, and policing, as outlined in the accompanying table. Each is addressed in the following subsections.

Questions Communities and Municipal Governments May Answer with Geospatial Information and Technology

• **Understanding your community**
  • What will that new development look like when it is done?
  • Where can we build XYZ so its visual impact will be minimized?
 	
• **Infrastructure**
  • Where is our infrastructure and what shape is it in?
  • What infrastructure needs to be replaced and in what order to maintain or improve service while minimizing costs and disruptions?
 	
• **Land use planning and zoning**
  • What is the ideal zoning of this land?
  • Where should building be allowed and where not?
 	
• **Transit, school bus routing and service delivery routing**
  • How should we route school buses to minimize distance travelled, cost, and the time smaller children spend on the bus?
  • How can we route transit buses to maximize the people using the system while minimizing the cost?
  • How should we plan the day's work travel of our maintenance crews to minimize delays and cost and to maximize their output?

- **School planning**
  - Where should the new school be located?
  - Which school should be closed?
- **Urban parks and recreation**
  - Where are the tourist areas and how can tourists best be told their location?
  - Where should the new park be located?
  - What is the condition of the vegetation in the parks and around the municipal buildings?
  - Is some form of corrective action needed for the vegetation and if so, what must be done and in what order?
  - What recreation needs are there in that community?
- **Policing and security**
  - In what area does the serial criminal likely live?
  - Where are the crime "hot-spots" that can benefit from increased community involvement?
  - Where do the most traffic accidents occur, and what in that locale can be changed to reduce accidents?
  - What threat will the police officer (or fireman or medical response team) face at that address? (And WHERE IS that address?)
- **Economic development**
  - How can we best sell our municipality's attributes and promote sustainable development?
  - Our town's major employer has closed down – how can we re-invent our town?
- **What geospatial technology should we buy?**

6.3.2. Towards a Better Understanding of Your Community and Its Geospatial Needs

Most of us have heard the adage that "a picture is worth a thousand words." A picture in the form of an aerial photograph (or satellite image in some cases) and annotated map – both displayed in a GIS map output – is a useful tool to understand the many relationships that exist in a municipality. Clearly one of the benefits of a GIS has been to show politicians and senior managers – as well as citizens – these relationships. What will that new development look like when it is done? Where can we build XYZ so its visual impact will be minimized? Perhaps the best tool to answer these questions is what is known as a view-shed image – in effect a "before and after" picture of a development is created and placed over a 3-D model of the terrain. This image shows the visual impact that a development will have on the surrounding area.

A GIS can be used to examine relationships between commercial districts and residential areas, and between parks and residential areas, schools and roads. In many large cities with subway systems the growth of large residential complexes spreads across the city in a regular fashion – regular in that they coincide with the subway stops. But a properly created map is far more than a means of understanding relationships that have to do with decision making. Such information contributes to day-to-day municipal management and should be considered essential in all municipalities, no matter their size. To be effective and useful, information needs to be selected, organized, and distributed to suit the needs of those who will use it. Municipalities, officials, and others will differ in the geospatial information content and format most useful to them.

6.3.3. Infrastructure Management, Maintenance and Planning

There are two words that are striking terror into the hearts of city managers everywhere: infrastructure decay. Whether it is the rusting sewage treatment system in the quaint village of Merrickville, Ontario, collapsing bridges in Quebec, or potholes in roads and leaking water supply pipes almost everywhere, the cost of replacing infrastructure has been estimated to be in the trillions of dollars in North America alone.

What is the role of geospatial in infrastructure? First off, geospatial data are part of the municipal infrastructure. In addition geospatial data can answer, or help to answer, several important questions. Where is our infrastructure and what shape is it in? And what infrastructure needs to be replaced, and in what order, to minimize costs and disruptions? Obviously geospatial technology can be used to create an inventory of infrastructure by type and condition. Specially equipped vans now patrol our highways collecting valuable information on the condition of the pavement. Geospatial technology can be used to track age and assess the infrastructure most at risk, and when brought together with information from other municipalities, it can make the case for support from higher levels of government. Such an inventory can also be useful in helping to plan repair, reconstruction, or abandonment. We are a visually oriented species. When imagery (of exposed re-enforcing iron on a crumbling bridge, for example) is coupled with maps, aerial photographic imagery, view-shed imagery (described above), and perhaps a few catastrophic examples of failure from some other jurisdiction, such information can provide a strong visual stimulus for the policy maker and decision-maker.

Infrastructure Renewal and Repair. Water system upgrade (left) and road maintenance (right) are just two of myriad infrastructure renewals that municipalities are facing worldwide. Source: S. Aronoff

In short, geospatial information can at the same time help make the case for infrastructure renewal, and help decision-makers make better decisions, while reducing risks and cost. A recent estimate in Canada made by the Geomatics Industry Association suggests that the geospatial information needed to support infrastructure renewal in Canada would cost approximately $50 million per year over each of the next five years.[218] In 2007, the Canadian Federation of Municipalities published a report titled *Danger Ahead: The Coming Collapse of Canada's Municipal Infrastructure* that concluded $123B needed to be spent on infrastructure to avoid effective collapse of the system. The $123-billion estimate includes expenditures in four key categories:

- Water and wastewater systems ($31 billion);
- Transportation ($21.7 billion), and transit ($22.8 billion);
- Solid-waste management ($7.7 billion); and
- Community, recreational, cultural and social infrastructure ($40.2 billion).

To plan infrastructure renewal the priority needs must be identified. They are usually related to some element of public safety. Questions that must be asked include:

- What renewal is needed most (and why);
- Where is the renewal needed;
- What will it cost (to either perform the renewal – or delay it, or not do it at all); and
- How does the renewal fit into the overall infrastructure system?

Geo-based renewal can lead to more efficient and more integrated planning – and more effective spending. We have all seen examples of a new road being dug up to install sewer or water mains. Why were these services not put in before the road surface was built? Or why did the heavy equipment needed to build some structure drive across (and ruin) the new athletic field – or recently planted crops or trees? With a GIS and intelligent cooperation and coordination between different levels of government, utilities, and departments within the same government, costly and frustrating errors can be reduced. For example, putting all of the planned infrastructure renewal projects on one digital map (or in one GIS) helps policy makers, and those planning temporary transportation rerouting, anticipate overlapping or conflicting tasks and coordinate the work more effectively, thereby helping to avoid such problems. Simply put, renewal and infrastructure construction are best done using a GIS to better visualize the situation, look at the infrastructure as a system, and better assess risks. This is so whether one is discussing transportation planning, sewers, water delivery, schools, or public buildings.

6.3.4. Land Use Planning and Zoning

Land use planning and zoning were among the earliest applications of geospatial information in and by communities. Indeed, scores of books have been written on the subject – meaning that we need not cover it here in great detail. Essentially geospatial information can be used to answer two fundamental planning questions:

- What is the ideal zoning of this land? and
- Where should building be allowed and where not?

GIS was developed largely with such planning applications in mind. Originally, and still today, different attributes (including former land uses) were overlaid to assess the best location for certain types of developments or land uses. This ensures compatibility between various adjacent land uses and the inherent capability of the land. This application continues to be important today. Picturing existing uses, and planned zoning on a map or view-shed, helps decision makers and the populace understand what is being suggested and what the result is likely to be. Geo-information is also useful in flood control, assessing and planning responses to climate change, economic development for a region, and the like. Urban and regional planning now depends on geospatial information and technology. Throughout the planning process developers, planners and engineers use maps of varying detail, historical information, old and new aerial photography, viewsheds, demographic and cultural information (including census data), as well as transportation information overlaid on a map base to plan, approve, and build virtually all new developments. Never again should we find new schools or housing developments on old landfills.

6.3.5. Transit, School Buses and Municipal Service Delivery

For many years more efficient transit system planning, school bus routing, and delivery services have all used route optimization applications (commonly implemented with a GIS). These help to minimize travel distances and therefore costs to service an area or to include a specific group of pickup locations. For the transit system, it minimizes the distance travelled by its buses, maximizes ridership, and reduces costs. For school buses, it seeks to minimize total distance travelled while setting routes that minimize how long young students must spend on the bus. The same sort of application is used for service delivery activities such as garbage removal, snow clearing, street cleaning, and park and city maintenance. The cost savings cited in municipal management include savings in fuel, employee's time, and vehicle maintenance that can amount to a 15 to 25% reduction in costs.

School Bus Route Optimization. Efficient bus routes reduce travel time for children and save transportation costs. Source: S. Aronoff

A similar routing-related question is how should we plan a day's travel for maintenance crews to minimize delays and cost and to maximize their output? Delivery of municipal services can also benefit from the intelligent application of geo-information and related technology.

Ottawa is one of the coldest capital cities in the world and exhibits wide variations in winter temperatures and snowfall conditions, contributing to often severe and unpredictable road icing. In November 2009, the City of Ottawa announced[219] that GPS-enabled material handling devices were being installed on trucks that, in winter, distribute salt to melt the ice on the city's roads. The GPS devices allow supervisors to instantaneously view the location and the rate at which each vehicle is spreading salt. The operators can then be instructed to adjust the spread rates, as required, while they are on the move. These devices, which cost $200,000 to install on the

Hazardous Weather Conditions – Planning and Management. In northern climates, snow storms often produce hazardous driving conditions causing deaths from collisions and exposure to extreme cold. Geospatial information is now integral to the prediction of extreme weather events, planning of snow removal, road treatment operations, and other mitigation measures, and the delivery of emergency response and recovery services. Source: R. Ryerson

city's 200 salt trucks, will cost under $90,000 per year to operate. The savings from more precise salt distribution to match road conditions are expected to be $1 million per year in a $15 million budget – and reduce salt use by thousands of tons. The saving from this one application will pay for itself not in one year, or even six months, but in a matter of a few winter storms. At the same time, it advances the city's environmental goal of reducing road salt usage because of its harmful effects on vegetation, aquatic life, and water quality.

6.3.6. School Planning

Census and other demographic information have long been used in school planning. Such data as the number of women of child-bearing age, population projections, income levels, and other demographic information are essential to plan school development as well as school closings. Key location issues are: Where is additional school capacity needed, where should new schools be located, and which schools should be closed?

In 1952, the first school Ryerson attended was able to successfully plan for the incoming baby-boom wave of first graders by using basic demographic information: number of women of child-bearing age, number of women who were pregnant, and the number who planned to have more children. Mr. Graham, a school trustee and local farmer, completed a door-to-door survey of Grimsby Park School's catchment area to determine the number of women who had children in 1946 and who were expecting. He carried out a similar study in 1947 and 1948. He then used a map of that information to convince the local authorities that an addition was required for the school building. That addition, including indoor plumbing for the first time in that rural school, was ready in time for the post-World War II baby-boom.

Unfortunately, as David Foot noted in his landmark work on Canadian demographics in 1996,[220] not all school boards in the past have used demographic information as well as did Mr. Graham. Indeed, geographic information has at times not been used at all. Some boards were still opening new elementary schools as the school age population was dropping after the baby boom generation. Although this demographic shift was readily foreseen with the demographic data available to them, many jurisdictions evidently did not make use of it and wasted millions of tax payer's dollars.

Choosing the locations for new schools is a sensitive issue. It not only affects student travel times to school and convenience for parents, it also affects the mix of neighbourhoods from which the students will be drawn, house values, traffic patterns, policing and other service needs. GIS analyses are valuable for optimizing school location decisions. They can communicate to all stakeholders the factors considered and the priorities used, and they allow alternative scenarios to be easily examined. This transparency not only substantiates and legitimizes the decision-making process, it also improves the result by providing a means to assess the consequences of alternative priorities.

Transparency in decision-making and the value of geo-information is not restricted to developed countries. School planning using geo-information and demographics is even more useful in less developed countries where there are scarce resources. In a study of the use of GIS in Bangladesh we learned that a head-man of a small group of villages welcomed the information provided by GIS. He already had a school in his village and when funds were made available for another school in the area, there was some disagreement among his many relatives in surrounding villages as to where the school should be located. A GIS was used with demographic information and the locations of existing schools to optimize the location of the new school. By so doing he was able to avoid confrontation with his relatives – he could say "the computer made the decision – not me."

As the school age population in an area increases and decreases (as it always does), it is surprising that more schools are not constructed to adjust to the changing demographics in a locale. Why not design schools to be quasi-temporary – convertible to other uses? If the design consisted of a central core with offices and gymnasium facilities, surrounded by extendable hallways and removable classrooms, then as enrolment changes space could be reduced or added as needed. If the enrolment decreased beyond a certain level, the classrooms could be removed and used elsewhere, and the facility converted, for example, to a community centre with offices, a gymnasium, or some meeting rooms, with an outdoor playing field and parking. Then if the school-age population grows again, the classrooms could be returned from locations where they were not needed and the facility converted back to a school. The idea is not so farfetched. It is common today for schools to install portable classroom units to augment their classroom needs, the concept of a "quasi-temporary" school simply allows for the inevitable changes that occur in the geography of school age population.

6.3.7. Urban Parks and Recreation Management

The use of geo-information in tourism has been growing. More and more use is being made of tourism maps, satellite image maps and the like. Indeed, ESRI Canada features such use in its 2010 calendar.

But the uses of the data and technology go far beyond showing where things are. The uses of geospatial information and technology for parks and recreation management has exploded over the past few years, at the same time that demand for parks and recreation resources and the cost of maintaining them has skyrocketed.

Uses range from planning parks (where should the new park be located?), to assessing vegetation health using specialized aerial imagery (what is the condition of the vegetation in the parks and on city property – and is some form of corrective action needed?), and for planning effective service delivery, once the decision has been made to take some action.

Parks and other recreation facilities not only serve as a source of rest and relaxation for a community's citizens that foster positive group and community wide activities, they build a spirit of cooperation and mutual responsibility for the community's well-being. Team sports and other youth activities engage teenagers and young adults, lessening the attraction of gang membership and other potentially disruptive or illegal activities. Parks that welcome family-friendly activities like community events and welcome afternoon picnics and barbeques not only make a community a friendlier more "fun" place to live, they make for a safer more secure place to live by reducing crime, vandalism, and other types of urban blight.

Monitoring Urban Development. Residents are well aware of property development projects in their neighbourhood (left). Online access to geospatial information such as satellite imagery (right), though it may be somewhat dated, is useful to appreciate the overall scope and speed with which a community is expanding. Source: S. Aronoff (left), Google, Inc. (right)

Geospatial analysis can help improve the management of urban parks and recreation to capture more of the benefits these facilities have to offer. Management activities include:

- Monitoring the catchment area and activities;
- Maintaining an inventory of assets;
- Communicating and informing to improve park use and satisfaction; and
- Improving maintenance effectiveness.

Monitoring the Catchment Area and Activities: A park or recreation facility attracts people from specific areas based on the amenities that are offered. Knowing the age groups of those in the area who would likely come to the park, and the types of activities they want to do are important, not only to get the most benefits for the community's money spent but also to anticipate and plan for the way these needs will change in the future. By defining the catchment area for a facility (i.e. the areas where the people who use a facility live,) other information can be obtained. For example, demographic information would show whether the number of young people in the area is increasing, decreasing, or remaining stable. Information about the cultural make-up of the community being served could help define the types of facilities of most value. For some cultural groups it is common for an extended family to spend a summer afternoon at a park – with children playing soccer informally with their cousins, babies being cared for by their mothers, grandmothers lying on blankets spread out under the trees, and the men chatting around a barbeque grilling lunch or dinner.

It's a scene common in many parks around our city – but principally those parks with the facilities to make this activity easy to do. They have permanent barbeques for community use, convenient parking and washrooms, shade, and safe space for young children to run around and be clearly visible at all times.

Other communities tend not to use park space in the same way. Their family gatherings don't take place in parks. In these communities the principle use of park land is for organized team sports like baseball, football, soccer, and hockey. To support these activities you need spaces easily divided and sub-divided into playing fields, with sizes varied to suit different ages of children and different sports.

Many municipalities have geographic information systems with most, if not all, of the data needed to perform the analyses. Creative use of geospatial data can improve the benefits the citizens obtain from park resources.

Maintaining an Inventory of Assets: Municipalities already maintain inventories of major fixed assets like buildings and equipment (e.g. trucks and tractors). Increasingly, municipalities are using their geographic information systems to track characteristics of their park and recreation facilities. For example, many inventory their trees and certain other vegetation types. With aerial photography for a background display, and data easily collected by city employees using hand held GPS-equipped computers, the condition of trees and other characteristics of a park are readily monitored. This expands the types of assets a municipality inventories and ties that information to specific locations. This information can then be made available to all city departments, some of whom may find other useful applications. Perhaps knowing where there are large patches of shrub and forest in quiet park areas would be useful in setting police night time patrol schedules.

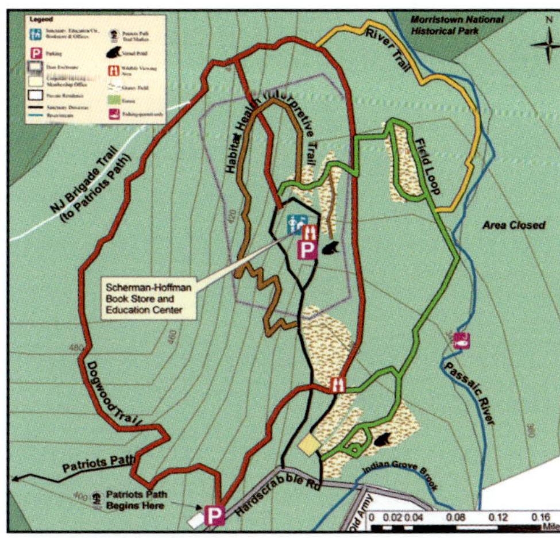

Innovative Trail Map Creation with GPS. Walking the trails with handheld GPS units recording their path, staff could quickly collect the digital geographic information to easily plot the trails for this visitor map for the Scherman-Hoffman Wildlife Sanctuary. Source: New Jersey Audubon Society

Communicating and Informing to Improve Park Use and Satisfaction: Parks and recreation facilities receive more use if people can easily find out where they are and what they offer. Most municipalities have web sites that offer basic information about these facilities and their location, often with links to maps. Geospatial technology can make it relatively quick and inexpensive to communicate more information and make it easier for visitors to plan their visit and enjoy the experience when they arrive. For example, the

New Jersey Audubon Society developed information to promote visits to their wildlife sanctuaries, many of which are close to or within the municipal boundaries of urban areas. To acquire data not available from existing sources, staff members walked the sanctuary trails carrying a handheld computer with GIS software and a GPS. They entered site-specific information such as scenic overlooks, trail benches, water falls and ponds, unique historical features, toilets, information displays and other useful visitor information. This data was then input to a GIS and combined with other GIS data layers such as a topographic map and aerial photography (to be more specific about certain landscape features). The final map product was then developed with supporting information about trail length, highlights, cautions, and other information.

Improving Maintenance Effectiveness: As noted above in Section 6.3.5, service delivery can be facilitated through the use of GPS to optimize routes from one place to another. Given the cost of equipment and the cost of staff, even a small saving in delivery costs can, over time, save millions of dollars in larger jurisdictions.

Service delivery – from mowing lawns to pruning trees and repairing athletic fields – is facilitated by geo-information. The uses range from planning the development of a facility to planning the day's work for a maintenance crew. Today it is common for service crews to have their day's routes plotted in advance to reduce down-time and traffic delays. While not always appreciated by employees, the same technology can also be used to determine if a truck (or group of trucks) all stop at a time and location not normally allocated for a break – perhaps for a coffee and donut. While seen by some as somewhat Orwellian, such monitoring saves time and money, leading to efficiencies in the use of staff and equipment. Such efficiencies inevitably lead to more secure futures – for the recreation facilities and the staff who maintain them.

6.3.8. Policing and Security

Policing and security operations have seen a tremendous growth in the use of geo-information – from its sophisticated and often surprising uses in the solution of crimes, to its more mundane application in planning policing intensity and coverage areas. For example, crime and accident statistics tied to location can be used to identify areas that would benefit most from increased police surveillance – areas where crime can be anticipated or accidents prevented. Police and security operations, such as those outlined below, currently benefit from the use of geo-information:

- Finding criminals;
- Reducing road accidents;
- Public dissemination of crime rate Statistics and locations; and
- Emergency response.

Finding Criminals: Most people have heard about the use of crime scene mapping and GIS to overlay various parameters which pinpoint the likely home base of perpetrators of serial crimes, and the projection as to where the next crime might take place. While this is the stuff of Hollywood legend, it is also a very real application. In one case in Toronto a senior officer had taken the specialized training to use this technology – and in his first instance of being asked to use it, he identified the most likely location and time for the next crime. While the officer told us that it was "just luck," his prediction was so accurate that the criminal was captured when and where projected. Needless to say, geographic information is an ongoing tool in the fight against crime in that city!

In another case that has received a great deal of attention,[221] relatively simple location information and technology were combined with dogged police work to provide sufficient information to convict a hit-and-run suspect. The basic information used was from cell phone records which gave the location of the cell phone relative to cell towers. When combined with surveillance photo records at a bank and gas station, it was sufficient to identify the car and driver and place them very close to the scene of the accident. In another recent case suspects in a murder case in Ottawa were apprehended within hours – they had in their possession a GPS-equipped cell phone that belonged to the victim.[222]

Reducing Road Accidents: As in all things, the spectacular application in policing tends to get the attention, while the true value of geospatial information is far more mundane. Where have the most road accidents occurred? What were the specific factors contributing to the accidents? What can be done to the routing of traffic or design of the roads and their access to reduce the carnage? And carnage it is. According to a study released in 2008 by the American Automobile Association, in the USA alone there are over 40,000 deaths on the roads each year. The cost of accidents is in excess of $160 billion for damage and health care.[223] This is far more than the costs attributed to congestion – which is also something that can be studied (if not always solved) through the use of geospatial information.

In 2004, the World Health Organization estimated that the world-wide cost of road traffic accidents was $518 billion per year. Furthermore, low- and middle-income countries lose more money to road accident and injury than they receive in development assistance. High-income countries lose 2% of their GDP to the cost of road accidents. Indeed, in 2004 the World Health Organization used World Health Day "to advocate a 'systems approach' to road safety, which takes into consideration the key aspects of the system: the road user, the vehicle and the infrastructure."[224] Obviously a key component would be the mapping and understanding of accidents and their causes.

Public Dissemination of Crime Rate Statistics and Locations: Crime rates in certain areas of the city have long been made public by police forces in cities such as Chicago and Winnipeg as an aid to community policing – to better engage the citizens in their own protection and to make them aware of issues as they arise. Indeed, a central part of the City of Chicago's crime prevention and community policing is its use of GIS.[225] Crime statistics organized by location tell us a lot about an area. For example, national and international statistics tell us that one of our clients, the Quebec Chaudiere and Appalache (QCA) Region of Quebec, is the safest of any city in North America.

In the city of Winnipeg, its "CrimeStat" web page is seen as sufficiently important that, at the time of writing, it is visible on the city's home page.[226] The CrimeStat web page can be used to map crimes by area of the city. On December 3, 2009, Winnipeg posted an alert on its CrimeStat page to ask citizens to help solve a series of crimes. There were composite drawings of the suspect, a map of the locations of the attacks, and media reports.[227] Such pages are limited to generalized information to protect the identity of individuals.

Protection of the identity of individuals is a growing concern to agencies involved in protecting our privacy. This is especially important given the ease with which information on the individual can be obtained by those familiar with the use of the Internet and the many data bases accessible on line. Geospatial information and maps have been specifically addressed in a number of meetings and presentations sponsored by, for example, the Privacy Commissioner of Canada.[228] At one such meeting Dr. Khaled El Emam, who holds the Canada Research Chair in Electronic Health Information at the University of Ottawa, illustrated how easy it is to use location information to identify people whose individual information (such as health) is supposed to be, and remain, confidential.[229]

Emergency Response: A less public use of crime information is by police agencies themselves, as they go about their normal duties responding to emergencies. Police officers need to know before reacting to a call what dangerous situation they may be facing. Has there been any previous history of violence at that location? Are there weapons registered there? What access points are there (and escape exits) from the location? What back-up is indicated? What weapons may be required by the police? By having a location-based data base with information tied to addresses, officers can approach a dangerous situation with more preparation and care. In a similar vein, fire brigades need to know if there are dangerous chemicals or flammable products in a building that is on fire. Proper use of geographic

Upgrading Municipal Labelling to Support Geospatially-Enabled Services. This highly reflective street number sign is required by local law to help emergency responders more easily locate a specific address. Source: R. Ryerson

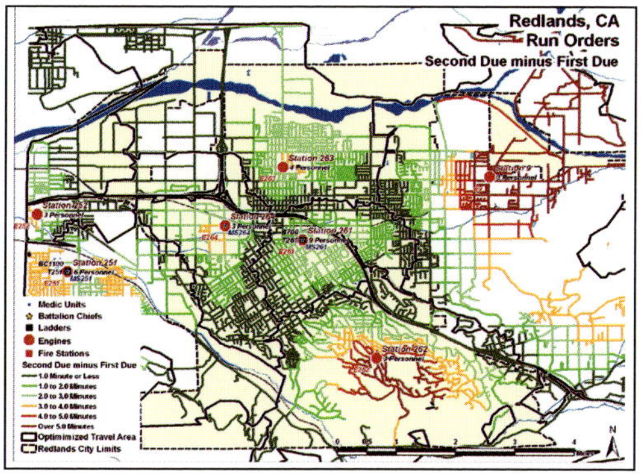

Customized Map Production with Geospatial Data. This map shows the time needed for a second fire fighting unit to arrive on scene. It is a custom map product requested by the Redlands, California Fire Department. With municipal data organized within a geospatial information system, the map was easily generated in short order. Source: Entrada/ San Juan Inc.

information can make an inherently dangerous job somewhat safer. Returning to the roots of the technology, such applications are not unlike the use of surveillance imagery by the military, as they plan their tactical advance on an enemy position.

Another security or safety-related application is the well known "911" emergency dispatch system. Where the system exists, all telephones are coded to an address. When someone calls from that address emergency vehicles are sent as required – police, fire or ambulance. It is for this reason that addresses must be consistent and regularized. Instances have been reported of ambulances being unable to find an address or being delayed because of difficulties in the address. Delays can result in death, slower recovery, or incomplete recovery. In many places each building is identified by a number, presented in a consistent manner on a road-side sign, which the emergency vehicles can easily find and read – even in the dark.

Much has recently been written about the use of geospatial information in disaster response – which in many jurisdictions is a responsibility of local government. The disasters can be anything from flooding, to landslides, forest fires, or other issues. When such issues become large enough they are often handled by state and national governments and are discussed in more detail in Section 6.5.

6.3.9. Economic Development

Geospatial information has been used in developed and developing countries alike in attempts to attract industry and development. We have already discussed the value of geo-information in the decision to locate a business in Quebec, in a case study reported in Section 3.4.2.3. In the mid-late1990s, the United Nations Development Program (UNDP) had a project that introduced GIS to improve government services and information in India. Ryerson was involved in evaluating the results. One test site was in the District of Bankura, in the state of West Bengal, India.[230] District Magistrate Sri. Tempo obtained outputs from this UNDP GIS activity to generate information for potential investors in the region's industry. She was able to show levels of education, availability of land, and the availability of a skilled workforce to potential investors. While the original focus of the test was on resource management (notably the alleviation of drought), this additional use helped to convince her of the potential importance of the technology in a much broader range of activities. Starting small and simple and working up to more complex activities has clearly been very successful in this instance – and provides a useful approach no matter the location or country.

Amenities Attract Talent. City amenities not only benefit existing residents, they make it easier to attract outside talent. The annual Winterlude Festival organized by the National Capital Commission and the City of Ottawa with broad government and corporate sponsorship offers skating on the Rideau Canal, the world's longest skating rink, and numerous public events during February each year that are enjoyed by the local population as well as a large number of tourists. Source: S. Aronoff

Another of the applications in development is the re-inventing of cities and regions (and even countries) through paying attention to their geography and the benefits that their geography – their "where" – gives them. For example, as noted previously, the town of Elliot Lake in Ontario Canada has re-invented itself from a mining town in a picturesque setting to a retirement town, in order to replace the departed uranium mine. Niagara-on-the-Lake, Ontario has developed its beautiful setting, historical architecture, a fruit and grape industry, a college with a focus on food and wine, as well as a world-class theatre into a high-value wine- and food-based tourism destination. Napa California has become synonymous with wine (and wine tourism) in California and the USA – yet it produces a tiny percentage of California's wine. Maps and explanations of location play a key role in each of these development projects.

In the preceding paragraphs we identified some uses of geospatial information in promoting industrial development. What perhaps may be of more direct relevance (at least for the cities and regions we name in the table below) are those cities or regions that have developed capabilities in the field of geospatial. These are the regions and cities of the world that stand to come out the best in the GeoEconomy because they will be driving the technology and services that underlie the GeoEconomy.

We evaluated over forty factors in deciding whether or not a city or region made our list. The factors we looked at included corporate activity, research, education, national and state government support and interest in geospatial, national, state and local government use of the technology, the national and local policy environment, market share, exports, employment, human resource capacity, sustainability, banking health, international recognition, links to national and international networks, links to the national military operations and R&D, business set-up costs, international investment, and the presence of a local champion or champions.[231]

In our opinion these cities and regions of the world have the potential to build on their geospatial or geomatics expertise to further assure their success in the GeoEconomy. Whether they do so – or not – will be interesting to watch. Surprisingly enough, to date only Quebec City, Lethbridge, Calgary and Wuhan seem to have recognized their unique potential in an official way. We assume that eventually at least some of the others will recognize their inherent strengths and become engaged in the necessary strategic planning, with appropriate advisors, so that they can build on what they have.

Clusters of Geospatial Expertise: Poised to Reap the Rewards of the GeoEconomy?		
Colorado, USA	Lethbridge, AB, Canada	Redmond, WA USA
Calgary, Canada	Melbourne, Australia	Tokyo, Japan
Bangalore, India	Mountainview, CA, USA	Toronto, Canada
Boston, MA USA	Munich, Germany	Stennis/ Slidell (MS/LA, USA)
Enschede, Netherlands	Montpellier, France	Wageningen, Netherlands
Fredericton, Canada	Ottawa/Gatineau, Canada	Washington, D.C. USA
Geneva, Switzerland	Perth, Australia	Wuhan, China
Houston, TX, USA	Quebec City, Canada	
Huntsville AL, USA	Redlands, CA, USA	

Source: Study by Kim Geomatics Corporation. If your City belongs on this list, or if you would like to take advantage of the advantages you have because you are already on this list, contact the senior author.

What may be sad to some is that some of those with an early lead in this field (such as the UK and Portugal) are no longer represented on this list of geospatial clusters. In our opinion this comes as a result of policies that ran contrary to development of the geospatial industry and the use of its products and services. What is exciting is that India and China have both earned spots on the list. With the rapid evolution of the field, we expect that movement on and off our list will become even quicker than it has been in the past. As has been repeated throughout history, leadership today in a particular geospatial sector may well not matter in the future.

6.3.10. Getting the Right Technology for the Job

As municipalities or communities embark down the path of developing geospatial data and geo-awareness there are several important considerations to ensure that the right technology is applied to the right problems:

1. First, those who sell hammers tend to see every problem as a nail. In other words, vendors of both data and technology will not necessarily have the best interests of the buyer at heart – they will want to sell their data and their tools. As with all purchasers of all goods and services – buyer beware. Of course the best way to overcome this problem is to have a very clear plan of what you want to do (see #4 below);

2. Second, there is often a tendency to want to develop home-grown systems. In larger cities information technology (IT) people may want to develop their own custom-built geospatial data handling system. In this respect we urge caution. Among the many vendors there is quite likely an off-the-shelf system available that can be easily modified to meet almost any need – at a fraction of the cost of what the IT people would rather do. Furthermore, no system should be used (or developed) that is not fully compliant with Open Geospatial Consortium standards – otherwise you are likely to find that your data and systems will not be able to talk with other data and other systems with which you will have to interact;

3. Third, it is usually a good idea to buy from an established vendor who will be around for some years to come. What happens if your vendor disappears? Simply stated, this often means that your investment disappears too;

4. Fourth, ensure that everyone who may benefit from the use of the system has their data and their needs built into the data and presentation tools. Otherwise, the full benefit and integrated management of the municipality will not be possible. Sometimes it is difficult to achieve this for the simple reason that some managers (or their staff) may not fully understand the benefits and may not wish to participate. Through consulting in this area we have developed a standard approach to ensure that almost everyone's data and needs are built into the system. The approach has worked in Thailand, Bermuda, and the Maldives, among others; [232]

5. Fifth, learn from others. No matter the situation, another community somewhere has probably used geospatial data in a manner similar to what you need. We suggest reviewing several years of the trade press to see what others have done; [233] and

6. Sixth, if you are implementing a new system, there will be changes to standard operating procedures. The transition is NOT about technology, it is all about people accepting change that may dramatically affect their working lives. A change management program should be implemented, and a workshop for staff should be organized to help them adapt to the new geospatial world in which they will be operating. In our consulting practice we have found that such workshops are a useful starting point for all concerned.

6.4. Companies

6.4.1. Introduction

We have discussed the importance of geospatial information for countries as part of a national infrastructure and so far in this chapter we have presented how the data can be used (and is being used) by communities and individuals. Geospatial information is also the foundation for much of the economic decision-making in industry. It is for that reason that we and many others advocate the open provision and dissemination of geospatial information by governments so that it can be used to benefit the economy as a whole, while at the same time allowing individuals the opportunity to study certain issues of interest and make better decisions for themselves.

Without geo-information the whole economy is less efficient, as has been amply demonstrated in a landmark study in Australia by ACIL Tasman.[234] While it is important that industry gain the potential efficiencies with respect to the use of geo-information, at the same time, it is important for individuals to know how that information is being used or could be used by those wishing to sell something to us.

This section then has a dual purpose – to inform those in industry what the data may be used for and to help individuals understand what is being done with geospatial data. While we are tempted to say that most large retailers and those most active in the financial field likely do not need any advice, our experience to date suggests that inattention to geospatial information is often the harbinger of failure in even the largest and best known retailers and financial institutions. The sub-prime mortgage fiasco of 2007 in the USA is the best, but not only, example of this.

The questions companies typically ask that draw on geospatial technology and information can be grouped under a number of categories such as business intelligence, marketing and risk assessment, consulting and service delivery, natural resource management, and environment. In some cases the questions are quite direct and in others they are open-ended.

Some of the specific questions for each category are detailed in the following table.

Questions Companies May Answer with Geospatial Information and Technology

- **Business Intelligence, Marketing and Risk Assessment**
  - Where should I locate my business?
  - Where are my competitors?
  - What are my competitors doing?
  - Where are my current clients?
  - What are their buying habits by location?
  - Where are those to whom I should try to sell?
  - What variations in risk are tied to location?
- **Consulting and Service Delivery**
  - What are the "where" questions that our clients need answers to?
  - How can we best answer these "where" questions?
  - What are the "where" questions we can ask ourselves to make us more efficient in delivering our consulting services?
- **Natural Resource Management**
  - Can maps, imagery or real-time location make us more efficient? If so, how?
  - What are the most efficient delivery routes?
  - What resources do we have?
  - Where are my drivers and vehicles now?
  - Can we reduce inventory and increase profit through the intelligent use of geospatial information and technology?
- **Environment**
  - Can we demonstrate through imagery or other geospatial tools and ground sampling (acquired before, during and after development) that we are paying attention to the sustainability and protection of the environment?
  - Are our business practices damaging the environment?
  - If so, where and how much?
  - Are there any activities in the vicinity of our facility that exhibit damage (to vegetation, water, etc) that might be frivolously blamed on us?
  - Are we monitoring our facilities for failures (dams, holding ponds, pipelines) that might lead to major problems?

Business intelligence and marketing are the application areas most associated with the private sector. From the above list of questions it is clear that there are many other areas of application – many of which have a low cost to implement but potentially a very high return in terms of reduced cost, reduced risk, and perhaps lower insurance premiums.

This section addresses the questions in the four categories in the table and closes with a look to the procurement of technology by industry. Most of these questions can be answered by the interpretation of imagery done by a competent consultant.[235] The specially acquired imagery may cost anywhere from a few thousand dollars to much more, depending upon the size of the area being investigated and how close it is to a company offering such services. Given the liabilities that can be involved, a few thousand dollars invested every few years to monitor changes (if any) would seem to be wise insurance. As we have said in many public lectures, one remote sensing image can be worth a thousand words, and many hours of a lawyer's time!

6.4.2. Business Intelligence, Marketing, and Risk Assessment

Business Week Research reported that **80% of all data held in data bases in US industry had a location component.**[236] Most business decisions are made with some reference to location – where to locate the bank or the retail service, where my clients are located, and where best to try to sell insurance – or anything else. These and similar questions fall under the title of business intelligence (BI) and geo-marketing. BI and geo-marketing are often said to be using location intelligence – much as the military uses military intelligence to plan an attack, industry uses location intelligence to attack markets.

While risk assessment may not seem to be linked to marketing, we have done so here because what geo-risk analysis is doing is not unlike what a marketer does – both are trying to predict and influence future behaviour. The marketer tries to influence that behaviour to get someone to buy, while insurers try to influence people to not locate their homes in flood zones, for example.

There are a number of companies that have based their success on the provision of geospatial information and tools to industry and others who wish to understand the buying population – or spending population in the case of charities. While hardly a household name, MapInfo (now officially known as Pitney Bowes Software) has made a fine income developing tools for companies to study their respective markets. These tools allow the retailer or bank or insurance company to segment the market based on map information (hence the name MapInfo) to allow their marketing to be highly targeted. After all, it makes as little sense to send a mail-out to someone who is highly unlikely to buy your product as it does to insure a building for flood damage when past history and local geography tells you that it will never flood.

In some jurisdictions market studies are facilitated by detailed postal codes. In Canada, for example, the postal code signifies a particular block face (i.e. the string of houses along a road between two other roads), an apartment building, or a series of mailboxes at the Post Office. When postal codes are combined in a GIS with census information by enumeration areas or sub-areas one can determine, for that localized area, the average level of income, demographic makeup (ages, ethnicity, education), and the like. Armed with this information retailers and others can target their marketing – or plan the location of their next store or branch. While this may seem like an expensive proposition, it is well within reach of larger retailers. Indeed, most large retail chains have GIS experts on staff.

Can the smaller retailer use such information? Of course! In Canada and most other developed countries most residents have been asked at a check-out at a grocery store or hardware store "what is your postal code?" Or "What street do you live on?" By responding, you will almost guarantee that you will receive advertising from that store – as will your neighbours. That "point-of-sale" information tells the store where to concentrate their advertising. Established stores use this information to identify the catchment area for the store – the area from which clients are drawn and where marketing should be targeted to be most effective.

Before a store is established, information on the population density and general demographics can be easily obtained. The locations of the competition can be identified by examining the local area (as easily as by cruising the streets AND the Internet). Plotting competitors' locations and the location of the target demographic on a map (even on a street map in "hard copy") is a useful beginning. Then the decision must be made – do I locate beside or close to my competition, or between my competition and the major market, or far away? Such decisions will depend on the nature of the product or service, the degree to which the competition is vulnerable, and other non-location factors. The main point to note, however, is that the START is based on location.

Banks and other financial institutions such as insurance companies are long-time major users of location information for marketing. Indeed, some of the brightest geographers, and some of the first to work with GIS, were hired by banks in Canada as early as the early-mid 1970s. Banks and insurance companies provide a wonderful, if often secretive, window on how geospatial data are used in sales. While we are quite certain that the banks need no advice in this domain, it may be instructive for their clients to know how they do what they do.

The concepts upon which BI and geo-marketing are based are simple in the extreme. The first assumption is that people who live near each other will have similar interests and "wants." (People buy what they want, not necessarily what they need!) The second assumption is that if some people in a given area buy a product, others from the same area are also likely to do so. If a bank (or telephone company, or insurance company) sees that it sells a certain product to a number of people in an area, it makes sense to try to sell the same product to others in the same area.

But with banks, there is an added complexity in marketing and sales that comes from the fact that the banks sell a myriad of services and products including credit card services, mortgages, car loans, home improvement loans, investment funds, retail and commercial banking, student loans, insurance, traveller's cheques, and foreign exchange services.

In selling all of these financial products, the bank has information on who buys what – and where they live – and they can combine all of this information in a GIS to better target marketing and track trends. They also have information on who has repaid loans – and who has not – and where they live. They not only try to sell a specific product to those who are not yet clients in the same neighbourhood, they use the information they have to sell existing clients additional services – services that mesh with what the bank knows about that client and the behaviour of other clients with a similar demographic profile.

At the same time as the marketing is done, banks usually factor risk into the decision to loan funds. One of the factors associated with risk is location. It is one of the factors that US banks did not adequately take into account which led to the sub-prime mortgage problem in 2007 and 2008. In addition to a different regulatory environment in Canada, it may also be that paying more attention to mortgage risks related to location helped Canadian banks avoid such a crisis. Canadian banks weathered the financial storm largely unscathed.

Sophisticated GIS applications are used to do market segmentation and targeting analyses. Not surprisingly successful banks and insurance companies have some of the most sophisticated systems in the world. The basis of the insurance industry is risk assessment – and risk varies from one place to another. In the United Kingdom the insurance industry has used maps created from airborne radar imagery that show flood-prone areas.[237] The insurance industry also uses spatial data to assess where the risk of serious illness, accident or crime is higher. The differences from one area to another can be seen in car insurance. Our friends who live near the foothills of the Rockies tell us that almost every car finishes the winter with a cracked windshield. There it is impossible to get damage insurance for windshields as we do in the east where less gravel is used on the roads.

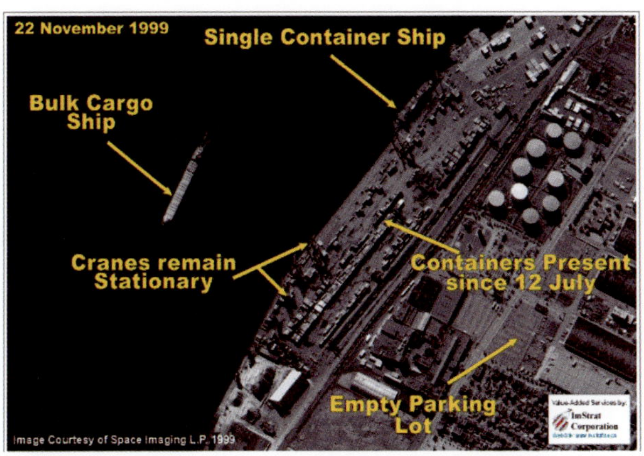

Business Intelligence from Satellite Imagery. A skilled analyst can infer valuable business information from the study of a manufacturing site using satellite images acquired over time. Annotated features of this industrial site compared with an earlier image suggest the plant has not been operating. Source: Imstrat Corporation and GeoEye satellite imagery

We have also found that those running GIS systems inside banks seem reluctant to talk about them, although there are some excellent summaries of what banks do with GIS.[238] Recently some major banks have not only been hiring people in the field (as they have for decades), senior GIS appointments have been widely publicized and GIS specialists have been seen as sufficiently important to join the Executive Committee of a major bank division in one well known international bank.[239]

Marketing and using location information is not restricted to banks, retailers and insurance companies. Not surprisingly, non-profit organizations and charities market in much the same fashion. They focus their attention where they have had previous success.

Others using the data for intelligence purposes include oil and mineral exploration companies. For example, a major early user of SPOT satellite imagery was a company offering the monitoring of who was doing what with their oil leases in Alberta. With the imagery it was possible to tell who was actively exploring by building access roads, drilling, etc. That information helped those with surrounding leases determine what they should consider doing.

Similarly, aerial photography has long been used in a variety of industries to determine what the competition is doing. Is the competing chemical company adding capacity and if so, of what type? Others have used historical imagery to make decisions about buying land or the assets of another company. No-one wants to buy a company only to find that it is sitting on polluted land leaving the purchaser with a huge liability, or that the forest resource has been damaged by disease.

The list of applications of geospatial data for any industry or company can be arrived at by answering two simple questions:

1. For remote sensing, what useful information can you gain through a view from above? and
2. For geospatial information, what useful information can you gain by plotting different sets or types of information on a map to visualize trends or differences over an area of interest?

6.4.3. Consulting and Service Delivery

Marketing and risk assessment applications have historically been major users of location information in the consulting field as stated previously. However, as societies become more attuned to the GeoEconomy, the use of geospatial technology and information is spreading across a broader spectrum of consulting clients and uses. There is growing application of location-based information in everyday business – to answer the "where" question in everything from management consulting areas such as strategic planning and change management, to environmental and engineering consulting, and health care delivery. Indeed, geospatial information should play a role (ranging from modest to critical) in almost every consulting field where professional advice is provided. This is especially so where advice is associated with construction, transportation, land use planning, tourism, delivery of services, and international development. Indeed, international development, in particular, can benefit from the intelligent application of geo-information and technology.

It should come as no surprise that geospatial information is routinely used by management, engineering, and environmental consulting firms across the entire range of professional services they offer to their clients. In order to help clients develop state-of-the-art business solutions, consulting firms must either hire professionals or sub-contract for the geospatial expertise they need. We have seen time and again that larger engineering consulting firms such as Golder Associates, AMEC, and Hexagon have bought smaller firms specializing in geospatial in order to have in-house capability to meet the consulting challenges they face in the GeoEconomy. We do not think it is coincidence that the most successful consulting firms seem to be those who have whole-heartedly adopted geospatial information and tools.

In our experience, many consulting contracts would benefit from considering the impacts of location on the specific assignment. Management and other consulting firms that cannot recognize when geospatial information needs to be part of a client solution, and provide a process to incorporate it effectively, will be at a progressively greater disadvantage.

Simply put, virtually everyone in the consulting industry should find that geospatial information can make them more efficient and help them in providing their clients with more effective advice.

6.4.4. Natural Resource Management

In natural resource management spatial data are used by companies involved in precision agriculture, fishing, forest management, oil and gas, and mineral exploration. Simply stated, spatial decisions impact almost every aspect of natural resource management today: "where" matters. They all use location-based information to some degree because they cannot use and properly manage resources unless they know how much is where. It is not surprising then that natural resource industries were among the first to adopt geospatial technologies – including aerial photography, satellite imagery, geographic information systems, and satellite positioning and navigation.

The first simple question that can be asked by those in the resource industries is: can maps or imagery or real-time location make us more efficient? The answer usually is yes, and that leads to the next even simpler question: how? While the question is simple, the response varies dramatically in its complexity. In some cases inputs are reduced, thus lowering costs. In other cases a completely new way of doing things develops. One of the interesting things we have found over our careers is that as imagery and information becomes more widely available, more and more unique applications are uncovered. Outside of the military, nowhere has imagery been more widely used than in natural resource management. This has resulted in a plethora of new applications. The breadth of applications and their importance are highlighted in the next two sub-sections, the first focused on the forestry, mineral and energy resources, and fishing, and the second focused on agriculture.

6.4.4.1. Natural Resource Industries

Today's highly competitive natural resource industries have incorporated geo-technologies into their operations and are dependent upon them. The more valuable resources, such as gold and diamonds, attract world class experts using the most advanced geospatial technologies available. For these industries the challenge is not to introduce geospatial technologies, it is to keep up with rapid developments and correctly assess which new advances will give them some competitive edge and improve their success, and which will be merely costly disruptions to their operations.

Mineral and oil exploration businesses are significant and long time users of geospatial information – ranging from government geological maps to specially acquired imagery from aircraft and spacecraft. While many scientists publish their findings, the exploration business is, if anything, more secretive than the banks in terms of sharing what they do with geospatial data (including imagery) and the results of their work. Given that a single mine can be worth hundreds of millions of dollars, it is not surprising that anything that provides an edge to one company over another is not widely shared. When Ryerson was selling imagery it was not uncommon for several prospectors working independently to buy the same imagery over the same "hot" area. In one case, it was only two partners who bought Landsat imagery over an area everyone else had given up on prospecting. But the two prospectors saw something and used Landsat data to provide an overview of the geological forces that operated on that area. That area, now known as the Hemlo Gold Field, ended up being one of the most valuable gold finds in history and the prospectors ended up making many millions of dollars.

The contribution of geospatial information and technologies goes far beyond the location, inventory, extraction, and processing of a resource. Today natural resource industries must also navigate a complex regulatory environment, manage investor and public expectations, and respond effectively to the unexpected – from accidents and disasters to incidents that galvanize the public's wrath. An Alberta oil sands company attracted national public criticism when over 1,600 ducks, that landed in a tailings pond in 2008 during a winter snowstorm, were said to have died. On March 1, 2010, the company began an eight-week defence against charges of failing to take adequate measures to divert waterfowl from its tailing ponds.[240]

These industries have a competitive imperative. They must upgrade and innovate in order to survive. They need to make their current operations more efficient and adopt new methods at least as fast as their competitors. The following discussion suggests four ways that natural resource industries can use geospatial information and technology to improve their effectiveness and stay competitive:

- Improve the assessment of the resource and management of operations;
- Reduce incidents that contravene regulations;

- Find opportunities to reduce environmental impacts and improve sustainability that also reduce costs and improve productivity; and
- Use geospatial information and technology to communicate more effectively and increase perceived transparency

Improve the assessment of the resource and management of operations: All of the geospatial technology we have discussed in this book is rapidly being improved, and new applications and improvements to existing ones are becoming available. Because the natural resource industry is already so heavily dependent on these technologies, there will be an ongoing stream of opportunities to upgrade. The challenge is to know which upgrades will actually deliver bankable benefits and which will simply cost money, disrupt personnel, and reduce production. Considerable expertise is needed to correctly assess these trade-offs for a particular operation and the fact that each situation is unique further complicates things. Simply ignoring new developments is not an option; the imperative of competition offers little redress to those who fall behind. So there will be a high demand for skilled geospatial professionals capable of advising on those decisions – decisions that can easily be multi-million dollar decisions for large resource companies.

The forestry community has been using imagery and maps for many decades to carry out inventories, plan access roads, and assess depletions from harvesting, insect damage, forest fires and the like. What has been added to the forester's repertoire is the ability to use higher resolution (i.e. more detailed) imagery from newer satellites, combined with the ability to overlay all of this information in a GIS. What this leads to is just-in-time harvesting – reducing the inventory of harvested trees. In effect, this reduces the inventory (a cost) while at the same time the value of the resource continues to appreciate – the trees continue to grow until the moment that they are needed – or products based on them, sold.

Reduce incidents that contravene regulations: In the past it was possible to flout regulations – or be unaware that regulations were not being adhered to. Today it is much more difficult to break the rules, in part because of geospatial technologies like remote sensing and easy access to the media via the Internet. As the previously mentioned dead ducks in the tailings pond issue illustrates, even incidents where there is no damage to human life or property can be costly to a company in court costs and, perhaps more importantly, in public relations.

Find opportunities to reduce environmental impacts and improve sustainability that also reduce costs and improve productivity: Reducing environmental impacts and improving sustainability are usually viewed as measures that increase costs and reduce production. This isn't necessarily the case. Although more difficult to find, sometimes these objectives can be attained with financial gains to the company. For example, more frequent monitoring of waste water might not only reduce the risk of toxic events that produce fish kills in nearby lakes, streams, and aquaculture sites, they might also provide earlier warning of process controls that need to be adjusted to maintain proper operations.

Use geospatial information and technology to communicate more effectively and increase perceived transparency: The public is now familiar with information related to location being presented as maps, air photos, and satellite images. The Internet makes delivering communication to the public both fast and efficient. Perhaps more importantly, using these media and the method of delivery itself sends a message. It gives an impression of immediacy and transparency, of "coming clean" so to speak. To some extent "the medium is the message" – a phrase Marshall McLuhan famously coined in his 1964 book *Understanding Media: The Extensions of Man* to describe how the medium of communication influences the perceived meaning of the message.

6.4.4.2. Agriculture

Agriculture can be big business or small business – from corporate entities with thousands of hectares and hundreds or even thousands of employees to a small family farm barely providing a subsistence living. The scale of an operation, the value of the product and the availability of suitably educated personnel, all will affect the degree to which geospatial technology can be profitably applied to farming operations.

Geospatial Technology for Viticulture – Napa Valley, California. GPS, remote sensing, and GIS are among the geospatial technologies commonly used in the management of high quality wine grape production. Source: R. Ryerson

One of the less common but growing applications is carried out by farmers and others in agribusiness. As noted in Section 4.3.9, *The Economist*[241] (Nov 7, 2009) highlighted the application of remote sensing and geospatial data in precision farming – a farming technique that makes use of geospatial technologies to modify treatments within a field to compensate for natural variations in soil chemistry, nutrients and the like. Remote sensing and GIS technology are used to detect, map, and display these variations. Then soil sampling is typically carried out to better understand the underlying causes of these variations. All of this information is then used to analyze the differences – what might cause them and what treatments or prescriptions should be applied, in what amounts, to different parts of the field.

The prescription is then used in farm machinery, equipped with a GPS-enabled computer and variable-rate spreading-equipment, which automatically adjusts the spread rate according to the need for each location. The system can be used for a range of treatments including fertilizers, herbicides, and pesticides. Optimizing the rate of fertilizer, herbicide or pesticide application in a farmer's field is similar to the application discussed in Section 6.3.5, where the right amount of salt is delivered to an icy roadway.

Optimizing Wine Quality with Airborne Multispectral Imagery. By choosing different band combinations and processing methods diverse information products useful in vineyard management and wine production are created. The normal color image (left) shows what an observer would see. For the vineyard area a vegetation index has been generated from selected image bands and color coded from low (green) to high (red) values (center). From the index values an image of vegetation vigour customized for the needs of the vineyard manager is produced (right) which the winemaker uses to guide field sampling and to in turn decide which grapes to harvest together in order to maximize production of the most profitable wines. Source: Jay Hutton, Gray Hawk Imaging, Napa, California. © 2010 Jay Hutton.

One of the interesting aspects of the GeoEconomy is how very different disciplines have borrowed from each other to develop new applications of geospatial technology and data. Another interesting aspect is that through the introduction of such efficiencies, there are environmental benefits as well. In the case of precision agriculture, for example, fewer chemicals are needed and there is both a cost saving and fewer chemicals are swept into our streams and rivers. Thus, there is the potential for precision agriculture to reduce environmental impacts, improve productivity, and save input costs. This value has been clearly demonstrated in the USA's Napa Valley by a commercial venture that uses imagery to help wineries improve management and increase production of more profitable reserve-quality wines. Every additional acre of high value wines produced leads to an increase in profit of as much as $50,000.[242]

6.4.5. Environment

The environment is, or should be, "top-of-mind" for every company president and owner. Aside from the moral obligation to protect the environment we leave for our children and grandchildren, there are liabilities easily measured in the tens of millions of dollars associated with contaminated land. If these two issues are not enough to gain the attention of the average business owner, consider the cost of frivolous or unfounded law suits or negative publicity.

As we have discussed in earlier sections, geospatial information can be used to detect environmental impacts, often earlier than would otherwise be the case. Caught early, they may be less expensive to handle and cause less damage to the environment, people's health, and property. As well, early detection can provide more options to mitigate or eliminate the damage, as illustrated in the case study below.

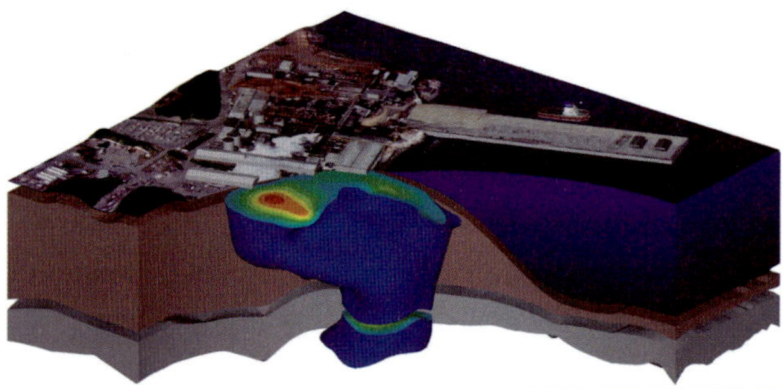

Modeling Sub-Surface Contamination. Geospatial data is one component used in sophisticated computer models of the spread of contaminants in the atmosphere, water, and soils. This image depicts the spread of a contaminant from a source within an industrial site into sub-surface soils and close to an adjacent water body. Contaminant concentrations are coded from high (red) to low (blue). This type of modelling analysis has been used in court to assign responsibility for pollution events and to assess damages. Source: Image provided courtesy of C Tech Development

---

Case Study:

On the night of April 25th of 1998, a catastrophic spillage of toxic mining waste upstream of the Doñana National Park in Spain contaminated a large area with acid material with high heavy-metal content. The spillage of 5,000,000 m^3 of acid water saturated with heavy metal was due to the break in the Aznalcóllar mine dam. It affected the Guadiamar basin, with serious impacts on the land, water, flora and fauna. Studies since[243] suggest that there was almost no warning. Studies just before the dam's failure seemed to have given no indication of what was to come. Today one must wonder whether the use of Radar Interferometry (to discover minor changes in elevation and movement) or other geodetic deformation measurements[244] might have led to an earlier identification of the problem and avoidance of the problem.

Geospatial technologies are also a means to regularly document conditions in order to rebut legitimate concerns that a particular problem is caused by a company's activities, when in fact it is not, or to counter frivolous claims, as demonstrated in the case study below.

Case Study:

A number of years ago we were doing field interviews with farmers across the province of Ontario. One day we visited a farm that was near a major farm chemical factory. We approached the farmhouse that was surrounded by fields of poor looking, stunted and obviously stressed tomatoes. The farmer was sound asleep on the porch in mid-morning – surrounded by empty bottles that, along with the odours emanating from the porch, suggested that he had been drinking whiskey. While we didn't try very hard, we could not wake him.

We went on to the next farm where the tomato fields were full of luxurious vegetation and ripening tomatoes. When we asked about the poor looking fields, the neighbour said "Oh yes, he was probably drunk again. He gets drunk and forgets that he sprayed the day before and then will give it another dose of herbicide the next day." A double dose of even the best selective herbicide is never good!

Much to our surprise a week or two later we saw the drinking neighbour, on television, all cleaned up, standing in front of his wretched looking fields, with the smoke stacks of the chemical factory in the background. He was lamenting the apparent impact of the factory's pollution on his fields. The story eventually made it to the national news. The factory management could do nothing but say: "It wasn't us." Had they acquired an image they could have simply said: "Look at all of the other fields surrounding our factory – they are all fine – just this farmer's fields are a problem. Don't blame us, he has to blame himself." They would have also avoided bad publicity and a drop in their stock price that came with the negative publicity. As has been said before, one picture can be worth a thousand words – or several hours of a lawyer's time, court costs and a law suit!

As mentioned earlier, it isn't enough to act responsibly, it is imperative for a company's "health and well-being" to be *perceived* as acting responsibly. This requires not only taking appropriate action but also communicating to regulators and the public the action the company has taken, when it detected a problem, how quickly it was handled, and what was done. Communicating these actions effectively and in a timely and easily accessible manner is at least as important as the substantive steps that were actually taken. Public opinion influences politicians and regulators. This affects not only which laws will be created but also how stringently they will be applied. It isn't enough to do the right thing – you have to get credit for it too! Geospatial technologies in general, and remote sensing in particular, are useful to show what has been done.

6.4.6. Choosing the Technology

One of the important decisions a company has to make is choosing what geospatial technology to use. There is often a tendency among larger organizations (be they in government or industry) to develop home-grown systems. For these larger companies we generally suggest buying off the shelf – but from well established vendors who are likely to be here over the lifetime of your system. Buying off the shelf is usually better and lower cost in the long run – in spite of what the IT specialists might want to do! Again, searching the web to see what others in a similar position have done and consulting with others in the same business area is not only recommended, but regarded as essential. In addition, one of the important things to understand is that very few of the software packages available are easy to use for the non-specialist. However they are designed to be customized so that an organization can provide very easy to use applications that perform specific tasks tailored to an organization's needs. For example, personnel reviewing property records could be presented with a display outlining the property boundaries with an airphoto background and a text box listing the relevant ownership information. Updating tools could be provided allowing changes depending on the level of authority the operator has been given.

Many standard business software applications have some mapping or map-like operations that may satisfy some needs. MapPoint in Microsoft Office is one example.

6.5. Countries and States

6.5.1. Introduction

Here we address the use of geo-information and technology by states, provinces and countries. We treat these different levels of government in the same section because while all can benefit from the use of geo-information, which use is appropriate at each level of government varies from country to country based on the country's history and constitution. For example, in some countries resource management is in the purview of a national government, while in others it is the responsibility of a state or province.

We have also assumed that appropriate data have been collected and kept up-to-date – ready to be used. What data are required (and how often they must be updated), will depend on the uses. Some advice on the way to approach building an appropriate geospatial data program is given in Section 6.5.10.

The number and range of potential applications of geospatial data and technology for states, provinces and countries is staggering. It has been estimated in a variety of studies and reports that on the order of 80% of decisions made by these jurisdictions have a geospatial component. While the number and range of applications is staggering, so too is the value to a national or regional economy of a well organized and well coordinated geospatial application program.

ACIL Tasman (2008) estimated that, in 2006-07, the impact on GDP of the spatial information industry in Australia ranged between $6.43 billion and $12.57 billion for the quantifiable lower bound and estimated realistic scenario respectively - equivalent to 0.6% and 1.2% of GDP respectively. Of these amounts, productivity improvements are responsible for $5.31 billion and $10.31 billion respectively. The results also show that real wages are estimated to have increased by between 0.60% and 1.12% as a result of the impact of spatial information systems on the economy. These are significant numbers for a national economy. Part of the value is derived from the fact that information, and decision making based on that information, can be better coordinated and managed when proper attention is paid to geospatial. The obviousness of this value proposition is, of course, one of the reasons that we say that we are in a new economy, the GeoEconomy.

The governments of any country are like a many-headed hydra. Multiple departments and divisions within departments are charged with pursuing their mandates which are diverse, often overlapping, and sometimes competing.

Here we focus on what we consider to be key uses for countries and their state or provincial governments. We have organized these uses by the key issues faced by governments everywhere. We have discussed all of these issues in earlier sections of this book. Typically each of these is addressed by one Minister or a committee of Ministers. These issues are:

- Environment and global change;
- Safety and security;
- Sovereignty;
- Infrastructure building and management;
- Resource management;
- International development assistance (both giving and receiving);
- National economic planning;
- Audit and management;
- Planning a governmental geospatial program; and
- Education and training.

The sort of questions that a country, state or province might ask and answer with geospatial data and technologies in these key issue areas (with the obvious exception of planning the geospatial program itself) are given in the following table. Uses in each of the areas identified in the table are discussed in the following subsections, as is planning a geospatial program. It should be noted that this is only a partial list. For each issue we could identify many more questions – and write a separate book. Indeed, for resource management dealing with agriculture and forestry we have already written a short book on the application of geo-information for the Food and Agricultural Organization of the United Nations![245]

Questions a National or State Government May Answer with Geospatial Information and Technology

- **Environment and global change**
  - What was the situation (with vegetation, sea level, climate, size of glacier, size of lake, etc) five (or ten or twenty) years ago and if it has changed, how?
  - If the climate changes, where will the new limits of critical commercial forest species habitats be?
  - If the climate changes, where will the new limits of agricultural production be?
  - Where can one expect major diseases (e.g. West Nile Virus) to impact people and animals?
  - Where did the disease originate and can it be stopped in the future?
  - Are environmental laws being broken (e.g. secondary containment, leakage of tailings ponds, clear-cutting too close to streams and rivers, etc)?
  - What is the extent of the drought?
  - Have any wetlands been illegally drained?
- **Safety and security**
  - Where are the internal threats to security? (Map locations.)
  - Where is our water supply most vulnerable to terrorist attack?
  - What is the extent of the damage from the flood (or ice storm, tsunami, hurricane, landslide, earthquake, fire, etc)?
- **Sovereignty**
  - Is the area under question (i.e. the disputed area) monitored by us? How?
  - Has the disputed area been mapped by us?
  - Do the maps show linkages from the bathymetric maps (water depth) to the mainland of the country (related to the Law of the Sea)?
  - Is the area under question (i.e. the disputed area) managed by us?
- **Infrastructure management**
  - Where is our infrastructure and what shape is it in?
  - What infrastructure needs to be replaced and in what order to minimize costs and disruptions?
  - Where should the new road/pipeline/railway/canal go?
  - Where should the new university be located?
  - Where should the new hospital be located?
- **Resource management**
  - Where are new parks required?
  - What is the state of forest depletions: where are they, how large, and what impact do they have on sustainability?
  - What is the extent of soil erosion? How much? Where?

- Where does the geological base mapping suggest that there may be economic deposits of minerals, oil or natural gas?
- Where should containment dams be built to recharge ground water?
- What is the extent of the drought?

- **International aid**
  - Where is our international aid being given? (or where does it come from?)
  - What is the regional economic impact of the aid? (Map it, relate to other variables in a GIS)

- **National economic planning**
  - Where are the lowest incomes? Lowest cost of living? Lowest level of education?
  - Where are the major centers of production? (And what do they produce.)

- **Audit and management**
  - Where do the problems occur with regard to misspending?
  - What is the average cost of providing service X, Y, or Z by location? If there are differences, why? If not, why not?

- **Education**
  - What important issues in history, geography, and the natural environment would benefit from a geographic or map presentation/explanation?
  - What geospatial education and research is required in our state or country to keep up?

Ultimately the value of geospatial lies in improving information and its visualization for better and more transparent decision making. This value can be derived by providing an inventory (determining and showing what is where), monitoring (determining how the situation is changing or not changing – and showing this by mapping it), and by using GIS, remote sensing, and associated technologies to build predictive models to assist in planning and evaluation. Obviously the information must be used and decisions must be made for the value to be realized, and one of the important factors in ensuring that the technology is used is to have adequate education in the geospatial field. In this way then, the last issue listed above has double importance. Geospatial information can be used in education, and education in its use is important.

In our work around the world we never cease to be amazed at the way in which some people in governments understand the technology and its potential. As noted elsewhere, this is especially so among politicians in democratic countries, including developing countries. And as is also discussed above, we suspect that this is because they know which polls voted for them and which didn't: that, to them, is a critical geospatial problem that they must understand if they wish to get elected!

What IS surprising is how often some of the senior levels of the bureaucracy of the larger more complex and more developed countries do not see the value of geospatial data, or see no need to update the information. We have a number of ideas as to why this is so – none of them flattering to those who hold what we consider to be backward views. To close on this painful subject, we believe that running a country based on out-of-date geospatial information is as ridiculous as trying to run your personal finances without knowing what your current income is, what your past expenditures have been, what your present obligations are, and how much money you have left in the bank.

Running a country based on out-of-date geospatial information is as ridiculous as trying to run your personal finances without knowing what your current income is, what your past expenditures have been, what your present obligations are, and how much money you have left in the bank.

In some jurisdictions what we have specified as being within the mandate of communities is actually (or also) under the authority of the state or national governments. Those covered under Section 6.3 are not repeated here.

Before getting into the details of the key issues faced by governments, we again reiterate that, before one can effectively use geospatial data at the national level, there should be a strategic analysis and the development of a coherent national approach to geospatial data policy and data renewal as we have described in several papers.[246] To do otherwise will lead to lost opportunities, unnecessary discord, and increased costs. This subject is considered in more depth in Section 6.5.10.

6.5.2. Environment and Global Change

Countries, and in some cases states and provinces set environmental policies and regulations to govern activities within their own borders. Whether national or international in scope, addressing environmental issues effectively requires a flow of information relevant to the issues that is sufficiently current and accurate to be trusted. Most of that information is tied to location – it is geospatial information.

There are myriad geospatial uses upon which countries should rely to set and implement environmental policies. New demands and the continuing introduction of new technology have together produced a stream of improvements and innovative uses. Many of these have already been discussed in this book. While environmental issues differ widely, and their information needs are diverse, there are five key information components required to support the decision making process. All of them depend on geospatial and governments at all levels should make use of them to some degree. We believe that their decisions can be improved by making more effective use of geospatial in general and especially in these areas. They are:

- Inventory and monitoring;
- Forecasting and predictive modeling;
- Communication;
- Ensuring data quality; and
- Enforcement.

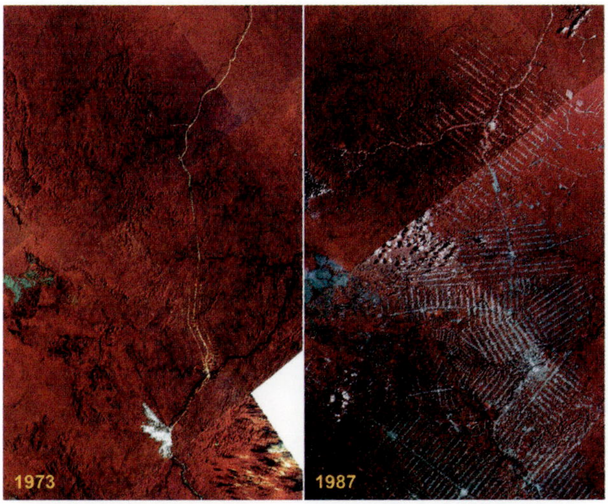

Landsat Imagery Documents Rainforest Loss - Rondônia, Brazil 1973 and 1987. Rainforest loss in the Amazon Basin was among the first uses of historical publicly-available satellite imagery to document and monitor major environmental changes caused by human activities that were previously difficult to quantify and were hidden from public scrutiny. Source: Instituto Nacional de Pesquisas Espaciais (INPE), Brazil

To develop environmental policy and take effective action that delivers the intended results at a reasonable cost, a country must have the information to understand its own environment. Whatever the specific issue, geospatial information should always be used to inventory, assess, and monitor environmental factors, as well as to identify trends. It should also be used to develop forecasts of future environmental conditions, to communicate proposals, policies and their justification, as well as to ensure the information being used and communicated is suitably accurate for its intended use. Lastly geo-information can be used to enforce whatever regulations a nation decides to enact.

This section will discuss the critical uses of geospatial information in these five information components.

Inventory and Monitoring: Geospatial information and technology can be used to monitor the environment and how it is changing. This requires baseline information documenting the status of relevant environmental factors (inventory) and periodic assessments of what changes have occurred (monitoring). Much of the change over time, such as changes in wetlands, size of a quarry or land fill, as well as larger impacts such as desert encroachment, the diminishing size of glaciers, or the decreasing size of lakes, can be monitored using satellite data as well as aerial photography – and much has been written on how to do it.[247]

Geospatial technologies have revolutionized inventory and monitoring methods that are critical to national level decision making for environmental issues. Another major application of geospatial technology is in the use of remote sensing for identifying and monitoring parks, wetlands and nature reserves. This tool has been applied around the world in both developed and developing countries.

GPS technology can be used to ensure that field measurements are accurately tied to location in areas where accurately finding your position on a map – in the middle of the ocean, a tropical rainforest, or Arctic tundra – would be practically impossible. GIS technology is used to allow the information from satellites and other sensors to be integrated with field observations and other data, to define the status and monitor changes that affect global weather patterns, carbon cycling, migration of fish stocks, and distribution of rainfall. Perhaps most dramatically, these technologies have made possible both national and global scale environmental analyses that previously could not be done or were prohibitively expensive. While global monitoring may not be an issue to every country, the results of global monitoring can certainly impact certain countries – island nations or those with extensive low-lying areas near the sea, or those whose activities are seen to contribute to climate change. For example, before satellites it was not possible to map deforestation in the Amazon Basin, or track sub-Saharan desert encroachment. Satellite imagery provides this essential data.

Airborne Radar Image of the Mere Bleu Wetland – Ottawa. This image, produced from several images captured on different dates, has been useful for monitoring environmental changes to this significant example of an unusual wetland feature with naturally high acidity. Source: Natural Resources Canada.

Programs such as the Global Information and Early Warning System (GIEWS) operated by the Food and Agriculture Organization (FAO)[248] and the Famine Early Warning System (FEWS) operated by US AID[249] are but two examples of the way global geospatial information is used to monitor global scale environmental processes.

Environment and climate change has been a subject on the minds of the leaders of many if not most of the world's countries, culminating in the conference in Copenhagen in December 2009. With the record provided by satellite remote sensing since the early 1970s, it is relatively easy to determine what changes have taken place in most regions of the world with respect to vegetation.[250] Have forested areas been cut over or burned, have agricultural areas been urbanized? What was the situation five (or ten or twenty) years ago, and if it has changed, how? Has the glacier really retreated, and if so, by how much and how fast?

Case Study:

We can use one of the case studies previously noted for corporations, as an example of what a country can do with data to monitor the environment.

We return to the April 1998 failure of the Aznalcóllar mine dam located at Sevilla, Spain, upstream from the famous Doñana National Park and wetland of the same name. While the most serious long term threat to the wetland is over-exploitation of regional aquifers – which reduces the input of freshwater, lowering the fresh water table and increasing salinity levels – the dam failure was a catastrophic event that was tied to the company operating the mine.

The accident created the need to develop more efficient controls and management tools in Doñana, taking into account the ecological concerns as well as public opinion. Remediation action was undertaken by the Spanish Government, but the long term effects of this event must be monitored and carefully assessed. As far as the question of remediation is concerned, different levels of the local administration (Junta de Andalucía, Ministerio de Medio Ambiente) have developed programs for the regeneration of the area. The development of tourism facilities was proposed, but these had to be carefully assessed to prevent adverse effects.

Spanish authorities have developed a Strategic Plan for the management of Doñana which includes efforts by all levels of government, and other stakeholders (Salathé, 2002). The overall objective is to slow and reverse the degradation of Mediterranean wetlands. This strategic plan calls for the production and sharing of information about the National Park and critical surrounding areas. In particular, the areal extent and condition of the wetlands are to be monitored on a periodic basis, using methodologies and information products which are harmonized across jurisdictions. It also calls for capacity building of institutions involved in the protection and development of Doñana. These activities have resulted in a long-term management plan which protects and enhances the ecological integrity of Doñana. Central to the success has been the use of geospatial information and technologies to monitor and analyze wetland conditions.

Forecasting and Predictive Modeling: We have mentioned the critical importance of using geospatial information and technology in modeling – predicting the future. Knowing the status of key environmental factors and monitoring change over time is critical to understanding current status and its relation to historical norms. That is sometimes all that is needed to make appropriate policy decisions. More commonly, we also want to know what is likely to occur, when it is likely to happen, and what we can do to change expected outcomes we'd rather not have.

Such an approach is being used to determine where the new limits of critical commercial forest species habitats will be under certain projected climate changes. This is done using a GIS to combine the knowledge of the climate and soil requirements of specific species with projections of the climate. This sort of planning is especially important given the time required for trees to mature. If you are planting a tree today that will not be harvestable for 50 or 75 years, it is important to ensure that over that time the climate will allow that tree to grow to maturity without being weakened by a new climate regime.

The same sort of projections related to trees and climate change can be used to identify the new limits to agricultural production and the potential for drought and its alleviation. Areas that will have more droughts, or more rain, can be identified and planning can be done to account for these changes. Similarly, as glaciers decrease and the potential exists for reduced water flow, changes can be made in irrigation planning, or alternatives will have to be found to supply areas that have typically relied on glacier melt-water.

While we have only noted forests, agriculture and water, the potential effects of climate change across a wide array of issues are, in most cases, capable of being modelled and visualized using geospatial information and a GIS. Clearly, climate change should be considered in all state and national planning processes.

Geospatial information and technology are critically important in modelling and predicting the future.

A growing and as yet not well developed field is the projection of the outbreak of major diseases (e.g. West Nile Virus) that impact people and animals. We reviewed the work done in the mid-1800's by John Snow in London in Section 4.3.7 above. While these tools have been used to understand patterns of disease, they are only beginning to be used to predict outbreaks and responses to disease. Geospatial technologies were used to understand the causes and source of avian influenza by Gilbert at al (2008). This information helped in the understanding of how the disease was transmitted. An interesting and related application was done much earlier in the province of Ontario where rabid foxes were a problem. GIS was used to identify where rabid foxes (and foxes in general) were located. Then treated bait was distributed to the appropriate locations using GIS and an early GPS system. Rabies in the fox population was virtually eliminated, and the same technique has been applied to other wildlife including skunks and racoons. **This work was done before GIS was routinely available in university labs.**[251]

Case Study:

The El Niño–Southern Oscillation (ENSO) cycle is a fluctuation between unusually warm (El Niño) and cold (La Niña) conditions in the tropical Pacific. It is the most prominent year-to-year climate variation on Earth. El Niño and La Niña typically recur every 2 to 7 years. It is unique among climate phenomena in its strength and global influence affecting weather conditions worldwide. Thanks to a network of satellite and Earth surface based sensors scientists have also made this climate phenomenon the most predictable. (McPhaden et al, 2006).
The impact of the ENSO cycle is significant. Some of the major effects occur in agriculture, power generation, fresh water resources, public health and safety, forestry, fisheries, transportation, tourism, and financial markets.

During the 1997-1998 El Niño, NOAA (Sponberg, 1999 and WMO, 1999) estimated the global impact to be:

Direct loss	US$34 billion
Mortality	24 thousand
Morbidity	533 thousand
Displaced persons	6 million
Affected persons	111 million
Land affected	56 million acres

For example, ENSO-related changes in temperature and precipitation promote the spread of some insect and water-borne diseases such as hantavirus, malaria, dengue fever, and cholera. Many of the 22,000 lives lost during the 1997–1998 El Niño were related to disease outbreaks in the developing world. There was an unprecedented scale of forest fires in Indonesia which caused persistent haze and smog throughout Southeast Asia during 1997–1998 causing widespread respiratory illness.

Drought as well as flooding can adversely affect both subsistence agriculture and the production of cash crops. ENSO-related shifts in precipitation patterns have affected the production of wheat, rice, maize, sugar cane, and other important crops in far-flung corners of the globe. Historically, drought related crop failures due to El Niño were a major contributor to famine that would periodically devastate vulnerable populations.

Predicting ENSO cycle events is far from perfect. However, models that depend on global scale geospatial data have been developed to forecast the occurrence, intensity, and duration of these events. This information has been used to mitigate impacts on public health, agriculture and food security, fresh water resources, and fisheries. Predicted storm impacts on shoreline erosion and accretion have been factored into land development and coastal zone management plans.

Judicious use of these forecasts has paid off. For example, California is estimated to have saved US $1 billion in 1997–1998 as a result of actions taken by individuals, businesses, and government in response to advance warning. Peru makes use of ENSO cycle forecasts to optimize the mix of crops to sow in order to maximize the overall yield. Two of the primary crops grown in northern Peru are rice and cotton – both very sensitive to the quantities and timing of rainfall. Rice thrives on wet conditions during the growing season followed by drier conditions during the ripening phase. Cotton, with its deeper root system, can tolerate drier weather. A forecast of El Niño produces wetter conditions in Peru, an incentive for farmers to plant more rice and less cotton. Australia, Brazil, Ethiopia, and India have taken similar initiatives.[252]

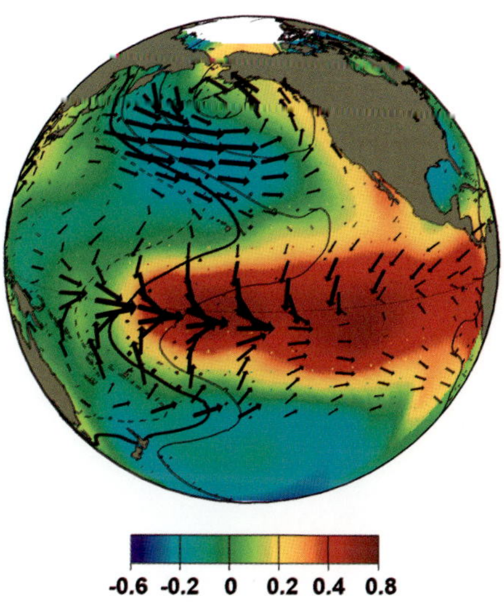

Sea surface temperature patterns derived from satellite data were used in this analysis of ocean currents during an El Niño Event. Source: Steven Hare, University of Washington

Communication: Governments can use geo-information to communicate in two ways. The first is to simply disseminate facts. The second is based on the fact that visual presentation of information can be powerfully persuasive – geo-information can be used to convey a point of view. Maps and images allow people to SEE what the information tells us about location relationships – from information about dangerous avalanche areas to avoid, to the environment.

There is nothing like an image of dead trees extending out downwind from a smelter to suggest perhaps that the smelter may be the cause of the dead and dying trees. However, such

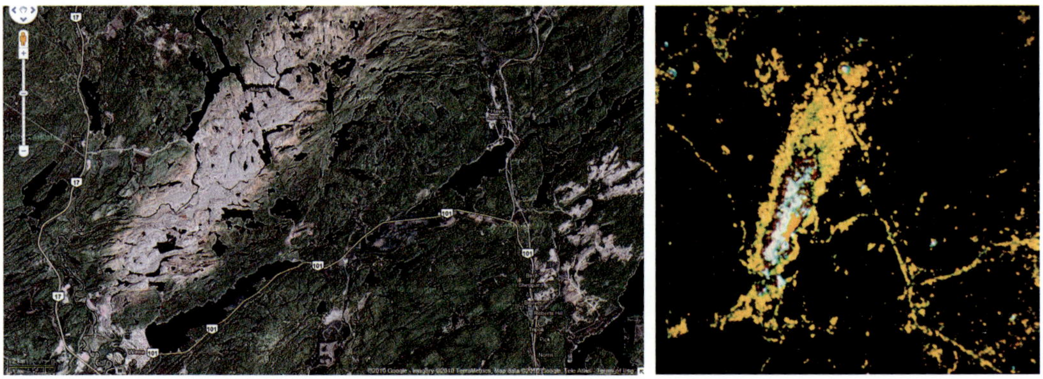

Vegetation Stress on Satellite Imagery – Wawa, Ontario. Stressed vegetation appears as a large grey area on satellite imagery (left). A display (right) from the original image analysis published in 1973 that together with field verification determined the cause of this vegetation stress to be sulphur dioxide emission from an industrial plant. Source: Google, Inc. (left image) P. Murtha (right image)

Fieldwork is Aided by Ground Photographs. This perspective image of pipeline construction in a forested area in the Rocky Mountain foothills provides detail that can augment aerial surveys. With today's geo technologies such imagery can be tied to a precise location on the ground and be included in a GIS. Source: R. Ryerson.

visual information must be supported by the necessary data such as field evaluation, to avoid drawing incorrect conclusions, like the problem of the tomato farmer mentioned in Section 6.4.5. It is a specific example of the more general issue of data quality.

Ensuring Data Quality: A critical factor in using geo-information is ensuring that the data are accurate. This is so since there is often substantial financial, political, or social incentive to influence national environmental policy decisions. There will be companies and regions within a country that could be significantly affected by these decisions. There will also be international players with a vested interest in the outcome.

There is always the possibility that what is being shown with geospatial information may have been modified, biasing the data to sway a decision. Environmental information can be presented selectively to suit one side or another. Data quality can be compromised by unsuitable collection methods. Data may also be corrupted due to technical problems or even sabotaged – i.e. intentionally altered, destroyed, or with access selectively denied. It should be noted that the environment, by its nature, is constantly changing, so accurate data can be rendered inaccurate simply with the passage of time. Environmental data must be sufficiently current for the intended analysis.

As we discussed in Section 2.8, one can lie with maps. For that reason the accuracy with which maps are made and information collected are both critical. As scientists we have spent a great deal of time and effort throughout our careers to ensure that the methods we used resulted in information that met the needs of our clients. Accuracy has long been important.[253] Put very simply, it makes little sense to us to provide information in such a way that leaves any doubt about the accuracy of the result.[254]

Imagery presents the same challenges. In assessing multi-year changes in wetland habitat imagery must be from the same season for comparisons to be valid: it makes little sense to compare the size of a wetland in a rainy season in one year to one acquired during the dry season in another year – unless you wish to lie with the imagery![255]

Because national scale environmental issues typically have major financial and political impacts there is added incentive to influence these decisions by any available means. So vigilance is needed not only to ensure the decision-making process is balanced and comprehensive, but also to ensure that the quality of the data and the analyses used are sufficiently accurate for the intended use and free of intentional bias.

Enforcing Environmental Regulations: Geospatial information can be used to enforce environmental regulations. It can be used to document conditions, integrate multiple information sources, and monitor compliance. Satellite imagery, aerial photography, and other remotely sensed image data are not only important as data sources, but also the knowledge that they will be used can be an important deterrent to inappropriate actions.

In the 1970s, one of the authors visited a Scandinavian country where a newly purchased radar sensor was being used to monitor coastal areas, to identify and charge ship captains who were dumping oil and contaminated water within the country's territorial waters. The technology had successfully been used in court to prosecute a case. However the court process was lengthy and expensive. What was found to have the most influence was the program of posting information about the radar sensor system in all the locations where ship captains and their crews would congregate. The posters highlighted that the technology could monitor and document their polluting activities at night and in bad weather. The budget for monitoring was so limited that the sensor could only be flown a few days each month. However the posters made ship captains sufficiently wary that these oil pollution events declined dramatically. Enforcement by publicity can be very cost effective when periodically reinforced by well-publicized arrests.

The enforcement of environmental regulations commonly falls to more than one department and may also involve multiple levels of government or cross international borders. While the types of remotely sensed data that may support enforcement efforts may differ with the specific case, geospatial data and tools, including GIS, are usually part of the integration process.

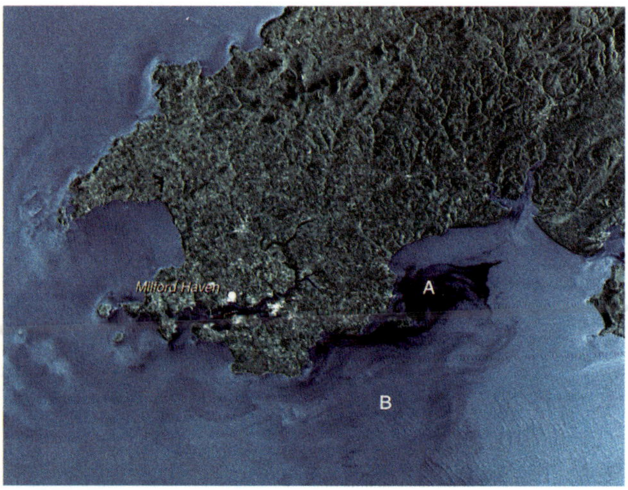

Oil Spill Detection with Satellite Radar Imagery – Wales, UK. Under certain wave conditions oil on water reduces wave heights enough to make water with an oil surface film appear darker (A) than unaffected areas (B). Source: Radarsat-1 Data © Canadian Space Agency/Agence spatiale canadienne (1996)

The most common use in enforcement sees remote sensing data from aircraft or spacecraft used to determine if there is cause for a more thorough ground verification of a problem. One early, and continuing, application was the use of aerial photography to determine if laws concerning secondary containment around oil and chemical storage tanks are being broken. Secondary containment for oil tanks, for example, consist of surrounding the tanks with berms or raised banks that would prevent the oil from escaping into the surrounding environment should the tank fail. Other uses include the assessment of wetlands that have been illegally drained or encroached upon, if tailings ponds appear to be leaking into natural waterways, or if trees have been cut down too close to a stream. We have discussed a number of these uses previously.

In this discussion we have suggested the use of geospatial information to support the regulation and enforcement of a jurisdiction's own laws. However, the globalization of geospatial information has dramatically increased the transparency of activities that produce environmental effects – be they positive or negative. This has led to the development of other important aspects of national scale environmental enforcement and regulation that originate outside of government. We have touched on these issues in our previous discussion of transparency earlier in this chapter and elsewhere in the book. One is the use of data by special interest groups both inside and outside a country that scrutinize a country's environmental record and publicize those actions of which they are critical. Their actions can in some cases influence the policies of their own governments to threaten sanctions against the other targeted country. This suggests that where there may be some question being raised by an external interest group, governments would be well advised to use geo-information to carry out their own monitoring. By so doing they will have a better idea of what the situation is and they will also be able to determine if the special interest group is, in fact, showing the "real" situation.

6.5.3. Safety and Security

It should come as no surprise that geospatial data are widely used to address safety and security issues. In many countries geospatial activities are managed by military agencies and advances in the field – from GPS to imaging satellites – have often been driven by military needs and military research. Indeed, much of the "dark" in the so-called "dark world" is the secrecy over the advanced nature of what the military can do and is doing with geospatial data and technology. While we will leave this dark world (into which we have had the odd glimpse) to the military and those managing its activities, there are still many areas of safety and security which can be discussed "in the clear."

The value of geospatial to safety and security is the same as for any other issue: it lies in improving information and its visualization for better and more transparent decision making. This value can be derived by providing an inventory (determining and showing what is where), monitoring (determining how the situation is changing or not changing – and showing this by mapping it), and by using GIS, remote sensing, and associated technologies to build predictive models to assist in planning and evaluation.

Disaster mitigation and management is a major safety and security application of geospatial data and technology.

One of the major areas of safety in which geospatial data and technologies are applied relates to disaster mitigation and management. A common question that can be answered includes what is the extent of the flood damage? (Or damage from the ice storm, tsunami, hurricane, landslide, earthquake, fire, etc.) Satellite imagery is so useful for such applications that there is an international agreement among the so-called "space-faring nations" to provide satellite imagery in the event of a major disaster.[256] Another disaster-related application is the prediction of where flooding might occur with a certain flood level of a river. Obviously to do this sort of prediction, especially in flat areas such as the Red River Valley in the USA and Canada, or the river systems in Bangladesh, detailed topographic information is required. To be accurate such predictions use information on the height of the land that is more accurate than what is provided by standard topographic maps. Today that information is often provided by airborne LiDAR sensors.

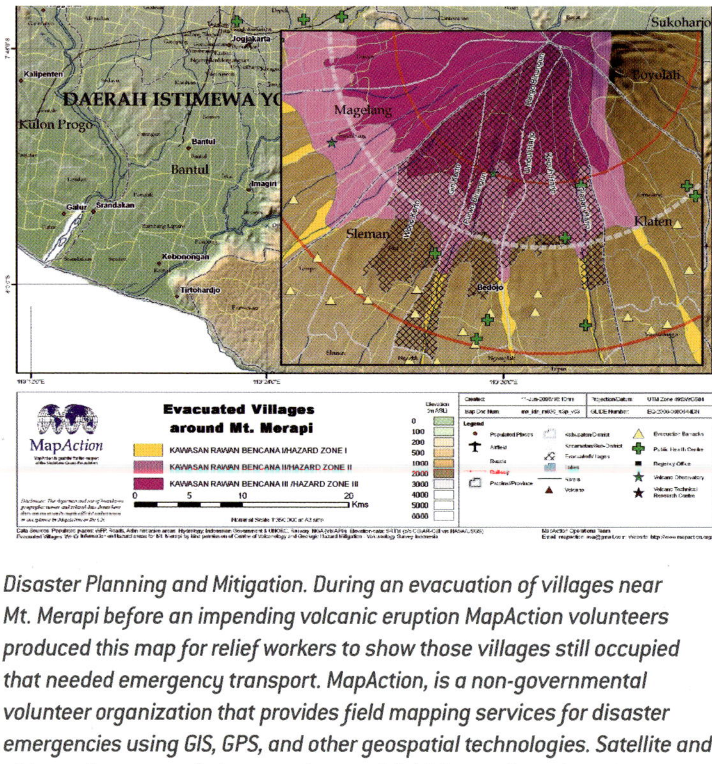

Disaster Planning and Mitigation. During an evacuation of villages near Mt. Merapi before an impending volcanic eruption MapAction volunteers produced this map for relief workers to show those villages still occupied that needed emergency transport. MapAction, is a non-governmental volunteer organization that provides field mapping services for disaster emergencies using GIS, GPS, and other geospatial technologies. Satellite and airborne imagery, existing map data, and field data collected on site are combined to produce up-to-date information products such as paper and electronic maps for use by emergency relief workers. Source: MapAction

The way in which different areas of science can be combined with geospatial information to arrive at potentially valuable information in the realm of safety and security (and almost any other field) can be illustrated by an idea that one of the authors came up with following flooding in Mozambique. Mozambique has gained some notoriety for being one of the most heavily land-mined countries of the world. It turns out that one type of land mine used in that country floats. We were told that land mines that were near the surface in areas that were flooded often "popped up" and floated away. There was some concern about how to locate where the land mines might end up. We suggested that by using Radarsat satellite data one could determine where the flooding occurred. By superimposing on this an indication of where land mines were thought to have been located before the floods, we could determine the potential sources of mines. We then suggested that one could use fluvial (i.e. river) geomorphology, river flow-models and deposition models using the weight and size of the land mines to estimate where the mines might have been deposited. While seemingly complex, the concept was simple and is indicative of the type of use to which geospatial information can be put by those who think spatially about the problems that face them.

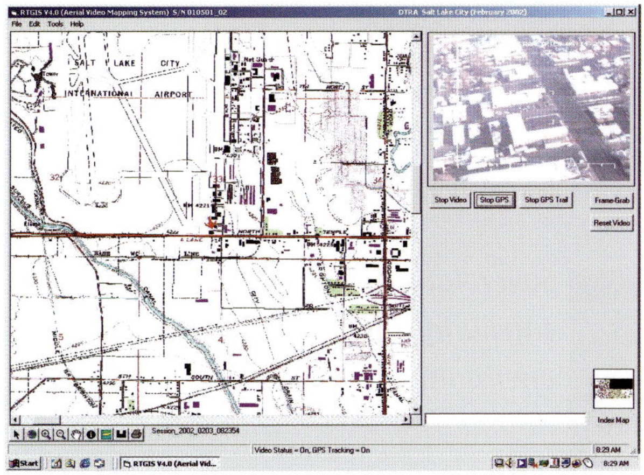

Geospatial Technology for Security Surveillance – Winter Olympics, Salt Lake City, Utah 2002. This GIS-based security and emergency response system used video imagery from a helicopter-based video camera and data on the camera's geographic position and orientation transmitted over a wireless link to provide real-time airborne video imagery and a map display of the camera position and orientation for an operator at a ground-based emergency control center. Source: Dr. Tim McCarthy.

There are a number of other security and policing uses that can be recommended, some of which we outlined in Section 6.3.8. One can identify and locate threats to security along borders (where the smugglers might bring contraband across an unprotected border), locate the activity areas of terrorists (and project future activities), and identify areas where there may be an increased threat to security forces, border patrols and drug interdiction agents, for example. Security at specific events can be improved through the use of video surveillance and GPS combined – such as for the 2002 Winter Olympics held in Salt Lake City. Thermal remote sensing has been used to detect heat-loss in buildings – but as explained in Section 3.2.3.2, it can also be used to pinpoint buildings that might be sites of marijuana grow-ops which are notorious for their use of energy and release of heat.

6.5.4. Sovereignty

Sovereignty is an important issue in many countries. Much of the issue surrounds the question of borders and control of resources and the environment. In previous sections we have discussed the importance of countries ensuring they have as much information about their resources and their environment as foreign players – be they foreign industrial or national organizations seeking access to a nation's valued assets. We focus here on the role geospatial can play in securing and maintaining its claim and control over the territory within its borders.

A border is first and foremost a geospatial manifestation of sovereignty. The case for ownership under the Law of the Sea and other international statutes is usually based on geologic linkages of underwater features to the mainland or a height of land, drainage basin, or river – all of which can be measured (and mapped) using geospatial information.

In the 1960s and 1970s there was a continuing border dispute off the east coast of Canada and the USA in the George's Bank area. For the first time, Canada used GIS to present its case to the International Court of Justice, and was generally successful in making its claim. Simply stated, geographic information was used to show what parts of the area were more closely linked by geology and other factors to Canada than to the USA.

> ***Sovereignty is an important issue in many countries. Much of the issue surrounds the question of borders and control of resources and the environment. Geospatial information is critical in determining land boundaries – both with respect to establishing the boundary and for maintaining it.***

Other geographic elements are important too. Is the area under question (i.e. the disputed area) monitored by us or by the other claimant? How? To illustrate how this argument plays out consider that Canada has long taken the position that the so-called "Northwest Passage" is Canadian territorial water. Others (most notably the USA) dispute that claim. With climate change suggesting a continuing warming of the region, the potential for time- and money-saving shipping in these waters is increasing. This in turn raises concerns for the potential for environmental damage to a sensitive ecosystem, as well as ownership of potential underwater resources.

Sovereignty in the Arctic. Canadian and American coast guard ships take part in a survey of the Arctic sea floor to map the extent of the continental shelf. The data will be used to resolve border disagreements that have become more important as a result of climate change. Warmer temperatures are therefore making formerly ice bound routes passable. Source: USGS

Long before the discussion of climate change really began, the concept of a radar satellite (a concept that later became Radarsat) was proposed as a means for Canada to monitor and project sovereignty over its north. The 1974 study[257] suggested that this monitoring could be done on a daily basis for the entire region – something not possible before the advent of radar satellites. When MacDonald Dettwiler was proposing the sale of its space division to a USA defence contractor in 2008, one of the major arguments against the sale was that it would put Canada's capability to monitor the north into the hands of the Americans. As noted previously, the sale was blocked under the Investment Canada Act of

1985 (which replaced the Foreign Investment Review Act of 1984, then known as FIRA).[258] It was the first time a sale has been prevented in the 23 years the act has been in place – previously, hundreds had been approved. Many of the legal analyses linked the blocking of the sale to issues of sovereignty, as well as to the sale of vital assets that had been paid for in large part by the Canadian taxpayer.

Many other countries and international agencies now routinely use geographic information to demarcate and gain agreement on boundaries. For maritime nations, the United Nations Convention on the Law of the Sea (UNCLOS) has been established to settle the question of boundaries. Such questions can become complex with reclamation of land – or a volcano that extends an island or establishes a new island. Most countries have ratified UNCLOS 1982 which explains how specific maritime boundaries are measured seaward, as well as the types and dimensions of claimable maritime zones, and the rules for maritime boundary delimitation. Reclamation, for example, can extend a boundary seaward, but only where no boundary currently exists with another country.[259]

While the international law for land boundaries is not as clear as UNCLOS would seem to provide, geospatial information is also important for determining land boundaries – both with respect to establishing the boundary and for maintaining it.

The establishment of a boundary draws on a range of special geospatial and political skills. For example, it was announced in October of 2008 that the UN peace-keeping mission in the Sudan donated a satellite imagery system to produce a border map, which was said to be "one of the key elements of the Comprehensive Peace Agreement (CPA) which ended the long-running north-south civil war in the ... African nation."[260] The equipment was said to be worth $600,000 and "would be used to create the map to delineate the 1 January 1956 border," one of the major benchmarks of the CPA. A final report, including the map, was to be presented to the President. Once endorsed, the Committee must proceed with "demarcation on the ground." Abdallah El Sadig Ali, the border team's chairman, called the UN's delivery of the satellite imagery system "historic." But the involvement of geospatial technology and the making of maps must still deal with disagreements on the ground, as the world's many simmering border disputes will attest. Indeed, there is so much controversy that many international organizations studiously avoid even attempting to identify borders in documents such as country profiles – all that is provided in FAO Country profiles are the names of bordering countries.[261]

The second role for geospatial in the area of boundaries is in boundary maintenance. To those unfamiliar with concept of boundary maintenance, it would seem to be a simple matter. A boundary is established and once that has been done, nothing more is needed. To those of us who have had the privilege of managing government geospatial programs, we soon learn that boundary maintenance is indeed a much more complex matter, even with countries that have enjoyed a long and generally positive relationship.

One of the best examples of boundary maintenance is provided by the International Boundary Commission that maintains the boundary between Canada and the United States. "Officially, the Commission's work is described as maintaining the boundary in an effective state of demarcation. This is done by inspecting it regularly; repairing, relocating or rebuilding damaged monuments or buoys; keeping the vista cleared; and erecting new boundary markers at such locations as new road crossings. The Commission also regulates, under the provisions of the 1960 International Boundary Commission Act, all construction within 3 metres or 10 feet of the boundary and is responsible for defining the boundary location in any legal situation involving the border. The commissioners report annually on the work done during the year and provide to both governments the latest data on the boundary monuments. All monuments along the boundary are located so that they tie in with the survey networks of both the United States and Canada through 1000 survey control stations established for this purpose near the border. The position of any monument may be re-determined at any time by the survey crews of the Commission, which are called on constantly to perform assorted survey duties along the boundary. These duties include improving survey connections to the control stations and establishing new monuments."[262] The annual cost of this activity for the 8,891 kilometre border is $4 million done on a fifteen year cycle. Although there is significant geospatial work carried out, the majority of the costs are for vegetation clearing and monument maintenance. More information on the 15 year plan and associated costs can be found in the annual reports.[263]

Landsat Image of the Canada-United States Border. This early Landsat Multispectral Scanner (MSS) image was originally created to illustrate improved image geometric correction quality. While the geography is the same on both sides of the border, the clearly demarcated border illustrates the effect of differences in land-use policies on either side. Source: Natural Resources Canada

6.5.5. Infrastructure Building and Management

Geospatial information and technology is inherently a part of a country's physical infrastructure. They are also important in what some call civil engineering infrastructure planning,[264] building, and management, whether one is speaking of the local, state or national level. The degree to which geospatial is integrated into a nation's overall information infrastructure directly affects the speed and effectiveness with which a country's full information resources can be harnessed to address complex issues.

For example, responding to a regional drought has diverse economic and social impacts. Farmers may demand subsidies to weather the financial hardships of dramatically lowered crop yields. Assessing the size of the affected areas, the expected level of financial hardship, the impact on food prices, the possible need to encourage foreign imports to make-up lost supply, the regional economic impacts, and so on involve the use of ever broader and more diverse data sets. The ease with which these data can be found and are sufficiently compatible to be used together is an indication of the effectiveness of a nation's information infrastructure – of which geospatial data is an important part. Nations having higher quality data and the infrastructure to find and combine those data sets to address diverse issues possess a significant advantage over those less well prepared. They can identify issues, develop responses, detect effectiveness, and make changes to improve results more quickly and accurately.

It is curious that geospatial, one of the major elements of a state or national infrastructure, is rarely accorded the importance it deserves. Geospatial information and the enabling technologies are, or should be, integral parts of the infrastructure, as the motion adopted by the UN in 1999 stated quite unequivocally.

What we find curious is that geospatial, one of the major elements of a state or national infrastructure, is often not accorded the importance it deserves. Geospatial information and the enabling technologies are, or should be, integral parts of the infrastructure, as the motion adopted by the UN in 1999 stated quite unequivocally. It is important to reinforce what we have been saying throughout this book: what is important in terms of geospatial information in one area may (or may not) be important elsewhere. For this reason, as one plans the approach to creating a geospatial data infrastructure, one must keep at top of mind the factors that are critical to the people and environment for which the infrastructure is being developed.

Geospatial information is so important, that we are tempted to repeat the words from section 6.3.3 where we addressed infrastructure from the municipal point of view – but we won't. However, we strongly recommend that those involved in state and national infrastructure planning, development and maintenance should review the discussion of infrastructure in the community context.

The municipal view usually encompasses things such as local roads, bridges, water mains, sewers, storm drains, and the like. At the state and national level some of these areas are also of concern (for national highways, major bridges, etc). But to that one also must add infrastructure such as railways, water transportation, energy distribution (pipelines, electrical transmission, for example), health, and education. Resource management is sufficiently important and sufficiently different in the tools and information used, that it is considered separately.

As noted above, the cost of replacing the built infrastructure (roads, bridges, sewers, and water mains) has been estimated to be in the trillions of dollars in North America alone. This figure of a "trillion dollars" is in itself instructive in terms of geospatial. One interesting description of what such an amount entails is given, not surprisingly, using map-like representations.[265] We find this way of describing what a trillion dollars looks like interesting, in part because the Centre for Globalization has chosen to do the representation much like a scaled map – with a human figure placed beside the representations of the various amounts for a scale. A million dollars in $100 dollar bills can be "stuff(ed) in a grocery bag," while a 100 million dollars "fits neatly on a standard (shipping) pallet," and a billion dollars would require "10 pallets," and a trillion would require 10,000 pallets. While a billion would take up about 1/5 of a standard 53 foot transport truck trailer, a trillion would require over 190 trailers. Interesting too is the fact that the repair and renewal of the geospatial infrastructure is only a very small fraction of that trillions of dollars – perhaps $1.5 to $2 billion for Canada, the USA and Mexico together. This $1.5 to $2 billion for the renewal of the geospatial infrastructure represents less than 1/1000 of the total civil engineering infrastructure replacement cost – and clearly would be a good investment!

Collapse of the Cypress Freeway Structure – Oakland, California 1989. The need to upgrade the freeway structure had been repeatedly recommended to the City of Oakland. Thirty two motorists died in the collapse.[266] Source: HJW Geospatial Inc.

For national and state governments geospatial information and technology can answer or help to answer several important questions – the same ones posed for communities. Where is our built infrastructure and what shape is it in, and what infrastructure needs to be replaced and in what order to minimize costs and disruptions? As noted in Section 6.3.3, geospatial technology can be used to create an inventory of infrastructure by type and condition. Geospatial technology can also be used to track age and assess the infrastructure most at risk. Such an inventory can also be useful in helping to plan repair, reconstruction, or abandonment.

Once work is being planned geospatial tools can help ensure that the work is done in a logical fashion, and that better decisions are made on the priorities for renewal and/or construction. In many cases roads are simply renewed, added to (i.e. widened) or built on existing rights of way – so the actual location of the road is not open to significant change. However, some roads and many other types of infrastructure are often new, and do not simply replace or improve upon existing routes. For example, construction of railroads, pipelines, electrical transmission lines and cell tower locations are today rarely embarked upon without at least a rudimentary set of geospatial information and significant analysis. This information is used to identify the best route or best location. In the case of a cell tower, it is usually to reach the maximum area – which is in large part determined by the elevation of the location and shadowing by other hills or features in the local environment. To find the best location, detailed elevation information is required and this is often provided by LiDAR systems. For railways and roads there is the desire to minimize the need for construction materials – so often a route is designed so that materials removed from hills are used to fill in the valleys.

While simple engineering concerns are important, for many linear routes (such as for pipelines, railways, and transmission lines) the "best location" is often defined by different criteria in different places. In some areas one might wish to avoid Leda clays (that can liquefy and lead to landslides). In others, one might want to avoid (or be close to!) population centres.

Satellite Imagery for Potato Crop Area Estimation. Potatoes (left) are an important cash crop in Canada. An analysis of Landsat satellite imagery classified potato fields (color coded red) and generated estimates of the area planted to potato for the outlined administrative districts. Source: R. Ryerson (left), Statistics Canada (right)

In still others, one must avoid areas of cultural, biologic or touristic importance. In all cases, local geographic knowledge is essential.

As an example of how important local concerns can be in terms of location decisions, consider the location of a pipeline in the St. John Valley in the province of New Brunswick, Canada. At the time, we were involved in developing satellite technology to estimate the potato acreage in that province. The main non-engineering criterion for the pipeline routing was to avoid areas where potatoes were grown. At first glance to an outsider this may appear to be a curious request. Why are potatoes seen to be so important? The answer is simple economics – and another manifestation of the GeoEconomy. Between 50,000 and 60,000 acres of potatoes are typically grown each year in the province and when primary, secondary and tertiary contributions of the crop are considered, the total value to the economy has been estimated at

$1.3 billion.[267] Thus the potato crop's value to the economy represents almost $1,750 for every man, woman and child in the province. In New Brunswick you don't "mess" with potatoes!

The infrastructure applications of geospatial data and technology do not stop with roads, bridges, pipelines and railways. Another related area is in water transportation and planning associated with lakes, rivers and canals. For example, water depth mapping to plan dredging uses LiDAR, while remote sensing is used in assessments of river ice and its possible impacts on everything from wharfs to cables for electricity and communications.

Moving farther afield from the built infrastructure, geospatial data and technology have both long been used in planning the location of schools and educational facilities as noted in Section 6.3.6, and at the state and national level can be used to better locate hospitals and universities to serve a region. Indeed, recent tests in Canada have seen better service and significant potential reduction of infections within hospitals through the use of geospatial technologies.[268]

> ***Geo-based renewal and planning of virtually any type of infrastructure will lead to more efficient and better integrated planning of the infrastructure – and more efficient spending on both development and construction.***

Geo-based renewal and planning for virtually any type of infrastructure will lead to more efficient and better integrated planning – and more efficient spending on both development and construction. Simply put, planning and carrying out renewal and infrastructure construction are best done using a GIS to better visualize the situation, look at the infrastructure as a system, plan development, better assess risks, and carry out the construction. This is the case no matter the type of infrastructure being planned, studied, built or repaired.

Obviously, for the rest of the infrastructure to be properly scoped, planned and renewed, the first stage of renewal must be the evaluation and renewal of the geospatial data infrastructure.

6.5.6. Resource Management

6.5.6.1. Introduction

Resource management is often one of the most important tasks of government: how well it is done can impact the state or national economy, sustainability of the environment, and ultimately, the quality of life and security of its citizens. Natural resource management can also have international repercussions, given the potential for trans-border movement of pollution, water, and civil unrest that often accompanies poor resource management.

In part because of its importance, and in part because of the need for the kind of information and analysis geospatial can provide, resource management has been transformed over the last two decades through the use of geospatial information. More than any other sector, resource management has come to depend on geospatial information and technology.

The questions that geospatial information can answer for government range from those that are general and apply to a range of resources, to more specific questions on the resources we use and manage – such as forests, crops, ground water, minerals, oil, and natural gas. Because of this complexity, we have subdivided resource management into the general and more specific. The general applications section deals with the general questions asked and information provided. The specific sections that follow address those dealing with each of the resource sectors we consider here: forestry, agriculture, water resources, and minerals, oil and gas. The fact that some of the same questions can be posed by those managing different resources highlights the inter-relationships that exist between them.

6.5.6.2. General Applications

For resource management in general, the value of geospatial can come from providing a wide range of information to answer a number of questions and meet a number of government's needs.

As we have stated previously, ultimately the value of geospatial lies in improving information and its visualization for better and more transparent decision making.

A GIS can be useful in facilitating collaboration among diverse stakeholders – even among those who disagree.

Both government officials and the political level can answer questions about government resource management issues and policies with a geographic information system. A GIS can also be useful in facilitating collaboration among diverse stakeholders. As noted above, a well designed geospatial data infrastructure enables a variety of data sets to be accurately combined. This in turn allows diverse stakeholders to show each other their concerns and present their data in a combined map display, so interactions can be explored and conflicting activities identified and addressed collaboratively. This is especially important today, since almost any management activity will affect multiple resources and multiple stakeholders – not just the targeted ones.

Looking at the same sort of issue from a slightly different perspective, one can use geospatial tools to answer the question "what are the inter-relationships and impacts of a (specific) natural resource management policy?" For example, policies that determine the size of allowable forest clear cuts will affect the rate of soil erosion and how rapidly rainfall enters the river network. This in turn affects water quality for flora and fauna, as well as for humans. Being able to see and explore these relationships is important for the resource managers and the citizens, both as represented by their governments, and as individuals or communities of interest.

Geospatial data can answer a number of key general questions concerning natural resource management. These include:

- What is where? (provision of an inventory);
- How is the situation changing or not changing? (monitoring);
- What will happen in the future? (Predictive modeling with a GIS can be used to assist in planning and evaluation related to what is likely to happen in the future); and
- What will the environmental impacts be of development activities?

With respect to the last of the four general questions posed above, it is important to note that environmental regulation is now an integral part of natural resource management. Furthermore, there has been a trend in both developed and developing countries to introduce more stringent environmental regulations. Making them more stringent is a reflection of changing societal goals: people world-wide want to both identify, and mitigate, unintended consequences that degrade the environment. Simply stated, without geospatial information and technology, these regulations could exist, but could never be enforced, nor their impact assessed.

Natural resource practices are frequent sources of international disputes and here there are many general (and specific) questions that one could pose and answer. For example, water use and polluting activities in one country can impact the quantity and quality of water available to users in an adjacent country. Repeated assessments over time can be used to judge the cross-border impact of water management or other natural resource management policies – whether involving vegetation removal and soil erosion, or removal of water for irrigation or almost any other resource management activity. With the advent of satellite remote sensing this can be done within one jurisdiction or across state or national borders, and need not require approval for aircraft over-flights as was the case with aerial photography.

6.5.6.3. Forestry

Government forestry departments have long been users of geospatial technology to carry out inventories, monitor forest health, and estimate potential timber volume or yield. Today virtually every major forest producing state or country has an on-going program to carry out inventory. These are well described in the literature and on many web sites.[269]

Predictive modeling is one of the newer applications of geospatial technology in forestry.

Automated Satellite-Based Forest Fire Detection. In August 2002 this image from NASA's satellite-borne MODIS sensor was used to monitor the progress of forest fires in the Klamath Mountains of Oregon as it burned over the state line into California. The Rapid Fire Response and Detection System implemented by the US Forest Service in conjunction with NASA and the University of Maryland uses daily MODIS imagery and a dedicated direct image reception and processing facility to detect forest fires throughout the United States. Sourse: NASA

A newer application of geospatial in forestry is found in predictive modeling. Foresters (and other resource managers) commonly use geospatial modeling analyses to assess what the status of a resource is expected to be some time in the future and how that outcome can be altered by changing management policies – or climate. This type of analysis makes use of various kinds of predictive modeling techniques, which combine multiple data sets of historical information to develop computational means of forecasting changes into the future. In forestry, historical inventory information, that includes data on tree heights, density of trees, and species composition for each forest type in a region together with the occurrence of losses to forest fires, hurricanes and other natural disasters, and regeneration rates, can be used to predict the quantity and characteristics that a forest stand would be expected to have at some point in the future. The effects of different management policies, such as increasing fire suppression, or reducing harvest rates, can be explored using this type of model.

Customizing Information Products to Suit the Recipient - Landsat TM Image of Burned Forest Areas. The value of an information product depends on its ability to communicate as well as on the quality of its information content. When false color infrared images were the standard for vegetation analysis, analysts were used to its depiction of healthy vegetation in shades of pink to red. The foresters for whom this information product was being prepared associated red with fire and green with healthy vegetation. To suit the needs of forestry practitioners Dr. Ahern created this image product in which trees were color coded green and burned areas red. Source: F. Ahern and Natural Resources Canada

Predictive modeling, implemented using geospatial technologies, has revolutionized complex planning projects across the resources sector. These technologies allow for the production of different scenarios which are prepared iteratively and quickly, adjusting parameters such as a forest harvest rate, climate, etc. and assessing the result. For example, a forest area used to produce timber can, during its growth period, be used for some forms of recreation. Tree thinning operations, a standard forest management procedure for many plantations, can be used to not only generate additional products, such as Christmas trees, but also to provide educational opportunities (such as interpretive programs), and recreational opportunities. The sorts of questions that can be asked about the forest resource is truly only limited by the imagination and need of the person posing the question.

A Sample of Geospatial or "Where" Questions in Forest Resource Management

- How much timber/wood will be produced by that particular block of land?
- What is the state of forest depletions: where are they, how large are they, and what impact do they have on sustainability?
- What is the impact of the recent fire (or insect attack)? Is there any salvageable timber and if so, where?
- How well have the replanted trees been growing following harvest? Where are the trees healthy and where are they not?
- Where should we site access roads to minimize erosion?
- What is the extent of soil erosion? How much? Where?
- Is there any evidence of illegal logging? Where? How much area?
- Is there any forest disease? If so, where is the disease? If the disease is spreading, at what rate is it spreading?
- In the planned harvest area are there any areas designated as habitat for endangered species?
- Where might there be alternative income streams from eco-tourism, grazing, recreation?
- Where have urban areas encroached on forested land? What sort (i.e. what land capability types for forestry) of land has been encroached upon?

6.5.6.4. Agriculture and Food

Food, like water, is essential for human survival. For this reason alone, agriculture would be important. But beyond that, agriculture is important in terms of the number of people it employs in many (especially developing) countries, and in terms of the oft-stated goal of countries to be self-sufficient in food production.

Food, like water, is essential for human survival and its safe, economic production is everywhere increasingly dependent on geospatial data and technology.

While geospatial data and technology is playing an increasingly important role in agriculture, it also plays (or can play) an important role in food safety – tracking sources of problems in food production by (and to) specific locations. One need only think of some of the more famous outbreaks of food poisoning and the role that geospatial information played in identifying the source of the problem.

We have discussed the value of geospatial information for very localized decision making in the private sector through precision agriculture. The role of the state or national government is usually, but not always, at a higher level. For example, the timely detection of poor crop condition or of poor growing conditions from factors such as bad weather or natural disasters, can allow action to be taken to mitigate the problem and address the consequences. During a major drought in Manitoba Canada in 1981, satellite imagery was used to categorize the impact of the drought on forage crops in order to allocate support based on the severity of the impact.[270] In an even earlier application, farm extension workers in Prince Edward Island determined that the lack of a winter cover crop tended to lead to increased soil erosion over the very important potato fields. Satellite imagery was used to determine where there was no cover crop and those farmers were visited to explain the value of the cover crop to productivity – and profit. In Bangladesh, where flooding can occur at virtually any time of year, a very localized program was developed in the 1990s that provided agricultural advisors with the information on what best to plant to maximize output on any given plot following the recession of flood waters. The advice varied depending on the soil type, time of year, and the length of the inundation. As might be expected, China has been active in monitoring rice with a variety of sensors.[271]

The Global Information and Early Warning System of the United Nations Food and Agricultural Organization (previously mentioned in Section 4.3.9 and in this chapter in Section 6.5.2) provides monitoring on a regional basis in sub-Saharan Africa. It can be expected that more regional programs will come into being as more countries of the world see the benefits of doing so.

Many government programs provide agricultural subsidies such as support to farmers suffering losses due to flooding, drought, fire or other calamities. Information obtained from remotely sensed imagery that documents the extent and severity of the calamity together with land ownership information, and conditions specific to each farm are routinely used to assess damages and calculate subsidy entitlements. Similar assessments may be used by insurance companies to verify losses. Such applications are anything but new. Many years ago imagery

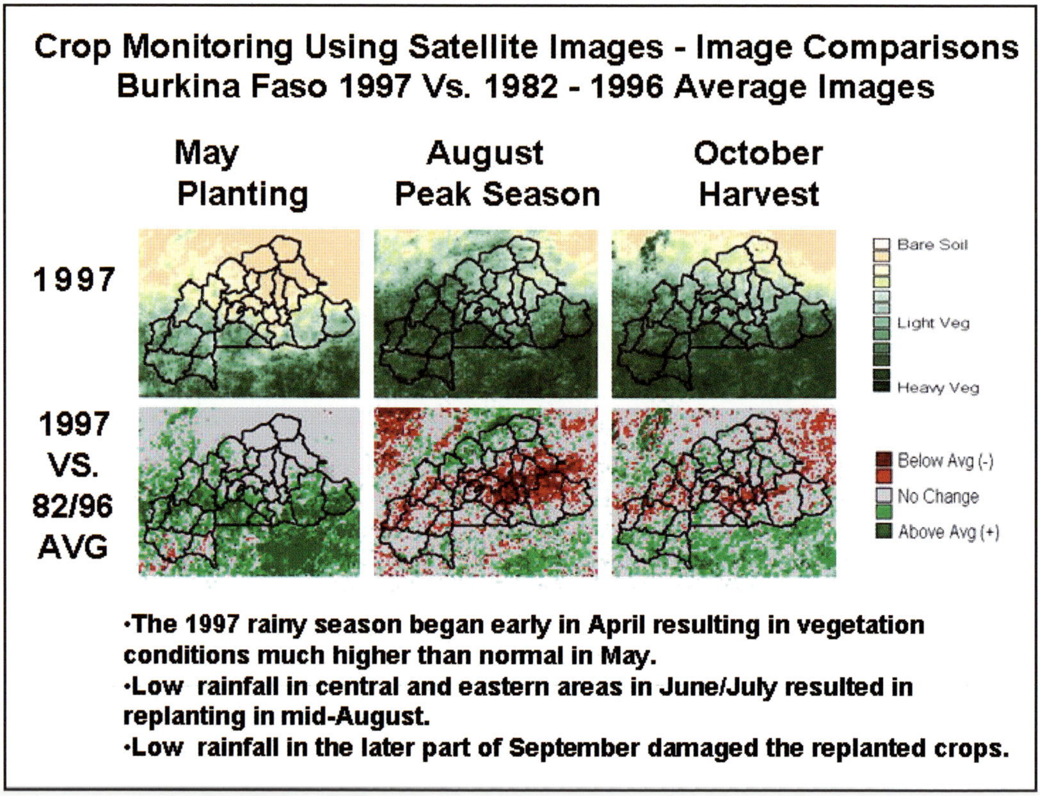

Geospatial Analysis to Anticipate Harvest Shortfalls. This poster illustrates the use of satellite image analysis to assess the condition of the 1997 crop compared with an average for the years 1982 through 1996. The red areas in the bottom set of images show that for large areas, crop condition was below average in August and October indicating food production was expected to be below average and food shortages would likely occur. Source: Food and Agriculture Organization of the United Nations

from aircraft was used to convince farmers that crop losses would be greatly reduced through the use of certified seed. The imagery showed in a convincing fashion that virtually the only significant crop losses in field beans were sustained by farmers who did not use certified seed: millions of dollars of crop losses were subsequently avoided as a result.[272]

Another major and long time application of geospatial data is in providing the base for both the census of agriculture and for the on-going estimation of production. The estimation of production helps to identify where there may be shortages of food or, in the case of commercial agriculture, gives an indication of what the market will be. In fact, the "Holy Grail" for satellite remote sensing has long been associated with answering this question concerning agricultural production estimation before harvest. Governments (and agribusiness) have wanted to predict what the crop production will be so that they can better forecast what the market will be – or how much crop they may have to buy. Literally hundreds of millions of dollars were spent on this activity in the 1970s and early 1980s in programs with catchy names like LACIE (Large Area Crop Inventory Experiment) and AgriSTARS (Agriculture and Resources Inventory Surveys Through Aerospace Remote Sensing) sponsored by agencies of the US Government. Europe has developed its own approach through its MARS (Monitoring Agricultural Resources) program sponsored by the European Union.

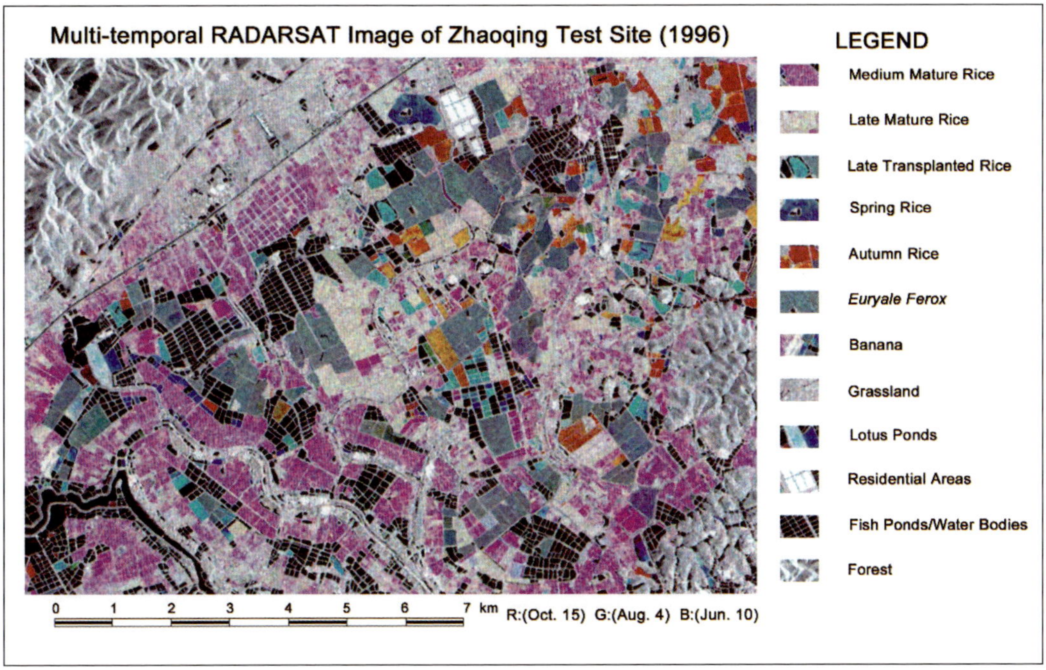

Satellite Radar Imagery to Map Rice Production in China. Three dates of Radarsat-1 imagery were combined to generate a composite image to map five different categories of rice as well as other land use and land cover types. Source: Analysis provided by Dr. Shao Yun, Institute of Remote Sensing Applications, Chinese Academy of Sciences, Beijing. Background image courtesy of Radarsat International © Canadian Space Agency/Agence spatiale canadienne (1996)

A Sample of Geospatial or "Where" Questions in Agricultural Resource Management

- Where will the crop yields be higher than usual and where will they be lower?
- Where will the crop yields be higher than usual and where will they be lower in neighboring countries?
- What will the impact be on crop production of the recent insect attack/tornado/hurricane/flood?
- What is the extent of soil erosion? How much? Where?
- Is there any evidence of illegal crops? Where? How much area? On whose land?
- Is there any disease? If so, where is the disease? If the disease is spreading, at what rate is it spreading?
- Is there salt accumulation in the irrigated area? If so, how much land is affected?
- Is water being used for irrigation on land that is not approved for irrigation?
- What areas of the state/country will open/close to crop production with global warming?
- Where are farmers leaving agriculture?
- Where (if anywhere) have wetlands been drained for agriculture?
- Where have urban areas encroached on agricultural land? What sort (i.e. what land capability types) of land has been encroached upon?
- To what extent have land mines reduced agricultural production? On what quality of land? What land should be de-mined first to maximize food production?
- Where is the locus or center of the recent outbreak of food poisoning/e coli? Is there a common geo-spatial denominator?

6.5.6.5. Water Resources

As we began the previous section, water, like food, is essential for survival. It is therefore not surprising that a great deal of attention has been paid to water in the geospatial world. From the mapping of stream courses on topographic maps, to the assessment of the impact of livestock on water pollution, to where to build catchment dams to replenish ground water, our access to fresh and clean water has benefited from the intelligent application of geospatial data and related technologies.

Geospatial data has been applied to find water, inventory water (what water is where), monitor the situation over time (both the quality of the water and the amount), model the future situation, and assess the environmental impacts of developments on water.

As with other resources, geospatial data has been and can be used to find water, inventory water (what water is where), monitor the situation over time (both the quality of the water and the amount), model the future situation, and assess the environmental impacts of developments on water. The first two elements are quite straightforward, and we have given a number of examples of how these first two steps are done. The last two – modeling and assessing impacts – are much more complex, and more interesting.

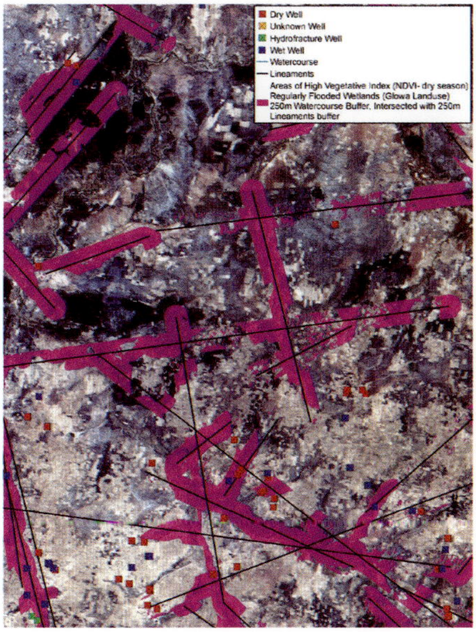

Finding New Groundwater Sources. An analysis of IKONOS satellite imagery together with local geologic and hydrological information were used to construct this map showing promising locations (purple areas) to drill new ground water wells. Source: Golder Associates Ltd.

Modeling is an analysis process that seeks to identify and quantify relationships among observed features. In addition to detection/identification and geographic extent, this analysis method also develops mathematical formulas to describe processes and relationships among different features and phenomena – such as the effect of a storm event on river water levels and flow rates. For example, hydrologic models can predict how soon the water level in a river will peak and how high it is likely to rise after a storm event of a given size. Such models typically include such geospatial information as the topography of the catchment basin, the extent and type of vegetation cover, and the extent of different types of soil. (The soil's absorption of water impacts how rapidly and to what extent water levels will increase after extreme storm events, and the incidence and severity of flooding that could damage infrastructure and affect food production.)

Monitoring Sedimentation. Sediment carried in the water runoff from industrial sites into lakes and rivers is a common contaminant easily seen on normal color aerial photography (A). Source: S. Aronoff

An important water source in many areas of the world is the ground. This has led to the complex task of modeling ground water to identify how and where it can be recharged

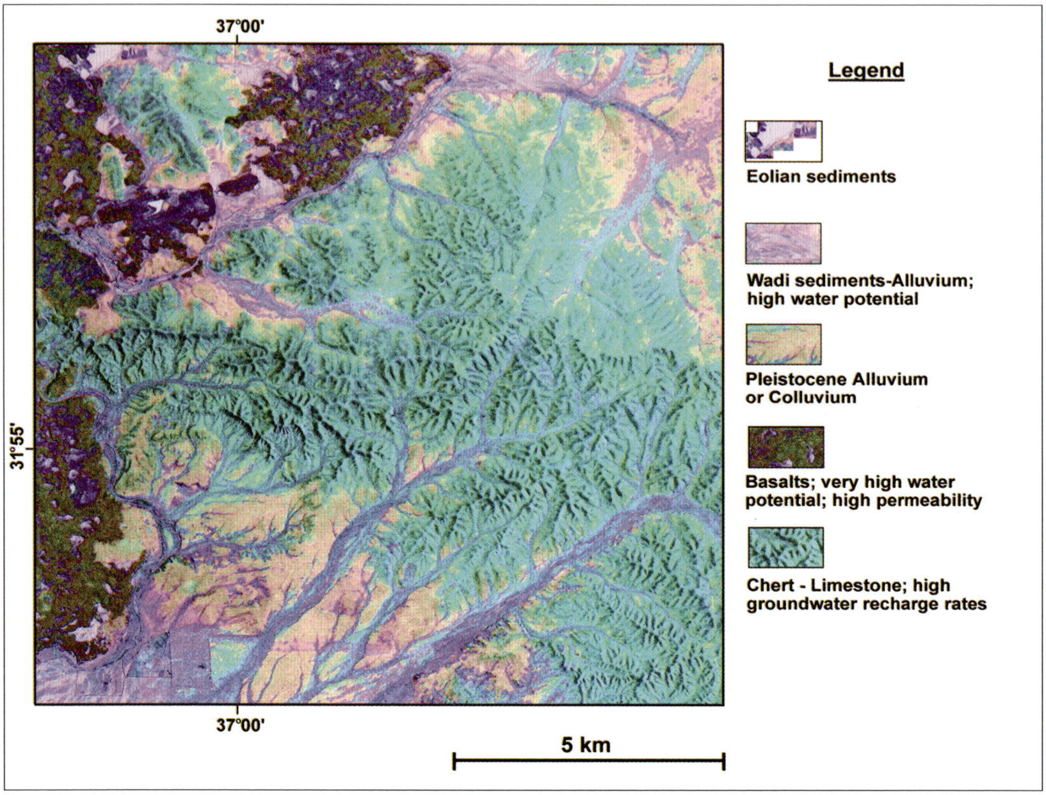

Hydrology of Azraq, Jordan. Landsat and synthetic aperture radar satellite imagery were analyzed to identify hydrologically important geologic features useful in the identification of ground water recharge zones from which sub-surface aquifers are replenished. Source: Dr. Vern Singhroy, Canada Centre for Remote Sensing, Natural Resources Canada.

to assure its continued availability. This is accomplished by the inventory and mapping of soil materials, topography, and hydrology to identify areas important for recharging ground water aquifers that should be protected or that might be suitable for locating containment dams to enhance ground water recharge. Remote sensing has been used to accumulate the information necessary to identify where the containment dams must be built to optimize the entry of surface water into the ground before it runs into an ocean or evaporates.

The second complex area of study related to water is that of environmental impact assessment. Water is important in two respects when one considers environmental impacts. Water is like the air we breathe: it is both something that can be polluted and, at the same time, can act as a carrier of pollution. Water is much more complex than air in that it can be polluted and carrying pollution both on the surface (in streams, rivers, lakes and oceans) and underground. Indeed, much of today's environmental legislation has as its ultimate aim the prevention of pollutants from entering water, either above or below ground.

A Sample of Geospatial or "Where" Questions in Water Resource Management
• How much has the lake shrunk and what impact does/will that have on wildlife/recreation/the fishery and on water for drinking/irrigation? • Where are the concentrations of livestock close to streams or rivers or ground water re-charge areas such that their waste may affect drinking water? • What is the extent of soil erosion? How much? Where is sediment from erosion entering the water? • What is the extent of the oil spill slick? • What is the temperature of the water? Where is warmer water entering the river/lake? • Is water being used for irrigation on land that is not approved for irrigation? • How close is the red tide to the fish farms on the coast? • Where is our water source vulnerable to terrorist attack? • How fast are the glaciers melting and what impact will that have on future water availability? • Are there adequate secondary containments surrounding the tank farm (i.e. group of oil or chemical tanks)? • Is there any evidence of oil or chemical pollutants entering the stream/river/lake? • Is there any evidence of soil staining or dead or dying vegetation near the stream/river/lake that may indicate the presence of threats to water quality? • Are the irrigation canals being monitored for water loss? What is the extent of leakage from the irrigation canals and where is this leakage? • How much has the land subsided over the area from which ground water has been extracted?

6.5.6.6. Minerals, Oil and Gas

Minerals, oil and gas are somewhat different from the other resources. First, they are regarded as non-renewable. Second, other resource applications are generally focused on determining the geographic extent, characterizing condition, or identifying variation such as species composition.

The major use of geospatial technology in minerals, oil and gas is in finding the resource.

For minerals and oil and gas resources the major application of geospatial technologies is actually in finding the resource. Indeed, just knowing the likelihood of finding a resource can be of great economic value. Governments have found that providing basic geological maps tends to spur further exploration by mining and oil and gas companies. That in turn leads to increases in local expenditures and a higher probability of actually finding something of economic importance, as is clearly shown in the study by ACIL Tasman (2008) and Lambert (1999). Given that a mine can generate billions of dollars of economic activity, it is not surprising that exploration is a highly competitive business with both significant risk and significant rewards.

THEOS Satellite Image of the Mir Diamond Mine, Russia. The scale of this open pit diamond mine, 525m deep and 1,200m wide which operated until 2004, is difficult to appreciate by the numbers alone. By showing the mine with the adjacent city, this satellite image provides the comparative context to appreciate the size of the mine in human terms. Source: © 2009 Geo Information and Space Technology Development Agency of Thailand

In the mining industry, for example, the exploration for high value minerals such as diamonds and precious metals involves the use of a range of sophisticated and expensive instruments. Hyperspectral remote sensing (discussed earlier), airborne measurement of gravity and magnetic fields, and other reconnaissance methods are used to survey large areas rapidly. These data are then analyzed to identify those few small areas where deposits are likely to be found – greatly reducing the area to which expensive field investigations are applied. The field work is done by highly trained professionals, and subsequent drilling (sometimes to great depths and at high cost) is done to extract samples.

Detecting the presence of conditions where the target minerals are more likely to occur can dramatically reduce exploration costs and improve success. Once the drilling and other exploratory work has been done, one can then define the size, location, and quality of a deposit – information needed to assess whether development of the deposit is economically viable. As mineral, oil and/or gas resources are developed there are further applications of geospatial – from further exploration on the same site to measuring the amount of material removed from an open pit mine, and on to monitoring tailing ponds and potential contamination of the soil. All of these steps use, and in some cases rely on, geospatial technologies.

A Sample of Geospatial or "Where" Questions in Mineral, Oil and Gas

- How do our geological maps compare to those of our immediate neighbors' maps? Are they as detailed, accurate, and useful?
- Are our policies for release of data for exploration companies as favorable as our neighboring states/countries?
- We assume that interfermometric SAR is being used to monitor how much the land has subsided over the oil/gas field. How frequently is this being done and what is the measured rate of subsidence?
- What is the extent of the oil spill slick? How fast is it spreading? How quickly can it be contained?
- How fast are the glaciers melting and what impact will that have on future water availability for the proposed mining operation?
- Are there adequate secondary containments surrounding the tank farm (i.e. group of oil or chemical tanks)?
- Is there any evidence of oil or chemical pollutants entering the stream/river/lake near the refinery/mine/smelter?
- Are the mine tailing pond dams being monitored and if so, how frequently and with what technology?
- Is there any evidence of soil staining or dead or dying vegetation near the mine/smelter/refinery?
- How much material (cubic meters) has been removed from the open pit and what, therefore, is the royalty/tax/environmental levy?

6.5.6.7. Conclusion

From its early development geospatial data, technology, and analysis methods have been widely used for natural resource applications. The questions that can be answered for national and state governments range from the mundane to the remarkable. Therefore it is not surprising that these applications are among the most diverse and widely accepted uses of geospatial information. Indeed, geospatial data and technology have become an essential component of natural resource management world-wide. Countries should include the acquisition and use of this information in their on-going planning, policy development and management activities. Doing so can make national governments more aware of developing issues sooner, allow them to respond more quickly, and recognize opportunities before others claim the advantage.

6.5.7. International Development Assistance

The use of geospatial data and technology in international development assistance provides an interesting window not on just its value in aid programs, but also on the perceived value of geospatial information and technologies in general, by both the donor and recipient countries. This in turn provides some clues as to how these countries will perform in the GeoEconomy. We believe that those countries and agencies with greater advantage in the GeoEconomy will be those who are doing one or (preferably) more of the following:

- Actively providing appropriate aid in geospatial; or
- Actively seeking aid in geospatial; and
- Organizing their bilateral aid programs (donor or recipient) with geospatial technology.

Geospatial and development assistance come together in one of three ways. First there can be appropriate assistance in providing the basic geospatial information and technology used to run a country. By appropriate we mean that the assistance offers a complete solution to a specific need of the recipient country as a whole. It is not just meeting the needs of vendors in the donor country, or only certain ministries in the recipient country. Second, one can find geospatial information or technology linked to a specific program – environmental monitoring of a sensitive eco-system tied to ecotourism, or a forest inventory, for example. The third approach sees the technology used to monitor the results of the assistance. Each of these, if done well, can contribute to the continuing development of a country, as well as to the growth of its geospatial infrastructure and capabilities.

> ***The use of geospatial data and technology in international development assistance provides an interesting window not on just its value in aid programs, but also on the perceived value of geospatial information and technologies in general by both the donor and recipient countries.***
> ***This in turn provides some clues as to how these countries will perform in the GeoEconomy.***

At one time, geospatial information in the form of topographic mapping was seen as a basic building block for development. Indeed, it was funding under the Colombo Plan and other aid activities in the 1950s and 1960s that led to the early rapid growth of the aerial survey industry in the USA, Canada, and Europe following World War II.[273] Provision of maps was seen as the basis of development – whether in Africa, Asia, or elsewhere in the developing world. That situation has changed in ways that are sometimes beneficial and other times not.

Where assistance has enabled geospatial to be fully integrated across the country and used to strategically link its many ministries, it has advanced a country's development. Where assistance has resulted in piecemeal geospatial implementations and was not viewed or developed as a truly national effort, important opportunities were missed. At best there was a waste of scarce resources, at worst a legacy that hampers future development may have been created.

Some aid agencies seem to have moved away from providing what we call direct geospatial and other infrastructure assistance and now focus on so-called "softer" aid issues (such as gender equality, governance, and AIDS). Parenthetically, these softer aid issues lead to results that are often difficult to measure – especially if there is no geospatial basis upon which to compare results from place to place! A cynic might suggest that this situation is a good one for those who do not want the effectiveness of their aid programs assessed.

The lack of understanding of geospatial and its value in the aid community can at times be almost frightening. One senior official in a major national aid agency told one of the authors that they had no interest in, or use for, the technology. Interestingly the letter with this statement in it was sent a few days before the tsunami hit Asia – and just a few weeks before this same aid agency was scrambling to find maps of the affected areas! This of course begs the question – how can that aid agency know where their money is spent, for what purpose and with what results in different locations? Are their efforts being focused on the areas in greatest need? Does the aid they deliver account for regional variations in the area they service? Likely not, if they are not using geospatial information to track and assess their work![274] One beacon of hope among aid agencies in this regard is US AID. They have embarked on an ambitious and largely successful program to use geospatial to track what they are doing, and with what success.

However, before anyone embarks on the provision of geospatial technology and data to a country, we advocate a strategic analysis of the country's needs and requirements. Without a strategic analysis the provision of direct geospatial assistance will at best lead to inefficiencies, and at worst to a complete waste of the money spent. The strategic analysis should involve the main users and suppliers of geospatial data in the country and an unbiased assessment of the major issues faced by the country. Both the donor and recipient must also ensure that the data will be updated and used in a consistent and sustainable fashion.

Another of the problems we have seen in a number of countries receiving assistance is the piecemeal introduction of geospatial. One technology will be introduced into the forestry department by one donor, another into the environment department by another donor, and yet another into a third department. Often the result is that the systems cannot communicate with each other and it is impossible to bring together various data sets for national planning.[275]

Policies affect how much we have and who gets to have it. Source: S. Aronoff

Approaches have been developed to ensure that countries integrate geospatial information into their on-going governmental management and avoid the problems noted here. One such program was recently undertaken in the Maldives, and other analyses have been performed in Thailand, Bermuda, and other countries.[276]

Before anyone embarks on the provision of geospatial technology and data in a country, we advocate a strategic analysis of the needs and requirements of the state or country. Without a strategic analysis the provision of direct geospatial assistance will at best lead to inefficiencies, and at worst to a complete waste of the money spent.

While there has been a shift to soft issues in some aid agencies, many have still continued to provide geospatial training, data and technology – including US-AID, and the Japanese, Swedish, Dutch, Australian and German aid agencies. Indeed, some of these are providing assistance using another country's technology.

It is interesting that in many ways the development banks and agencies of the United Nations seem to be leading the way in geospatial and international assistance. This is most often done with a focus on planning national geospatial programs. This sees geospatial applied to topics like property rights, property ownership, and three areas of management: environmental (including climate change), natural resources (agriculture, forestry and water), and urban.[277] These various applications of geospatial data and technology are easily understood and answer many of the typical "where" questions posed in development assistance. Some examples of the sorts of questions that one can ask in these areas are given in the following table.

A Sample of Geospatial or "Where" Questions in Development Assistance

- Where is the AIDS infection level higher? Why?
- Where are the wetlands being destroyed?
- Where are there settling ponds or potential sources of pollution close to the river/lake?
- Where can additional tourism facilities be located without impacting the environment?
- Where are the Millenium Goals' Targets being met and not met?
- What is the range of the endangered animal that must be protected to foster eco-tourism?
- Where is the logging taking place? Is the annual cut on that forest sustainable?
- Where is the source of the pollution entering the river/lake/bay?
- Where is the new school/hospital/well required?
- Are there regional differences in the gender program? If so, how much and where?
- Where might small land holders benefit from small loans?
- Where should one locate small weirs on the stream course to recharge the ground water supply?
- Where is the environment in worse condition now than 10 years ago?
- Where was the flood? Where is temporary shelter required?
- Where is the drought? Where is food assistance required?
- What crops will be grown where with 20 cm less rain and a 2 degree increase in temperature?
- Where are the glaciers retreating and at what rate per year now compared to twenty to thirty years ago?
- Where can the new transit system be located to minimize the impact of housing and the environment?

What is perhaps even more exciting than the delivery of geospatial programs through development assistance, and the use of geospatial data to answer questions related to development assistance, is the use of geospatial technology to MANAGE development assistance. Simple but critical questions can be asked and answered where geospatial technology is properly employed in a monitoring role:

- Where is our international assistance being given?
- Where does our international assistance come from?
- What is the regional economic impact of the assistance?
- Does the impact of the assistance vary from place to place?
- What sort of assistance is being given in each region of the country?

The graphical presentation of the answers to such questions are frequently displayed on the web page of US-AID, one of the apparent leaders in the use of geospatial information as a management and communications tool in international aid.[278]

6.5.8. National Economic Planning

The value of geographic information for national level planning has been known, to at least some, for more than a century. In the early 1900s the Government of Canada saw the value of a National Atlas to show people (especially potential immigrants) the richness of Canada's west, as it sought to populate that largely empty region of the country. The 100th anniversary of the founding of that Atlas was marked by mapping being depicted on the back of the Canadian $100 banknote.[279]

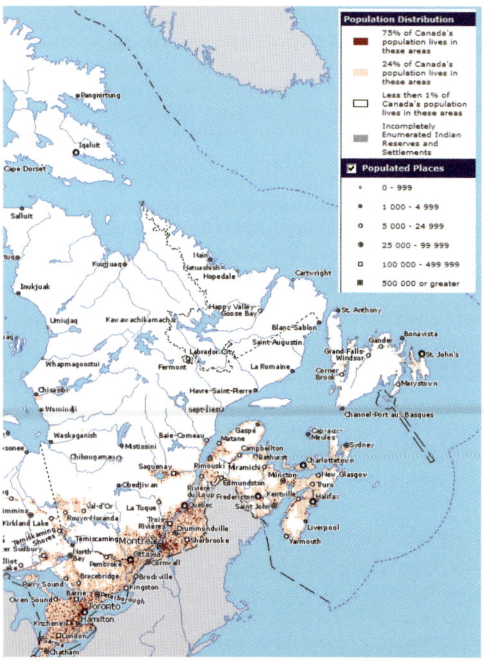

National Mapping – An Essential Public Service. Online access to information organized using a geographic map base has become common for disseminating basic national-level information to citizens and the world. Online offerings typically allow users to customize their search and produce map products tailored to their specific interests, such as this map of the population of Eastern Canada in 2006 created on the Atlas of Canada website. Source: Natural Resources Canada

The whole concept of GIS has been attributed to activities of the Government of Canada in the 1960s to map land capability, as it sought to provide the information needed to properly plan for development of a large country, featuring what is at best a harsh and fragile environment.[280] This is not to say that everything geospatial is wonderful in Canada. One of the problems that some might see in Canada is that the country has been resting on its geospatial laurels, so to speak. Renewal on a national level has been called for by a variety of groups, including industry.[281] But much the same can be said in the USA, the United Kingdom, India, China, Australia and New Zealand among others: all of these countries have been studying what is needed to effectively run a modern developing (or developed) country. As noted previously, Australia was the first country to commission a complete economic study of the value of geospatial to a national economy. That study, done by ACIL Tasman, cost under $200,000.[282]

The most important thing a consistent geospatial data set and framework provides is the confidence that across the country you are comparing things on the same basis. It is for this reason that statistical agencies, which are notorious for wanting to be accurate, have often developed geographic capabilities so that they can better respond to the "where" questions asked of them and their data bases. One can then begin to ask questions similar to many of those asked in the preceding section, as well as more general ones, such as are in the following table. Once the "where" questions are asked, one can better answer the next questions – "why?" and "what can we do about it?"

The most important thing a consistent geospatial data set and framework provides is the confidence that across the country you are comparing things on the same basis.

A Sample of Geospatial or "Where" Questions in State or National Level Planning
• Where is the lowest cost of living? Where is the highest cost of living? • Where is the incidence of cancer/heart disease/etc higher/lower? • Where are the lowest incomes? • Where are the lowest levels of education? What is the average time it takes for children to travel to school? • Where are the major centers of production? What do they produce? • Where are public funds being spent? How much per capita? On what?
As with the other lists of the applications of geospatial, the only real limitation on this list is the space allowed for it – and the imagination of the author.

As these questions are answered, one may well uncover information that will assist in the next area of application, audit and evaluation.

6.5.9. Audit and Management

"Where" questions are often asked by auditors and those evaluating how money and resources are being spent in both the public and private sectors. By putting various elements of spending (or sales, or results) against a geospatial backdrop, one can see patterns emerging that may assist auditors and evaluators in identifying and then tracking problems. While we have already made the case for using geospatial as a basis for evaluating aid projects, we would argue that managers in all sectors can also benefit from using geospatial data in their management of spending, people, and assets. Recognizing and solving problems before they grow into a serious issue is, in effect, one of a manager's primary responsibilities. All too often situations are looked at and evaluated without paying due attention to the spatial dimension. In the GeoEconomy, this can have dire consequences.

By putting various elements of spending (or sales, or results) against a geospatial backdrop, one can see patterns emerging that may assist auditors and evaluators in identifying and then tracking problems.

One simple question brings home the potential value of geospatial in the realm of audit and asset management: **Would the financial institutions that bought up the US-based securities based on sub-prime adjustable mortgages have done so if they had seen that much of the collateral was mortgages in low-income areas of US cities?**

The answer, we would hope, is a resounding no!

The following table presents a sample of the "where" questions one can ask in audit, evaluation and management, as well as some of the follow-up questions auditors would then typically ask.

A Sample of Geospatial or "Where" Questions in Audit and Management
• Where is the lowest cost of delivery of the service/product in the company or agency? Why? • Where is the highest cost of delivery of the service/product in the company or agency? Why? • Where is the incidence of misspending highest? Why? • What is the average cost of providing service X, Y, or Z by location? If there are differences, why? If not, why not? • Which region seems to have the highest expenses? Why in this territory, and not others? • Are there more service or product complaints from one or more regions? Why this/these region/s? • Do different parts of the organization spend different amounts on expenses in the same region? Why? • Are staff offices the same size and cost in different regions? If not, why? • Is the return on investment the same for research centres in different locations? If not, why not? • Do different parts of the organization in different locations have the same ratios of support staff? If not, why not? • Do different parts of the organization in different locations have the same overhead levels? If not, why not? • Are the risks in the investment the same at all locations? If not, is this as would be expected? If not, where are they worse and which ones should be avoided?

6.5.10. Planning a Municipal, State, or National Geospatial Program

6.5.10.1. Introduction

This section introduces an approach to planning government geospatial programs based on the experience accumulated in preparing such programs for both developed and developing countries.

National, and in some cases, state or provincial governments have a long history as the primary source of certain types of geospatial information. For example, they generate and maintain national or province-wide coverage of topographic maps, maps that support industries – such as geology maps used for mineral and oil and gas exploration, soils maps for agriculture, and various public information products such as population distribution. Providing and being responsible for this geospatial information are part of what asserts a country's sovereignty and are essential for it to defend its borders, deliver services to its citizens, and understand the conditions of its society and the forces affecting it.

Similarly the acquisition and maintenance of air photo coverage, usually in conjunction with topographic mapping, is universally accepted among developed countries and most developing countries as an essential responsibility of government. In Canada, for example, government acquisition and public availability of aerial photography began in the 1920s. In most developed countries, air photo coverage is readily and inexpensively available for public use for virtually all of the country. Among developing countries it is common to restrict public access to imagery for large parts, if not all, of the country and the military commonly controls public access to imagery directly or indirectly.

6.5.10.2. Current Drivers Affecting Government Geospatial Programs

There are several factors that have recently come to influence governmental geospatial programs. As we've mentioned previously, the impact of Google and Microsoft providing powerful and free online access to maps, satellite imagery, and aerial photography has educated the public (or at least the computer-literate segment of the public) to expect that location-related information is provided in a geospatial format such as maps or imagery.

At the same time, and perhaps as important, has been the opportunity for geospatial information to be used in the communication of all kinds of material that previously was either not provided to the public at all, or not in graphic form (e.g. because there was no means to disseminate the information quickly enough to be useful). Thus, for national governments, there is both a push from within to disseminate information that informs and influences their citizenry and the world, and a pull from the public for more current, more detailed, and more diverse geospatial information.

In responding to these demands, national governments typically developed a patchwork of geospatial databases and systems. While this technology often satisfied the needs of individual groups or departments, information could not easily be shared among them. In part this may have been due to technical incompatibilities. In other cases the desire to control their data led departments to make information sharing difficult or practically not possible.

The question of information sharing leads to another of the factors that influence government geospatial programs. Our research has shown that those countries and governments that share their data within government and with the population are more successful, and will likely realize more of the benefits of the GeoEconomy. These are the countries that have recognized that the needs of the government as a whole must somehow be imposed over the inherent competitiveness of individual departments and groups. In the private sector the ability to control your information, and withhold it from competitors to gain an advantage, imparts considerable power to those who can exercise it. It is also true that government organizations compete for funding, and hence must compete for power and influence over the decision-making process on which their budgets and personnel allocations depend. The extent to which national governments successfully manage this competition, as it affects the sharing of geospatial information, determines the economic, political, and quality of life benefits that the country can obtain from its valuable geospatial information assets.

In the private sector, guarding the organization's information, or an asset it has purchased or created, is part of healthy competition. Within a national government such an approach is all too common but generally inappropriate. Although a government department may have created it, and may believe that it "owns" this information, it actually belongs to the country, not to the bureaucrats placed in the position of authority.

It is telling that to overcome these obstacles of internal politics it has generally necessitated directives from the highest political levels requiring not only that data be shared, but that the necessary technical and administrative mechanisms be implemented. A directive that is not actionable may be good for publicity, but of no value to compete in the GeoEconomy.

It is within the context of these drivers that we have developed an approach to developing geospatial programs in government.

6.5.10.3. Developing a Geospatial Program in Government: A Nine Step Approach

One would think that planning a municipal, state or national geospatial program is, or should be, a simple process – you buy the technology and install it. The same can be said for developing a geospatial program in industry. But it isn't. It is definitely NOT simply a matter of buying technology and installing it! The experience we have gained over the past ten years in developing national geospatial programs has led to the development of an approach that has, with minor modifications, been applied in both developed and developing countries.

Some of the steps that have proven to be essential to implement the desired data sharing have been the development of a national geospatial data infrastructure, the definition of minimum requirements that the administrators of government geospatial databases must meet, and, and the solution of legal issues related to information sharing (such as limiting the liabilities of data providers if their information is improperly applied by outside users).

Perhaps the most important thing we have learned is that this sort of activity requires a number of different areas of expertise that are not found in one individual. For each project we have used experts in:

- Cross-cultural survey design;
- Cross-cultural meeting facilitation;
- Strategic analysis;
- Government organization;
- Geospatial data policy;
- High level (i.e. Senior management) geomatics expertise;
- High level (i.e. Senior management) government expertise;
- Technical system architecture design; and
- Change management.

What follows is our prescription for doing such work based on our experience.

In some of our work the program has focused on just one aspect of geospatial such as remote sensing,[283] some have focused on the special needs of geographically distinct environments such as small island states,[284] and still others have looked at national programs.[285] From this experience we have developed a systematic approach to the development of geospatial programs that are tailored to each situation.

Our approach focuses on five tasks: adequate preparation including collection of relevant background information, strategic analysis, identification of the relevant data policy issues that need to be addressed, the use of best practices (borrowing from the experience of others), and the use of appropriate planning tools for system architecture design.

In all cases we begin by securing the agreement and support of the client, that whatever their jurisdiction, the success of the program depends on it being designed to meet the needs and advance the goals of the organization (or state or country) as a whole. This objective must take precedence over the specific interests of any one group within the organization be it a municipality, state or country. Said in another way, the approach depends on a willingness of the players involved to agree to put the municipality, state or national government ahead of the individual agencies from which the players come. Once this agreement is reached and the organization's overall support is confirmed (often through or by the political level minister or the president), a number of preparatory assessments are made. This preparatory information is most often collected through surveys administered on-line and analyzed before the on-site consultation team arrives. The survey guides the consultation process by providing specific information that can be summarized as follows:

- Determine the information needs – who needs what information, for what purpose, and how often;
- Determine what geospatial information is available – what it is, how old it is and how it compares to what is needed;
- Determine who knows what about geospatial, their level of training, and the needed expertise that is likely missing. (This is especially important in the development phase of the project); and
- Determine the approaches that have been used in comparable jurisdictions (to avoid re-inventing the wheel).

From this preparatory information an understanding of the client's geospatial capabilities, experience, and objectives is developed. Then a Strengths, Weaknesses, Opportunities and Threats (SWOT) Analysis is prepared. The SWOT analysis provides information that feeds into further analysis and design.

The strategic analysis is carried out in the form of a workshop[286] at which the major players are gathered to agree on core values, define strategic drivers, assess the preliminary SWOT analysis, and arrive at a vision for the program and a mission statement. This is followed by identification of the organization's current state (i.e. the services currently provided and the degree to which current geospatial capabilities support them) and the desired future state (the type of services the organization would like to deliver). With this information in hand, a well-informed consulting team can then build a strategic framework that will outline specific required actions. A strategic plan is then developed and reviewed with the client. Concurrently with the development of the strategic framework, a data policy must also be developed.

Collaborative Geospatial Strategy Development. This workshop in the Maldives, facilitated by an expert in strategic planning, reached decisions on a comprehensive approach to develop a national GIS. Source: R. Ryerson

With the various strategic elements in place, along with the data policy and technical details arising from the requirements, an experienced designer develops a system proposal using relevant planning tools and a comprehensive knowledge of the best practices used elsewhere to address similar situations.[287] This proposal is then reviewed with the client and adjustments made to the size and scope of the proposed system. Trade-offs are negotiated among the stakeholders to choose capabilities that would be delivered first against those that would be added at a later date. These trade-offs need to take into account such factors as the organization's priorities and budget timing. At the same time, the institutional arrangements must be specified and agreed upon. Once these arrangements have been finalized, an implementation plan can be prepared and executed.

The result is a set of policies, a strategic plan, institutional arrangements, and a common view of where geospatial will take the government in question.

6.5.11. Education and Training

It has long been accepted that literacy and numeracy are important skills that one must possess to make one's way in this ever more complicated world. We believe that as important as these skills are, to make important decisions and to better understand the important issues of the day, one must also be geo-literate. We would therefore be remiss if we did not at least touch on one of the important factors in having success in the GeoEconomy: the development of geo-literacy among the population at large and the specialized training in geospatial of people who are often generically referred to as Highly Qualified Personnel (HQP).

> ***Literacy and numeracy are important skills that one must possess to make one's way in this ever more complicated world.***
> ***We believe that as important as these skills are, to make important decisions and to better understand the important issues of the day, one must also be geo-literate.***

Geo-literacy begins with geo-awareness in the schools and is often first developed through the study of maps, atlases (on-line and hard copy), and images under the banner of geography, or social studies. School children in the United Kingdom have long been trained in the understanding of maps and have even participated in their preparation. In the 1930s Sir Laurence Dudley Stamp used school children and volunteers to map land use in England – work credited with helping England avoid starvation during WW II.[288] While the Scout and Girl Guide movements have also long championed map reading, we would like to differentiate map reading skills associated with topographic maps from the more advanced understanding of geographic relationships that geo-awareness implies. Programs to develop what we call geo-awareness in the schools at the primary level began in several locations with the arrival of Landsat satellite imagery.

School children can easily be taught simple interpretation of Landsat-like images, and from that can be developed an understanding of how various elements of a region's geography interact and depend on each other.

To truly develop geo-awareness requires that the student understands the importance of geospatial in fields as diverse as the study of the environment, biology, civics, and history. In the same way that students are taught and expected to use calculators, computer spreadsheets and word processors in their work, they should be as familiar and comfortable using geospatial information. While it is easy to create maps, (most common office software has some capability to generate a map and there are any number of online resources to do so,) that is not the same as understanding how to bring geographic and spatial reasoning to how an issue is framed and addressed.

Geospatial reasoning is much more than displaying information in a map-like form to show objects' relative locations. Geospatial reasoning offers an added dimension to the way we think about a problem and the way we strategize about how to solve it. Displaying information on a map can be more than an effective means of communication. It can also represent a point of view. The information chosen to be represented and that which is left out, the relationships that are made easier to recognize (whether visually or by computational means) and those that are ignored, all affect the knowledge that will be seen by the person viewing the map. "Mapping is a way of thinking about abstract things in a graphic and practical way."(Koch, 2005)

In this book we have made the case that geospatial information is becoming increasingly important. For this reason alone, we can assume that more and more information will be presented in map form. Yet our education system provides little background on how to assess information received in this way. Maps are commonly presumed to be representative of an objective world. Yet a map results as much from the information that is left out as the information that is included. For a map to be readable, most of the information that could potentially be included must be left out. What is presented is the information the map maker decided was relevant to its intended use. What the map is to be used for determines what goes into it and what is excluded from it. It is a selective presentation of observations that may or may not be biased and its accuracy may or may not suit the use someone else may choose to make of it.

That the information on a map is a selection of what could be presented, and that this information may be biased or could be untrue or less accurate than needed, is the same whether you are using information from a book, or a statistical table, or any other source material. The difference is that the education system generally teaches so little about maps, let alone other geospatial tools, that most students graduate university never having learned this means of communicating information. It is as naïve as believing everything that is written in a book or seen in a commercial message.

The education system generally teaches so little about maps, let alone other geospatial tools, that most students graduate university never having learned this means of communicating information.
It is as naïve as believing everything that is written in a book or seen in a commercial message.

Teaching geospatial reasoning is not about learning to make maps, it is about learning to find, use, and generate geospatial information effectively for the various aspects of a person's life. To repeat, geospatial skills should be seen as fundamental to a person's education as arithmetic and report writing.

Instead of asking why geospatial skills should be taught, the better question policy makers and educators should be asking is how best can these skills be taught and incorporated into students' overall education program? Reading, writing, and arithmetic are taught as separate subjects but students use these skills in many, if not all, of their other subjects and, in fact, in their daily lives.

In the GeoEconomy geospatial information is now part of daily life. It is the way information is increasingly presented to the public, in business, and in government. Wherever the generation growing up today works and plays, geospatial information will become as ubiquitous as the printed word.

Wherever the generation growing up today works and plays, geospatial information will become as ubiquitous as the printed word.

As valuable as the printed word is to convey information, those who do not understand the subtleties of the printed word can easily fall victim to misleading advertising, unscrupulous promoters, and propaganda. In the same way, geospatial information, as valuable as it is, can be used to misrepresent, mislead, and, at worst, lie. Lacking the skill to intelligently use geospatial information will be an increasing disadvantage, just as being illiterate became an increasingly difficult handicap a century ago. We should be focusing on finding or developing the teaching aids needed to promote geo-awareness in our class-rooms.

A basic level of geospatial reasoning skills should be part of a modern education. There is also an ever growing need for highly qualified personnel (HQP) to work in the various geospatial fields. The development of HQP in geospatial has followed a number of different models depending on (not surprisingly) location. The training of geodesists and others in equally esoteric aspects of geospatial tends to be carried out in a few highly specialized schools in engineering or physics. Surveying is also typically found in engineering schools. Cartography or mapping was often part of geography programs in many universities, although much of the more practical training in mapping after the 1930s occurred in military mapping organizations. In many locations the early growth in the use of satellite data was also in engineering schools or departments of geography.[289]

As the field has grown in both its popularity and utility, there was a corresponding growth across a number of areas of academe. Indeed, in many places there developed whole departments dedicated to the field – such as at the University of New Brunswick in Canada or RMIT University in Melbourne, Australia. In others there were whole educational institutions developed in geospatial such as the ITC in the Netherlands and Wuhan in China. Interestingly enough, both Wuhan and ITC have been absorbed into other institutions at the same time as the field of geospatial is becoming even more pervasive. Perhaps this absorption into the broader academic community in larger institutions is a sign of the broader acceptance of the field and more convergence – as well as the understanding that geospatial should not stand alone. In any case, today virtually every resource or environment-oriented university program requires both theoretical and practical understanding of the use of geospatial data, and thus almost all of these programs offer courses in GIS, remote sensing and other geospatial technology.

What has occasionally held back development in the geospatial field in academe is the argument that it is not a "true" academic field worthy of the awarding of higher-level degrees such as the Ph.D. – geospatial is really "only" a technology. We disagree with that view – as do the many universities around the world now awarding degrees in these areas. The science behind geospatial technologies is complex and the applications can be sophisticated, drawing as they do knowledge of diverse scientific fields, as well as a range of social sciences too diversified to easily catalogue in this short summary. Some of the academic work has seen the development of effective solutions to complex spatial analysis problems, spatial accuracy assessment, and the propagation of spatial accuracy, as data sets are combined. These, as well as the integration of diverse data sets, have also required the development of new information theory, structures, and the invention of new sensors. Many more examples of the high level of academic achievement needed in geospatial work could be identified. The term *geographic information science* is sometimes used to describe a scientific discipline that would include the diverse body of knowledge that geospatial encompasses.

Whatever geospatial may be, and whatever academic purists might say, without sufficient understanding of geospatial, national economies will falter. Those investing wisely in geospatial research, and development of HQP through mechanisms like the previously mentioned GEOIDE Network in Canada, will have a much greater chance of success in the GeoEconomy.

For a country to have ready access to leading expertise and to develop a pool of geospatially trained personnel it needs to attract experts doing leading edge research in the field. Our question for the policy maker in education is: What geospatial education and research is required in our state or country to keep up? Not surprisingly, the answer will depend on the policy maker's knowledge of geospatial. Those that "get it" will recognize and champion the importance and value of increases in teaching, research funding, and technology transfer mechanisms. They will give their students a better opportunity to succeed in the GeoEconomy. Those that don't get it will leave their students at a relative disadvantage. They will likely lag behind those better prepared for the GeoEconomy.

We hope that this book will help people everywhere "get it."

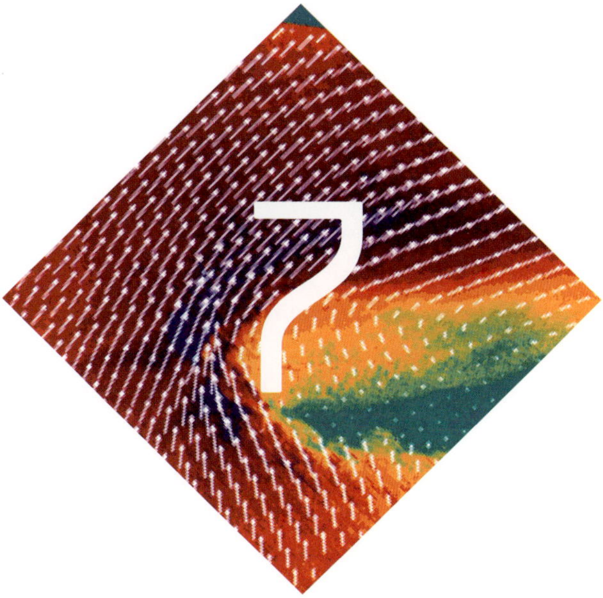

CONCLUSION

1. Welcome to the GeoEconomy.

We began by welcoming you to the GeoEconomy – the economy that is increasingly being driven by, and dependent on, geospatial or geographic information. In writing this book we have become even more convinced of the importance of geospatial information and hope that you, the reader, will agree with us from whatever perspective you may have had at the start.

We have argued that geospatial or geographic information is dramatically increasing in importance for individuals, businesses, communities and governments – be they local, state or national.

What has been both encouraging and at times perplexing as we have reviewed the field and the geospatial successes and failures of companies, cities, and countries, is that successes and failures are found in both the developed and developing world. It is not only the wealthy nations that are benefiting, there are developing countries benefiting as well. It isn't just about the quantity of funding, it is also about being smart about how those funds are spent. Those countries that have long embraced geospatial information seem to be the ones that are gaining ground, while those losing ground seem to be those that have seen it as less important.

We have explored the history of geographic knowledge and its use (Chapter 2), the technologies being used, the sort of data we now have access to, and how (and why) the need for geographic information is increasing (Chapter 3). We have reviewed a number of converging technologies, policies, and disciplines that influence geo-spatial technologies and data (Chapter 4), leading to an assessment, in general terms, of the current impacts of geospatial data on our society. We then introduced what we see as the primary future issues faced by our species, our environment and our society, and the important role that geographic information plays in addressing them (Chapter 5). In Chapter 6 we identified how the individual, communities, companies and countries can and should benefit from the intelligent use of geospatial or geographic information.

If we leave but one message, it is that very often one can see that the reasons for the successes – and the failures – in everything from our environmental protection to our economic decisions, are directly tied to how well we use geospatial or geographic information.

We have shown that the development of geospatial knowledge has had major impacts on the speed and course of human history.

But if we leave but one message it is that the reasons for the successes – and the failures – in everything we do from our environmental protection to our economic decisions, are directly tied to how well we use geospatial or geographic information.

Simply put: those who "get it," those who understand the GeoEconomy and how to use it to their advantage, have done and will do well. Those who do not will depend on geo-luck. Luck, geo or otherwise, like hope, is not a strategy.

REFERENCES

The following are published and presented papers, books, and articles referred to in the endnotes that follow. In some cases, specific articles from the popular press cited more than once in the endnotes (for example, B. Healy from *The New York Times*) are also listed.

ACIL Tasman. 2008. *The value of spatial information.* Cooperative Research Centre for Spatial Information, Melbourne, Australia. www.crcsi.com.au/pages/publications.aspx.

Army Space Journal. 2008. Corona codename "Discoverer" America's first space surveillance satellite. *Army Space Journal* 7 (1): 42-43. www.armyspace.army.mil/Pic_Archive/ASJ_PDFs/ASJ_VOL_7_NO_1_ WINTER_ 2008_11_Corona.pdf.

Aronoff, S. 1985. The minimum accuracy value as an index of classification accuracy. *Photogrammetric Engineering and Remote Sensing* 51(1): 99-111.

----------. 2005. A brief history of remote sensing for earth observation. In *Remote sensing for GIS managers,* ed. S. Aronoff, 9-52. Redlands: ESRI Press.

Atwood, D., D. Mayall, and R.A. Ryerson. 2009. Towards cooperative spatial data infrastructure in small island nations: The experience in Bermuda. *Proceedings American Society for Photogrammetry and Remote Sensing Annual Meeting.* Baltimore, Md., March 13-19.

Baker, J.C., et al. 2004. *Mapping the risks: Assessing the homeland security implications of publicly available geospatial information prepared for the National Geospatial-Intelligence Agency.* Santa Monica: RAND Corporation.

Banks, B. 1989. *Satellite images: Photographs of Canada from space.* Camden East: Camden House.

Biasi, F. 2007. New Conservation GeoPortal taps into a world of maps and data. *ArcNews* 29 (3): 35.

Black, J. 1997. *Maps and politics.* Chicago: University of Chicago Press.

Bourge, C. 2006. Europe's global positioning system. *Military Geospatial Technology Online Archives* 4 (1).

Clancy, T., with C. Horner. 1999. *Every man a tiger: The Gulf War air campaign.* New York: Berkley Books.

Colwell, R.N., ed. 1960. *Manual of photographic interpretation. Falls Church, Va.:* American Society for Photogrammetry.

----------, ed. 1983. *Manual of remote sensing. Vol. 1. Falls Church, Va.:* American Society for Photogrammetry.

Cornwall, C. 2010. Anatomy of a hit and run. *Reader's Digest.* (Feb.): 52-59.

de Crespigny, R.R. 1980. Two maps from Mawangdui. *Cartography* 11(4): 211-22.

de Soto, H. 2000. *The mystery of capital: Why capitalism triumphs in the West and fails everywhere else.* New York: Basic Books; London: Bantam Press/Random House.

----------. 2001. The mystery of capital. *Finance and development: A quarterly magazine of the IMF* 38 (1). www.imf.org/external/pubs/ft/fandd/2001/03/desoto.htm.

Diamond, J. 1999. *Guns, germs, and steel: The fates of human societies.* New York: W.W. Norton and Company.

----------. 2005. *Collapse: How societies choose to fail or succeed.* New York: Penguin Books.

Doldirina, C. 2009. Are intellectual property laws an impediment to the development of collaborative earth observation missions? Presentation at the *International Astronautical Congress* Paper 09-B1.1.7. October 2009 Daejeon, Republic of Korea.

Dorling, D., and D. Fairbairn. 1997. *Mapping: Ways of representing the world.* London: Longman.

Eberhart-Phillips, J.E., et al. 1994. Profile of mortality from the 1989 Loma Prieta earthquake using coroner and medical examiner reports. *Disasters* 18 (2): 160-70.

Economy, R., et al. 1974. Classification accuracy of the Image 100. In *Proceedings of the Second Canadian Symposium on Remote Sensing,* 227-88. Guelph, On., April 29-May 1. Eriksson, N., and P. Adamek. 2000. The tailings pond failure at the Aznalcóllar mine, Spain. Presentation at the *Sixth International Symposium in Environmental Issues and Waste Management in Energy and Mineral Production.* Calgary, Ab., May 30-June 2. www.mineralresourcesforum.org/workshops/regulators/2000/docs/LosFrailes.pdf.

ESRI. 2007. Rerouting application helps Minneapolis cope with bridge disaster. *Arc News* 29(3): 16.

Ferguson, J. 2009. Shaping Canada's landscape: Towards a national geomatics strategy leveraging geomatics to optimize Canada's infrastructure renewal. Presentation at the *GIAC Leaders' Forum. Ottawa, On.,* June 16-17. http://74.125.93.132/search?q=cache:6AX7Reqoy4UJ:www.giac.ca/lib/db2file.asp%3Ffileid%3D200+giac+infrastructure&cd=1&hl=en&ct=clnk&gl=ca.

Foot, D. 1996. *Boom, bust and echo: How to profit from the coming demographic shift.* Toronto: Macfarlane Walter & Ross.

Franklin, S. 2001. *Remote sensing for sustainable forest management. Boca Raton:* CRC Press.

Gabrynowicz, J.I. 1999. Expanding global remote sensing services: Three fundamental considerations. In *Proceedings of the Workshop on Space Law in the Twenty-First Century,* 97-117. UNISPACE-III Technical Forum, UN Doc. ST/SPACE/2 July 1999.

Gilbert, M., et al. 2008. Mapping H5N1 highly pathogenic avian influenza risk in Southeast Asia. In *Proceedings of the National Academy of Sciences of the United States of America,* 105 (12): 4769-74.

Godby, E.A., and R.A. Ryerson. 1988. Remote sensing. In *The Canadian Encyclopaedia,* 1855. Edmonton: Hurtig Publishers.

Goodenough, D.G. 1976. IMAGE 100 classification methods for ERTS scanner data. *Canadian Journal of Remote Sensing* 2 (1): 18-29.

----------. 1979. The Image Analysis System (CIAS) at the Canada Centre for Remote Sensing. *Canadian Journal of Remote Sensing* 5: 3-17.

Gower, J., ed. 2006. *Remote sensing of the marine environment. Vol. 6 of Manual of remote sensing,* 3rd ed., ed. A. Rencz. Bethseda, Md.: American Society for Photogrammetry and Remote Sensing; New York: John Wiley and Sons.

Haggett, P. 1965. *Locational analysis in human geography.* London: Edward Arnold.

Harley, J.B., and D. Woodward. 1987. Preface. In *The history of cartography, Volume 1: Cartography in prehistoric, ancient, and medieval Europe and the Mediterranean,* ed. J.B. Harley and D. Woodward, xv-xxi. Chicago: University of Chicago Press.

Harris, R. 2003. Current policy issues in remote sensing: Report by the International Policy Advisory Committee of the ISPRS. *Space Policy* 19 (4): 293-96.

Healy, B. 2004. Recreational sports sees more use of GPS. *Gizmos for Geeks; The New York Times.* www.gizmosforgeeks.com/2004/08/22/recreational-sports-sees-more-use-of-gps.

Henderson, F.M., and A.J. Lewis. 1998. *Principles and applications of imaging radar. Vol. 2 of Manual of remote sensing,* 3rd ed., ed. R.A. Ryerson. Bethseda, Md.: American Society for Photogrammetry and Remote Sensing; New York: John Wiley and Sons.

Hodder, I. 2005. *Catalhöyük: Excavations of a neolithic Anatolian Höyük, 2005 Archive Report.* Global Heritage Fund. Palo Alto, Ca. www.globalheritagefund.org/where/catalhoyuk.html.

Jin, Y-C. 1982. *China's achievements in surveying and mapping techniques during the Han Dynasty as reflected by the silk maps unearthed from the Han Tomb at Mawangdui, Changsha.* Beijing: Research Institute of Surveying and Mapping, Beijing.

Keegan, J. 2004. *The Iraq War.* Toronto: Key Porter Books.

Koch, T. 2005. *Cartographies of disease: Maps, mapping, and medicine.* Redlands: ESRI Press.

Lambert, I. 1999. Sustaining economic benefits from mineral resources: Government investment in geoscience. *AusIMM Bulletin* No.3 (April/May): 82-87.

MacDonald, E.H. 2001. GIS in banking: Evaluation of Canadian bank mergers. *Canadian Journal of Regional Science/Revue canadienne des sciences régionales* 24 (3): 419-42. www.lib.unb.ca/Texts/CJRS/Autumn01/pdf/macdonald.pdf.

MacInnes, C.D., et al. 2001. Elimination of rabies from red foxes in eastern Ontario. *Journal of Wildlife Diseases* 37(1): 119-32.

McGrath, G., and L. Sebert, eds. 1999. *Mapping a northern land. Montreal and Kingston:* McGill-Queen's University Press.

McHarg, I. 1969. *Design with nature.* Garden City, NY: The Natural History Press.

Monmonier, M. 1996. *How to lie with maps.* 2nd ed. Chicago: University of Chicago Press.

Ohlemacher, R. 2003. The earth observation summit: Heralding a new era for international cooperation. *Space Policy* 19 (4): 277-81.

Page, R. 2003. *Hope is not a strategy: The 6 keys to winning the complex sale. Columbus:* McGraw-Hill Professional.

Palmer, C.J. 2005. Boosting survival rates. *Occupational Health and Safety* 74 (6). www.ohsonline.com/articles/44867.

Philipson, W., ed. 1997. *Manual of photographic interpretation. 2nd ed. Bethseda, Md.:* American Society for Photogrammetry and Remote Sensing.

Pickles, J. 2004. *A history of spaces: Cartographic reason, mapping and the geo-coded world.* London: Routledge.

Pomfret, K. 2010. Location: Its time has come. *Space News* (Feb. 15): 19, 21.

Rencz, A., ed. 1999. *Remote sensing for the earth sciences. Vol. 3 of Manual of remote sensing,* 3rd ed., ed. R.A. Ryerson. Bethseda, Md.: American Society for Photogrammetry and Remote Sensing; New York: John Wiley and Sons.

Richardson, D.E., and W.A. Mackaness. 1999. Computational processes for map generalization. Cartography and Geographic Information Science 26 (1): 3-5.

Richelson, J.T. 1999. U.S. satellite imagery, 1960-1999. *National Security Archive Electronic Briefing Book No. 13.* National Security Archive. The George Washington University. Washington, D.C. www.gwu.edu/~nsarchiv/NSAEBB/NSAEBB186/index.htm.

Ridd, M.K., and J.D. Hipple, eds. 2006. *Remote sensing of human settlements.* Vol. 5 of *Manual of remote sensing,* 3rd ed., ed. A. Rencz. Bethseda, Md.: American Society for Photogrammetry and Remote Sensing; New York: John Wiley and Sons.

Robinson, A., and B. Petchenik. 1976. *The nature of maps: Essays toward understanding maps and mapping.* Chicago: University of Chicago Press.

Robinson, A., et al. 1984. *Elements of cartography.* 5th ed. New York: John Wiley and Sons.

Ryerson, R.A., and R. Arnason. 1981. Visual analysis of 1:250,000 Landsat data for forage assessment during the 1980 drought in western Manitoba. *In Proceedings of the Seventh Canadian Symposium on Remote Sensing,* 83-87. Winnipeg, Mb., September 8-11. Ottawa: Canadian Aeronautics and Space Institute.

Ryerson, R.A., and R.J. Batterham. 2000. An approach to the development of a sustainable national geomatics infrastructure. Photogrammetric Engineering and Remote Sensing 66 (1): 17-28.

Ryerson, R.A., and E. Quiroga. 2000. Taking remote sensing from development projects to operational use: Some common attributes of successful projects. Keynote presentation at the *Sixth Annual GEO Asia Pacific Conference.* Bangkok, Thailand, October 2-4. Presentation at the Twenty-third Canadian Symposium on Remote Sensing. Ste. Foy, Qc., August 2001.

Ryerson, R.A., Z. Kalensky, and C. Gosselin. 2003. Geo-information for agricultural development. *Working paper (FAO) No. 14.* Environment and Natural Resources. Rome: Food and Agricultural Organization of the United Nations.

Ryerson, R.A., et al. 2008. The commercialization of remote sensing and GIS for vineyard management: A simple but powerful application of change detection. *Proceedings of American Society for Photogrammetry and Remote Sensing Annual Meeting.* Portland, Or., April 28-May 2.

Ryerson, R.A., and C. Peanvijarnpong. 2007. Government policy, applied research, commercialization: Comparative analysis of Thailand's successful national program in remote sensing. Presented as an Opening Plenary Session Paper *Asian Conference on Remote Sensing.* Kuala Lumpur, Malaysia, 2007. Also published in 2008 in *Coordinates: A monthly magazine on positioning, navigation and beyond.* 4 (6). http://mycoordinates.org/government-policy-applied-research-commercialization.

Ryerson, R.A. 2009. Economic panic is a good thing – for the geospatial industry. Presentation at Geodiffusion. Montreal, Qc., May 13-14.

Salathé, T. 2002. Ramsar advisory mission to Doñana in Spain. The Ramsar Convention on Wetlands. www.ramsar.org/cda/en/ramsar-documents-rams-ramsar-advisory-mission/main/ramsar/1-31-112%5E17488_4000_0.

Sanders, R., et al. 2005. National flood modelling for insurance purposes: Using IFSAR for flood risk mapping in Europe. Hydrology and Earth System Sciences 9 (4): 449-56.

Seaman, V. 1798. An inquiry into the cause of the prevalence of the yellow fever in New-York. *Medical Repository* 1 (3): 314 32.

Sears, G. 2001. Geospatial data policy study. Prepared for GeoConnections Policy Advisory Node. March 28, Ref. #03-3425. Ottawa, On.

Seed, P. 2008. The cone of Africa ... took shape in Lisbon. *Humanities* 29 (6). www.neh.gov/news/humanities/2008-11/ConeOfAfrica.html.

Shafee, M., R. Ryerson, K. Lim, I. Sadiq, and A. Rasheed. 2010. Designing a GIS to meet real needs in the Maldives. *Proceedings of the American Society for Photogrammetry and Remote Sensing Annual Meeting.* San Diego, Ca., April 26-30.

Shao, Y., et al. 2001. Rice monitoring and production estimation using multitemporal RADARSA. *Remote Sensing of Environment* 76 (3): 310-25.

Shwiff, S., et al. 2009. Benefit cost scenarios of potential oral rabies vaccination for skunks in California. *Journal of Wildlife Diseases* 45(1): 227-33.

Smith, D.C. 1987. Cartography in the prehistoric period in the old world: Europe, the Middle East, and North Africa. In *The history of cartography, Volume 1: Cartography in prehistoric, ancient, and medieval Europe and the Mediterranean,* ed. J.B. Harley and D. Woodward. Chicago: University of Chicago Press.

Snow, John. 1855. *On the mode of communication of cholera.* 2nd ed. London: John Churchill. www.ph.ucla.edu/epi/snow/snowbook.html.

Stoker, J., D. Harding, and J. Parrish. 2008. The need for a national LiDAR dataset. *Photogrammetric Engineering and Remote Sensing* 74 (9): 1066-68.

Szostak-Chrzanowski, A., M. Massie´ra, and A. Chrzanowski. 2007. Analysis of deformations of large earth dams. *Journal of Applied Geodesy* 1 (2): 81-89.

Tomlinson, R.F. 1987. Current and potential uses of geographical information systems: The North American experience. *International Journal of Geographical Information Systems* 1(3): 203-18.

----------. 2003. *Thinking about GIS: Geographic information system planning for managers*. Redlands: ESRI Press.

UN-ESCAP. 1994. *Proceedings: Ministerial conference on space applications for development in Asia and the Pacific, 19-24 September 1994, Beijing, China*. Information, Communication and Space Technology Division, United Nations Economic and Social Commission for Asia and the Pacific. UN-ESCAP Reference No. ST/ESCAP/1459.

UN-ESCAP. 1995. *Ministerial conference on space applications for development in Asia and the Pacific, 19-24 September 1994, Beijing: Beijing Declaration on space technology applications for environmentally sound and sustainable development in Asia and the Pacific*. Information, Communication and Space Technology Division, United Nations Economic and Social Commission for Asia and the Pacific. UN-ESCAP Reference No. ST/ESCAP/1451.

UN-ESCAP. 2000a. *Second Ministerial conference on space applications for sustainable development in Asia and the Pacific,15-20 November 1999, New Delhi: Proceedings*. Information, Communication and Space Technology Division, United Nations Economic and Social Commission for Asia and the Pacific. UN-ESCAP Reference No. ST/ ESCAP/2062.

UN-ESCAP. 2000b. *Delhi Declaration and strategy and action plan on space technology applications in Asia and the Pacific for the new millennium*. Information, Communication and Space Technology Division, United Nations Economic and Social Commission for Asia and the Pacific. UN-ESCAP Reference No. ST/ESCAP/2059.

Ustin, S.L. 2004. *Remote sensing for natural resource management and environmental monitoring*. Vol.4 of *Manual of remote sensing*, 3rd ed., ed. A. Rencz. Bethseda, Md.: American Society for Photogrammetry and Remote Sensing; New York: John Wiley and Sons.

Wallen, V.R., and H.R. Jackson. 1971. Aerial photography as a survey technique for the assessment of bacterial blight of field beans. *Canadian Plant Disease Survey* 51 (4): 163-69.

Weaver, K.F. 1969. New eyes to see the world. *National Geographic* 135 (1): 46-74.

Wilford, J.N. 1981. *The mapmakers*. New York: Alfred A. Knopf.

Withee, G.W., et al. 2004. Progress in multilateral earth observation cooperation: CEOS, IGOS and the ad hoc Group on Earth Observations. *Space Policy* 20 (1): 37-43.

World Bank 2009. ***World Development Report 2009: Reshaping Economic Geography.*** Washington, D.C.

NOTES

Chapter 1

1. A paradigm is defined in the *Canadian Oxford Dictionary* (1998) as a "mode of viewing the world which underlies the theories and methodology of science etc. in a particular period of history." The GeoEconomy clearly qualifies as a new economic paradigm.

2. The phrase "hope is not a strategy" was popularized by Page, R. 2003.

3. Diamond, J. 1999; Diamond, J. 2005; World Bank 2009.

Chapter 2

4. Diamond, J. 1999.

5. Robinson, A., and B. Petchenik 1976.

6. Robinson, A., et al. 1984.

7. Pickles, J. 2004.

8. A more formal definition of a map provided in Harley, J.B., and D. Woodward, eds. 1987 is "Maps are graphic representations that facilitate a spatial understanding of things, concepts, conditions, processes, or events in the human world."

9. A picture map, common in prehistoric art, is characterized as having some images in plan (as viewed from above) and some in elevation or profile (as viewed horizontally). Smith, D.C. 1987.

10. Smith, D.C. 1987.

11. The Catalhöyük Global Heritage Fund Site. Global Heritage Fund. Palo Alto, California. www.globalheritagefund.org/sites/emea/europe/catalhoyuk_scroller.html.

12. Hodder, I. 2005.

13. Wilford, J.N. 1981.

14. Jin, Y-C. 1982 and de Crespigny,R.R. 1980.

15. Wilford, J.N. 1981.

16. Dorling, D., and D. Fairbairn 1997; Wilford, J.N. 1981.

17. Black, J. 1997.

18. Ibid., 19.

19. Ibid., 43.

20. Paraphrase of comments made by A. Brandenberger, consultant to the United Nations and professor, Laval University. See archives of the International Society for Photogrammetry, Lausanne, 1968; and Photogrammetric Engineering. April 1969.

21. Black, J. 1997, 137.

22. Europe came into its own as the ecologically fragile environment of the Fertile Crescent and Eastern Mediterranean was undermined by overuse. Diamond, J. 1999, 411.

23. Black, J. 1997, 27.

24. Diamond, J. 1999, 412.

25. Dr. He eventually joined the United Nations where he has headed a number of geospatially oriented agencies in the UN Headquarters in Bangkok and at the FAO in Rome. It was Dr. He who, as the head of the FAO in Asia, came up with the idea of tracking avian influenza using GIS. See Gilbert, M., et al. 2008.

26. Koch, T. 2005.

27. Monmonier, M. 1996.

28. For a thorough but introductory treatment see Aronoff, S. 2005, *Chapter 2 A brief history of remote sensing for earth observation.*

29. Evelyn Pruitt is credited with coining the term "remote sensing" in 1958. A research geographer with the United States Navy, active from the 1940s to the 1970s, her work advanced the study of coastal environments and the use of remote sensing in geography. Sources: Canada Centre for Remote Sensing Glossary of Remote Sensing Terms, Natural Resources Canada. www.ccrs.nrcan.gc.ca/glossary/index_e.php?id=486; N.M. Short, Sr., Remote Sensing Tutorial, NASA. http://rst.gsfc.nasa.gov/Intro/Part2_1.html; The Society of Woman Geographers Evelyn L. Pruitt National Fellowship for Dissertation Research Application Guidelines. Department of Geography and Regional Planning, Indiana University of Pennsylvania. www.iswg.org/SWGPruittDissertationApplGuidelines08-09.10-12-07W.doc.

30. Relative value of a US dollar as calculated using the Consumer Price Index by the Measuring Worth calculator based on S.H. Williamson (2008), Five Ways to Compute the Relative Value of a US Dollar Amount, 1790 to Present. www.measuringworth.com/uscompare.

31. Accuracy of the elevation of hills and the like on maps made with aerial photography was regarded as acceptable. Newer technologies such as LiDAR discussed in Section 3.2.3.3 have re-defined the meaning of accuracy, however.

32. Corona Program. NASA Information Document. Mission and Spacecraft Library. http://msl.jpl.nasa.gov/Programs/corona.html. Accessed March 2008.

33. See various publications by the Federation of American Scientists, for example. An interesting activity is to track what was once secret (for example, the Corona series of satellites) and then compare the conjecture made at that time as to what the capabilities were to what the capabilities have since been reported to be.

34. Aronoff, S. 2005.

35. McHarg, I. 1969.

Chapter 3

36. The key figures involved in the Workshop on Developing Indigenous Earth Observation Industrial Capabilities in Developing Countries (full reference is given in the next note) included Dr. Changchui He, now the Deputy Director General, Operations, of the FAO; Dr. Peter Adeniyi now Vice Chancellor, Federal University of Technology, Akure Nigeria; Dr. Sridhara Murthi, then Scientific Secretary of ISRO, India; Dr. Ade Abiodun, then employed by the United Nations Office of Space Applications; Mr. Mac Evans, who was President, Canadian Space Agency, Canada; Dr. Mahdi Kartasasmita, Deputy Chairman of LAPAN (Indonesian Space Agency), Jakarta, Indonesia. Ryerson organized the meeting and was Rapporteur. In addition to the speakers, there were senior people from a number of countries, including China, Mongolia, Iran, India, Indonesia, several African countries, and one from Latin America. In addition there were government people from Great Britain, France, and Germany. The biggest commercial player in Austria attended, as did someone from Boeing (USA), and a consultant from Spain. The Boeing representative later noted that the industry people who did not attend "missed out on learning some very useful information" from the meeting.

37. UNISPACE 1999. The Third United Nations Conference on the Exploration and Peaceful Uses of Outer Space, Vienna, July 19-30. Committee II, Agenda item 11. Economic and Societal Benefits, July 27. A/CONF.184/C.2/L.9.

38. The White House, Office of the Press Secretary. Statement by the President Regarding the United States' Decision to Stop Degrading Global Positioning System Accuracy, May 1, 2000. www.ngs.noaa.gov/FGCS/info/sans_SA/docs/statement.html.

39. www.military-geospatial-technology.com/article.cfm?DocID=1379; Bourge, C. 2006.

40. See for example: Keegan, J. 2004, 79; Clancy, T. with C. Horner 1999. The latter has four references to Global Positioning Satellites in the index that describe its use to direct cruise missiles, ground attacks, and the advantages the systems give in an attack.

41. See www.avis.co.za/main.asp?ID=1194 for rental information in southern Africa.

42. www.gpsnorthamerica.com/anti_theft.asp.

43. www.guardianmobility.com.

44. U.S. House of Representatives, 1997.

45. Ryerson, R.A., et al. 2008.

46. de Soto, H. 2001; de Soto, H. 2000.

47. See www.trimble.com/index.aspx. Trimble also has a detailed and useful summary of the applications of their technology on their web site at www.trimble.com/about_industry.shtml.

48. Healy, B. 2004.

49. Alan A. Varghese, senior director of semiconductor research at ABI Research in Oyster Bay, NY, quoted by Healy in Healy, B. 2004.

50. GPS Market Update, 2006. RNCOS E-Services Pvt. Ltd, New Delhi, India. The fact that this is but one of many market studies on the subject that range in cost from $1000 (in this case) to over $5000 tells us something about the market size. See www.the-infoshop.com/study/rnc39537-gps-market.html and www.electronics.ca/reports/gps/ for examples of the many studies.

51. Quoted in Healy, B. 2004.

52. See, for example, the book whose text was prepared by Brian Banks (1989) and the imagery selected and described by Jean Game of the Canada Centre for Remote Sensing.

53. One anecdote whose veracity we cannot verify is that when the Soviet Union started selling data from its satellites to try to obtain foreign currency, all data was found to be over US cities or areas of military significance – except for one area in an isolated part of the western US. The area was not known as being used for anything. According to what might well be another urban legend, when checked out the area turned out to be a secret military base.

54. Weaver, K.F. 1969.

55. Colwell, R.N., ed. 1983.

56. Godby, E.A., and R.A. Ryerson 1988.

57. Aronoff, S. 2005, 2.

58. Ibid., 70.

59. Ibid.

60. A number of books have been written on these various topics and many of them are listed in the references section.

61. See Google Earth at www.earth.google.com; Microsoft's Virtual Earth at www.microsoft.com/virtualearth; and NASA's Visible Earth at: http://visibleearth.nasa.gov.

62. For a comprehensive summary of the history of the use of photography see Aronoff, S. 2005 and Philipson, W. 1997.

63. Colwell, R.N., ed. 1960.

64. The Image 100 System, purchased by a team headed by Dr. David Goodenough, now resides in Canada's National Museum of Science and Technology. Research on that system led to the creation of a vibrant industry that still includes PCI Geomatics. See Goodenough, D.G. 1976; Goodenough, D.G. 1979; and Economy, R., et al. 1974.

65. See Philipson, W. 1997 for a comprehensive treatment of the use of aerial photography.

66. See Section 2.6 in Ridd, M.K., and J.D. Hipple, eds. 2006.

67. In 2004 the Supreme Court of Canada overturned the decision of the Court of Appeal of Ontario which in effect said that the state did not have the right to use forward-looking airborne infrared technology to look for marijuana grow operations for to do so was an invasion of privacy. Her Majesty the Queen v. Walter Tessling, Supreme Court of Canada Docket 29670: http://csc.lexum.umontreal.ca/en/2004/2004scc67/2004scc67.html.

68. Ryerson, R.A., et al. 2008.

69. Army Space Journal 2008; Richelson, J.T. 1999.

70. Forecast International, 2006. The Market for Civil & Commercial Remote Sensing Satellites. www.spacemart.com/reports/Commercial_Remote_Sensing_Satellite_Market_Stabilizing.html.

71. These costs are guidelines provided at the time of writing by one major commercial supplier. Costs to government agencies and educational institutions are less – as little as $1000.

72. See www.itres.com/Home.

73. A simple discussion of the principles is given in Ridd, M.K., and J.D. Hipple, eds. 2006. A more comprehensive and complete background to radar can be found in the 866-page manual by Henderson, F.M., and A.J. Lewis 1998.

74. See www.intermap.com.

75. The Federation of American Scientists maintains interesting historical background and technical details on many satellites – both military and civilian – world-wide. See their site: www.fas.org. More specifically on ALMAZ see: www.fas.org/spp/guide/russia/earth/almaz.htm.

76. See the JPL web page: www2.jpl.nasa.gov/srtm.

77. www.esa.int/esaEO/SEMWYN2VQUD_index_0_m.html.

78. www.dlr.de/tsx/main/mission_en.htm for more details.

79. See article in *The Hindu*, April 23, 2009. www.thehindu.com/2009/04/23/stories/2009042352431000.htm.

80. See, for example, the lists at both the TeraSAR-X and RADARSAT-2 web sites: www.dlr.de/tsx/main/applications_en.htm and www.radarsat2.info.

81. www.dlr.de/tsx/main/applications_en.htm.

82. See www.optech.ca.

83. With total area coverage, as compared to ship-borne water depth measurement systems that tend to provide a sample, LiDAR systems tend to identify isolated underwater rocks and shoals that can be missed with systems that sample.

84. Quoting Stewart Walker's response in SWOT Analysis: State of the Remote Sensing Industry, *Earth Imaging Journal* (Jan-Feb 2008).

85. Optech announced in May 2007 that it had sold 100 of its ALTM systems. www.optech.ca/press_100-ALTM.htm. In addition, Optech sells a system called SHOALS that can accurately map water depth.

86. See: Rencz, A., ed. 1999; Ustin, S.L. 2004; Ridd, M.K., and J.D. Hipple, eds. 2006; Gower, J., ed. 2006.

87. Ridd, M.K. and J.D. Hipple, eds. 2006.

88. An interesting paper on the use of commercial satellite imagery to monitor North Korea can be found at http://docs.nrdc.org/nuclear/files/nuc_04100401a_238.pdf.

89. See, for example, the United States Geological Survey web site for satellite image data for much of the world. http://edcsns17.cr.usgs.gov/EarthExplorer.

90. Ryerson, R.A., et al. 2008.

91. We are not lawyers and we are not offering legal advice here or anywhere else in this book.

92. If a company does not exercise "due diligence" (and cannot prove that they are exercising "due diligence") in some jurisdictions in protecting the environment, its senior management and directors face severe fines and, in some cases, jail.

93. Certified Mapping Scientist is a professional designation approved under an Act of Congress of the United States of America administered by the American Society for Photogrammetry and Remote Sensing. Co-author R.A. Ryerson is a Certified Mapping Scientist (Remote Sensing), holder of Certificate #R115RS.

94. See Holz et al. (1997), "Structures and Cultural Features," in Philipson, W. 1997.

95. A Canadian company based in the Ottawa region, Imstrat Corporation has developed a business providing consulting services and training in how to use imagery for intelligence applications. www.imstrat.on.ca.

96. See Section 6.0 of the Geomatics Accord at the Government of Canada's InterAgency Committee on Geomatics web site: www.iacg-cmoig.org/home/tor_e.php.

97. Pomfret, K. 2010.

98. UN-ESCAP 1994.

99. See www.mapquest.com.

100. See http://ww1.aaa.com/Scripts/WebObjects.dll/AAAOnline?association=caa&club=290&page=ITTServices.

101. Nokia bought Navteq for $8 billion in late 2007. See www.navteq.com/webapps/NewsUserServlet?action=NewsDetail&newsId=592&lang=en&englishonly=false.

102. At one time Pizza Pizza was said to have the best street map of Bangkok – to allow fast delivery by their pizza delivery motorcycles.

103. See Stamp (1940) and the more recent news report on the mapping that Stamp was reporting on: The Maps That Saved Britain from Starvation Win an Online Refuge www.port.ac.uk/research/gbhgis/mediaresources/freearticles/filetodownload,23045,en.pdf.

104. At http://erg.usgs.gov/isb/pubs/factsheets/fs07499.html it was announced that "using print-on-demand technology, the USGS is implementing map-on-demand (MOD) printing for certain infrequently requested maps. By providing MOD, the USGS can offer an alternative to traditional, large-volume printing and can improve its responsiveness to customers by giving them greater access to USGS scientific data in a format that otherwise might not be available. The first use of MOD technology in the USGS will be confined primarily to selected thematic maps that are of scientific value but rarely requested. It is anticipated that additional USGS mapping titles will be available through print-on-demand means in the future. The paper and inks used in the MOD process are not as durable as those used in traditional USGS lithography. Although the inks are ultraviolet resistant, they are not water stable. These products must be protected from high humidity, wet conditions, and prolonged and consistent exposure to the sun and other radiation sources. The MOD maps can be written on with either ink or pencil with no smudging. The maps can be heat laminated with no bleeding of the colors. Maps selected for sale using MOD printing will only be available in this format. Each map produced in this way will be evaluated periodically to determine if the demand warrants a lithographic printing. If lithographic prints are made

available, the price will be adjusted to the standard price for printed maps. Because the on-demand product is printed at the request of the customer, it cannot be returned or exchanged. The USGS is required by law to recover all costs associated with printing and distribution. The price of MOD products has been established to recover the cost of labor and material required to deliver the product to the customer. Because on-demand printing produces a map when a request for one is made, the full cost to prepare this map must be recovered at the time of sale. On-demand, unlike lithographic printing, does not have the economic value of volume printing (the more you print, the cheaper it is). There are no discounts on MOD products. MOD prices will be re-evaluated yearly to comply with the mandate to recover costs for printing and distribution."

105. Jin, Y-C. 1982.

106. See www.ordnancesurvey.co.uk/oswebsite for the Ordnance Survey web site.

107. Some might also argue that it is good to see mapping in the military where budgets are sometimes less well scrutinized than they are for their civilian cousins, while others would argue that the secrecy and costs can negatively influence the usefulness of the data.

108. This concept of levels of abstraction was advanced in Haggett, P. 1965.

109. In one case we have heard about, the senior managers were paid on the basis of their production and printing of maps. Since maps of remote areas have less information on them, they are easier to produce and less expensive to print. The result is predictable. The mapping agency has large piles of maps of remote areas, and fewer maps of the areas in which they are needed, and storage costs are high. This of course relates back to a fundamental rule of management: reward only that behaviour you wish to encourage. Perhaps the last telling statement in this sad story is that in the end the maps were cut up and bound together as note pads – to save the government the cost of buying note-paper for its employees.

110. See details at www.iacg-cmoig.org/home/tor_e.php.

111. Ryerson, R.A., and C. Peanvijarnpong 2008; Atwood, D., D. Mayall, and R.A. Ryerson 2009.

112. See, for example, www.worldsat.ca. This company prepares many of the satellite image maps used in *National Geographic*.

113. See http://maps.nrcan.gc.ca/topo101/faq_e.php#1.

114. Much has been written about this subject. See Aronoff, S. 2005, 37-38, for a brief synopsis, or almost any basic text book on photogrammetry or remote sensing.

115. In a sense the difference between the map a cartographer produces and the representation to which we refer is really a matter of degree. In both cases rules are used to decide how the world as it is perceived will be represented. The cartographer's rules may in some respect be less flexible – but there are still rules about how to represent geospatial information to suit certain uses. To be even more philosophical, the means we use to perceive the world affect what we know of it. Imagine if you were unlucky enough to be born with no color vision and saw only in black and white. For you, colour as others know it would not exist. How would someone even explain what a colour looked like? So, if you think of reality as what exists independent of its perception, then any information we have about reality is a representation of the real world. There is more to it than can be perceived. So, in the end what we consider "reality" has more to do with whether the information we have is suitable for our

intended use. That is, if we apply what we know about something will it give us the results we would like? If it does then we consider that what we know is reality and if it doesn't then we consider what we know to not be reality.

116. See http://nationalmap.usgs.gov.

117. See the text box in Section 3.2.4.3.

118. See www.cubewerx.com.

119. Ibid.

120. Doldirina, C. 2009.

121. The complexity of the map generalization process has been the topic of many papers and not a few Ph.D. theses. See for example: Richardson, D.E., and W.A. Mackaness 1999.

122. See www.iso.org/iso/iso_catalogue/catalogue_tc/catalogue_tc_browse.htm?commid=54904.

123. The Open Geospatial Consortium, Inc. (OGC) is a non-profit, international, voluntary consensus standards organization that is leading the development of standards for geospatial and location-based services. We note the word "voluntary": not all data are OGC-compliant. See www.opengeospatial.org.

124. For a more complete discussion of geospatial data and the challenges in developing countries see Ryerson, R.A., and R.J. Batterham 2000.

125. One might argue that military considerations may require that the ability to print maps should be maintained. If three thousand troops are to be mobilized, you may want to have sufficient maps for each soldier to have maps of the area of operations. Printing such maps on-demand would be problematic. While hardened laptops can carry map information into battle, we expect that hard copy maps will be used for the foreseeable future by the individual soldier – although it may well be that there will be more use of image maps.

126. BusinessWeek Research Services 2006. Location Intelligence: The New Geography of Business. http://mapinfo.landinghost.com/BusinessWeek.WP.deeperbluewebready6.28.06.pdf.

127. The concept of data being well-maintained is not one that is easily understood. Implied is that the data are current, accurate, and that metadata (that describe the data's characteristics) are available.

128. Open Geospatial Consortium 2005. The importance of going open. OGC White Paper. www.opengeospatial.org/pressroom/papers.

129. Mass market geomatics providers are those that provide geospatial information to the general public at no or low cost. Google Earth, Microsoft Virtual Earth, and MapQuest are examples.

130. This famous legend has it that Wal-Mart discovered that beer and diapers were often sold to the same customers and that when they were co-located the sales of both increased. The truth is somewhat less exciting. "In 1992, Thomas Blischok, manager of a retail consulting group at Teradata, and his staff prepared an analysis of 1.2 million market baskets from about 25 Osco Drug stores. Database queries were developed to identify affinities. The analysis "did discover that between 5:00 and 7:00 pm consumers bought beer and diapers. Osco managers

did NOT exploit the beer and diapers relationship by moving the products closer together on the shelves. This decision support study was conducted using query tools to find an association. The true story is very bland compared to the legend." Power, D.J., ed. 2002, *DSS News* 3 (23): Nov 10. www.dssresources.com/newsletters/66.php.

131. From CGDI Framework Data Definition. www.geoconnections.org/opportunities/rfp/framework/rfp_sd_definition_E.pdf.

132. See for example: Atwood, D., D. Mayall, and R.A. Ryerson 2009; Shafee, M., R. Ryerson, K. Lim, I. Sadiq, and A. Rasheed 2010.

133. Postal codes have very different meanings. In some jurisdictions they simply refer to a city or part of a city. In Canada a unique postal code is assigned to almost every block face and apartment building or set of floors in a building. This is the sort of precision required in business geographics. (Indeed, one single family dwelling we are aware of in the southern end of Ottawa has a unique postal code since it is the only house on that block face.)

134. A useful reference is an interview with Michael Goodchild. See: Michael F. Goodchild Talks about the Role of Volunteered Geographic Information in a Postmodern GIS World. www.esri.com/news/arcwatch/1208/goodchild-talks.html.

135. See www.mapinfo.com.

136. See www.goldcorpchallenge.com/challenge1/winnerslist/chaleng2.pdf and www.goldcorp.com.

137. We offer a workshop on data policy that delves into the subject in some detail. Much of what follows is from that workshop and our reports and published papers. We (and the reader wherever they may be) owe a debt of gratitude to two Canadian public servants in particular, Jeff Labonte and Jean-Marc Chouinard, whose work on policy has focused on the greater good of the public they serve – they "get it."

138. Sears, G. 2001. www.geoconnections.org/publications/policyDocs/keyDocs/KPMG/KPMG_E.pdf. (R.A. Ryerson put together the study team to do this work and provided the geospatial expertise.)

139. The District Magistrate (or Deputy Commissioner or District Collector) is the public servant in charge of administration and revenue collection for a district which is an administrative division of an Indian state or territory. There are 610 districts in India.

140. The web page of The National Air Photo Library (NAPL) of Natural Resources Canada states that it archives over six million aerial photographs covering all of Canada, some of which date back to the 1920s. NAPL indexes and stores federal aerial photography for Canada, and maintains a comprehensive historical archive and public reference centre. NAPL On-Line allows clients to search and retrieve metadata for over three million air photos using several criteria, including official place name, geographical coordinate, National Topographic System map number, or Roll and Photo number. http://airphotos.nrcan.gc.ca/index_e.php.

141. See Ryerson, R.A., and C. Peanvijarnpong 2008.

142. Sears, G. 2001.

143. These are Natural Resources Canada, Environment Canada and the Canadian Space Agency.

144. See details at www.iacg-cmoig.org/home/tor_e.php.

145. See the discussion in Gabrynowicz, J.I. 1999.

146. See Harris, R. 2003.

147. See the report (discussed further in one of the following notes) by Baker, J.C., et al. 2004.

148. See www.unece.org/env/pp.

149. www.ceos.org.

150. Withee, G.W., et al. 2004.

151. See the various UN reports: UN-ESCAP (2000a); UN-ESCAP (2000b); and UN-ESCAP (1995).

152. See www.earthobservationsummit.gov/terms_of_reference.html and Ohlemacher, R. 2003.

153. See www.mapps.org.

154. www.giac.ca/home.asp?lang=en.

155. See article in *The Hindu*, April 23, 2009. www.thehindu.com/2009/04/23/stories/2009042352431000.htm.

156. Ryerson, R.A., and C. Peanvijarnpong 2007. This is based on work by Kim Geomatics Corporation.

157. See the information on the Market Development Fund at the GEOIDE Network of Excellence in Geomatics. www.geoide.ulaval.ca/precom_EN.html.

158. Scc www.geoide.ulaval.ca/home.aspx.

159. The mining company was subsequently bought by another company and we are happy to report that the new owners seem to have developed a much more pro-active viewpoint of monitoring things such as mine waste.

160. Ryerson, R.A., and E. Quiroga 2000.

161. A more complete analysis is presented in a paper by Lambert, I. 1999. This document was the basis for Geoscience Australia's submission to the Australian Parliament's House of Representatives Standing Committee of Inquiry into Resources Exploration Impediments. The Committee tabled its report, *Exploring Australia's Future*, on September 15, 2003 in the House of Representatives. The Committee's web site includes links to the report and submissions; Chapter 4 is of special interest. www.aph.gov.au/house/committee/isr/resexp/contents.htm. For the submission of Geoscience Australia see: www.aph.gov.au/house/committee/isr/resexp/subs/sub53.pdf. (See Part 5.2, 18.)

162. See ACIL Tasman 2008.

163. See the web page of the Charter at www.disasterscharter.org. Member organizations at the time of writing provide access to over twenty satellites. Those involved include the following space agencies: European Space Agency (ESA); Centre national d'études spatiales (CNES); National Space Organization, Republic of China (NSPO); Canadian Space Agency (CSA); Indian Space Research Organisation (ISRO); National Oceanic and Atmospheric Administration (NOAA); Argentina's Comisión Nacional de Actividades Espaciales (CONAE); Japan Aerospace Exploration Agency (JAXA); United States Geological Survey

(USGS); DMC International Imaging (DMC); Centre National des Techniques Spatiales (Algeria); National Space Research and Development (Nigeria); Tübitak-BILTEN (Turkey); BNSC/Surrey Satellite Technology Limited (UK); China National Space Administration (CNSA).

164. See, for example, Stoker, J., D. Harding, and J. Parrish 2008.

165. Statement made by Mr. Xu DeMing, Vice Minister of Land and Resources and Director General of the State Bureau of Surveying and Mapping in China during the opening ceremony of the 30th Asian Conference on Remote Sensing in Beijing, China, October 19, 2009.

166. There have been examples of aid going to build inappropriate factories in inappropriate places. For example a rice processing factory was built where there was no water to grow rice, and a loan has been approved for a paper manufacturing plant when there were insufficient trees to sustainably supply the factory. More thorough analyses using satellite data (and a large dollop of common sense) could have prevented these terrible investments.

Chapter 4

167. See *Wired Magazine*, June 26, 2007.

168. As of February 10, 2009.

169. See http://europa.eu/rapid/pressReleasesAction.do?reference=IP/08/742.

170. See www.reuters.com/article/idUSWEA868520090224.

171. http://ess.nrcan.gc.ca/intl/intl_activities/brochure/pdf/30093_gis_english.pdf.359 Notes.

172. (2004). The issue of national security of geospatial information has been the topic of considerable study since the terrorist attacks in the USA of September 11, and the growth of other terrorist activities around the world. At the beginning of the study of geospatial data policy it was assumed that public geospatial information available on the Internet would be a useful tool for terrorists and enemies of the state. As has been discussed in some detail in Baker et al 2004, facts have not proven this to be the case. Of 629 US federally funded geospatial data bases evaluated, fewer than 6% were deemed to be of any use to an attacker, and less than 1% was deemed to contain information that was both critical and unique. In other words, most information available was either available from other sources (be they tourist-books, city guides, or other sources), or were not critical in planning or carrying out an attack. (Those data bases that were judged to be both critical and unique have since had their access modified.) Indeed, studies have suggested that limiting the accessibility of these more than 600 data bases may impede emergency first responders, not the enemies of the state. Our company is currently engaged in developing approaches to judge security concerns and risk related to geospatial data.

173. Ibid.

174. One of the concerns of the military is that they do not want so-called "secret" imagery falling into the hands of the enemy. For this reason some of the world's advanced military put civilian imagery whose characteristics are widely known into the hands of soldiers on the battlefield. If it falls into the hands of the enemy, no secrets are at stake. The amount of this imagery used by the military can be staggering – and provides a solid "base client" for a "civilian" imagery supplier.

175. See www.epa.gov/oecaagct/ttan.html.

176. New Scientist. www.newscientist.com/article/dn9931-facts-and-figures-asian-tsunami-disaster-.html.

177. ACIL Tasman 2008.

178. Such debates have been around geospatial for some years. At one time there were debates about whether it was cost-effective to have a GIS for urban planning. Over time the debate shifted to how best to implement a GIS.

179. For a non-technical summary of these two projects, see *Maclean's*, January 18, 2010, 44-48.

180. Biasi, F. 2007.

181. www.amnesty.ca/resource_centre/news/view.php?load=arcview&article=3495&c=Resource+Centre+News.

182. Snow, J. 1855. www.ph.ucla.edu/epi/snow.html. This site contains examples of the maps used.

183. Palmer, C.J. 2005.

184. ESRI 2007.

185. Crop Condition Assessment Program. In *The Daily*, 10 June 2008. Statistics Canada. www.statcan.ca/Daily/English/080610/d080610d.htm.

186. Aronoff, S. 2005, 339. See also the GIEWS web site at www.fao.org/giews/ENGLISH/index.htm.

187. See www26.statcan.ca/ccap-peec/start-debut-eng.jsp and Aronoff, S. 2005, 341.

188. Several years ago a potential client was horrified that we might be able to advise him of a pollution problem on his property. He didn't WANT to know. If he did, then he would have to initiate what might be a multi-million dollar clean-up! (He has subsequently left that company and the company has become a leader in environmental monitoring of their mine and processing sites.

189. Federal Geographic Data Committee. www.fgdc.gov/policyandplanning/executive_order.

Chapter 5

190. How green is their growth, *The Economist*, January 26, 2008, 58.

191. Ibid.

192. A ravenous dragon, *The Economist*, March 15, 2008, 3.

193. Waking up and catching up, *The Economist*, January 27, 2007, 24.

194. See Ryerson, R.A., Z. Kalensky, and C. Gosselin 2003.

195. We refer here to Dr. Changchui He, formerly Deputy Director General and Regional Representative for Asia and more recently, Deputy Director General Operations. Dr. He has a Ph.D. in remote sensing and is widely respected in the international community. It is also interesting to note how many people with geospatial backgrounds go on to be deans, vice presidents and presidents at universities.

196. How green is their growth, *The Economist*, January 26, 2008, 58.

197. A large black cloud, *The Economist*, March 15, 2008, 18.

198. The parable of the sower, *The Economist*, November 21, 2009, 71-73.

199. The new colonists, *The Economist*, March 15, 2008, 13.

200. Iron rations, *The Economist*, March 15, 2008, 4.

201. Niagara Falls Geology Facts & Figures. Niagara Parks Commission, Ontario Ministry of Tourism and Culture. www.niagaraparks.com/media/geology-facts-figures.html; Buttle, J. Muir, T. and Frain, J. 2004 Economic impacts of climate change on the Canadian Great Lakes hydro-electric power producers: a supply analysis; Canadian Water Resources Journal, v. 29, no. 2, p. 89 –110. Quoted in the report From Impacts to Adaptation: Canada in a Changing Climate 2007. Chapter 6 – Ontario Photograph by S. Aronoff.

202. For example satellite data were used to monitor old growth forests where the spotted owl was under threat from forest operations. After the initial maps were produced, there was considerable discussion on the approach and accuracy. In 2001 Dr. Steven Franklin, now President of Trent University in Ontario, wrote a book published by CRC Press titled *Remote sensing for sustainable forest management*. In this book he outlines approaches that have proven effective in mapping, including related to the spotted owl.

203. Five of these were from the USA Landsat series while the sixth was the French SPOT satellite.

204. See www.spacemart.com/reports/Commercial_Remote_Sensing_Satellite_Market_Stabilizing.html.

205. MacDonald Dettwiler (MDA) built, owns, and operates the Radarsat-2 satellite with major funding by the Canadian government, as well as long-term image purchase agreements. Radarsat-2 provides the highest resolution radar imagery available for civilian use. Unlike most other earth resource satellites, the orbit of Radarsat-2 was specifically designed to provide complete daily coverage of Canada's Arctic region – more northerly areas can be imaged twice or more each day. Since the radar can penetrate ("see through") clouds and has its own illumination, it is not subject to the vagaries of either climate or months of darkness. The radar imagery is used for a plethora of applications worldwide providing information used in oil and gas exploration, offshore drilling, fishing, environmental monitoring, disaster management, and of course for mapping. Two applications, of critical importance to Canada, involve monitoring Canada's northern territory. The Canadian Ice Service, a federal government service, depends on Radarsat data to produce charts and bulletins used by commercial ships to navigate ice-encumbered Canadian waters. The other is surveillance of Canada's vast Arctic and offshore territory – an application critical to asserting Canada's sovereignty of the region which has received more publicity as that sovereignty over some regions has been challenged. A storm of controversy ensued with the proposed sale of the system to a US corporation. This is because its operation might have been subject to laws that give the US government the right to preferential access and tasking of any satellite operated by a US-licensed commercial provider. This would mean that the US government would have the power to decide not to image an area or delay access to imagery if it was in their national interest – an interest that might not coincide with Canada's. The fact that the US law exists is an indication of the importance the US places on control of this kind of geospatial data. The fact that Canada did not allow the sale suggests the government of Canada was also aware of

this importance. In the GeoEconomy, not knowing the value of geospatial information will be a costly disadvantage.

206. How green is their growth, *The Economist*, January 26, 2008, 58.

207. Rushing on by road, rail, and air, *The Economist*, February 16, 2008, 30.

208. American Society of Civil Engineers, 2005 Report Card for America's Infrastructure. www.asce.org/reportcard/2005/index.cfm. Updated 2008.

Chapter 6

209. See Atwood, D., D. Mayall, and R.A. Ryerson 2009.

210. In some cases one might find that it is the good schools that close – as schools come closer and closer to the lower population limit to be sustainable classes become smaller and such schools tend to be older – with more experienced teachers – both factors would seem to tend to contribute to better education.

211. See Google Maps Street View or other providers. http://maps.google.com/maps?f=q&hl=en&ie=UTF8&ll=37.09024,-95.712891&spn=47.167389,92.8125&z=4&om=1&layer=c&utm_campaign=en&utm_source=en-ha-na-us-google-svn&utm_medium=ha.

212. Ryerson had a neighbor (named Peter) who sold his house in a quiet residential neighborhood and built another house right beside where a fire station was planned. He couldn't have been happier. He was (and still is) a volunteer firefighter. But not everyone wants to hear sirens at all times of the day and night!

213. See www.sinkholelawyer.com. The legal firm has a toll free number 1-877-SINK-LAW.

214. See http://en.wikipedia.org/wiki/Erin_Brockovich or www.imdb.com/title/tt0195685.

215. For example, the Chicago Police Force maintains a web site that shows crime incidents occurring in the city. Users can construct crime maps of their choice, selecting the city area of interest, the time range, and the types of crimes (see http://gis.chicagopolice.org/CLEARMap_crime_sums/startPage.htm). The City of Winnipeg in Manitoba, Canada provides a similar service directly off its main web page – click on the Crime Stat button at www.winnipeg.ca/interhom.

216. See, for example, www.retireelliotlake.com.

217. It is interesting to note that the wealthy people who live on the waterfront in Phuket tend to live in houses that are quite high off the beach. Lower level homes on or near the beach tend to be less expensive. One must ask if this location decision is based on some remnant of historical geospatial knowledge that resulted from an earlier tsunami.

218. See Ferguson, J. 2009.

219. See: www.ottawacitizen.com/news/Gadget+could+save+city+salt+costs/2260381/story.html; www.metronews.ca/ottawa/local/article/378241--new-technology-will-help-save-salt-resources-and-the-environment-this-winter; www.webtechwireless.com/data/news/webtech_wireless_interfleet_salt_spreader_solution_will_help_city_of_ottawa_monitor_road_salt_use_to_16_595.php.

220. See the original work by Foot, D. 1996 and several sequels.

221. See Cornwall, C. 2010.

222. Andrew Seymour and Meghan Hurley, Cell phone led police to suspects: GPS device tracked to trio sought in shooting death, *Ottawa Citizen*, February 24, 2010, D1.

223. See http://money.cnn.com/2008/03/05/news/economy/AAA_study.

224. See www.who.int/world-health-day/previous/2004/en/index.html.

225. See for example https://portal.chicagopolice.org/portal/page/portal/ClearPath which describes the community policing initiative and http://gis.chicagopolicc.org which shows how one can use the police GIS to determine information about crime and problems by area of the city. The first topic that comes up is the location of sex offenders.

226. See http://winnipeg.ca/CrimeStat.

227. This alert was posted at http://winnipeg.ca/crimestat/alerts/20080620LinkedSexualAssaults.stm. Accessed December 3, 2009.

228. See various speeches listed at www.priv.gc.ca/speech/2009/sp-d_09_e.cfm.

229. Dr. El Emam's work is summarized at www.chairs-chaires.gc.ca/chairholders-titulaires/profile-eng.aspx?profileID=1690.

230. For the location of this area see www.mapsofindia.com/maps/westbengal/districts/bankura.htm.

231. Over 40 factors were studied. The list here is but a sampling of key points.

232. Atwood, D., D. Mayall, and R.A. Ryerson 2009; Shafee, M., R. Ryerson, K. Lim, I. Sadiq, and A. Rasheed 2010; Ryerson, R.A., and C. Peanvijarnpong 2007.

233. We recommend looking at the following three publications for ideas and examples of applications – no matter where you are based: GIM International based in Europe, www.gim-international.com; the especially well-written *Position* magazine (Australia/Pacific) www.positionmag.com.au/POS/pos_frame.html; GeoWorld and Geoplace (USA) www.geoplace.com/ME2/Default.asp.

234. ACIL Tasman 2008.

235. There is a professional designation administered by the American Society for Photogrammetry and Remote Sensing called "Certified Mapping Scientist (Remote Sensing)." See www.asprs.org.

236. Business Week Research Services 2006. Location Intelligence: The New Geography of Business. http://mapinfo.landinghost.com/BusinessWeek.WP.deeperbluewebready6.28.06.pdf.

237. Sanders, R., et al. 2005.

238. See: a system vendors summary at www.esri.com/industries/banking/index.html; its use in a developing country such as Bangladesh, http://ezinearticles.com/?Banking-GIS---A-New-Approach-in-Bangladesh&id=332573; and on how mergers can be evaluated, MacDonald, E.H. 2001.

239. See the November 2009 announcement that Deutsche Bank Private Wealth Management appointed a head of GIS. www.bankingtimes.co.uk/30112009-deutsche-bank-pwm-appoints-gis-head.

240. *Edmonton Journal*, March 1, 2010 www.edmontonjournal.com/business/Ducks+deaths+case+Syncrude+tailing+pond+goes+court/2627576/story.html.

241. *The Economist*, Nov. 7, 2009.

242. See Ryerson, R.A., et al. 2008.

243. Eriksson, N., and P. Adamek 2000.

244. See, for example, Szostak-Chrzanowski, A., M. Massie´ra, and A. Chrzanowski 2007.

245. Ryerson, R.A., Z. Kalensky, and C. Gosselin 2003.

246. Atwood, D., D. Mayall, and R.A. Ryerson 2009.

247. We have already referenced the *Manual of remote sensing* series as well as the 2001 book by Dr. Steve Franklin published by CRC Press.

248. See www.fao.org/giews/ENGLISH/index.htm.

249. See www.fews.net/Pages/default.aspx.

250. The major problem is in areas that are perpetually cloudy. Such areas do not tend to have good and frequent coverage.

251. This work was pioneered by Dr. R. Tinline at Queen's University. See http://geog.queensu.ca/faculty/tinline.asp/ and MacInnes, C.D., et al. 2001. The work has since been applied to rabid skunks in California (see Shwiff, S., et al. 2009) and other wildlife.

252. See Global Consequences of El Niño, NOAA Tropical Atmosphere Ocean Project. www.pmel.noaa.gov/tao/elnino/impacts.html.

253. See the many papers by R.A. Ryerson cited here, as well as Aronoff, S. 1985.

254. Here we can identify a number of problems using more technical language and a few simple examples. First, one might be measuring changes between one time and another in land cover (forest vs. agriculture vs. urban vs. water, for example). One might be quite happy with an average accuracy of 92% using a common classification or information extraction algorithm. But here one might be wise to ask several questions: what is the class of most interest? Is it urban? Is it the change from agriculture to urban? What is the stated area in each class (forest, agriculture, urban and water? And what is the accuracy for each? If the movement from agriculture to urban is the change of most interest, we might have a completely useless map even if the accuracy was 92%. Consider the following example modeled after an actual study done in the early 1970s in upper New York State. Let us assume that the area is 40% water (which is mapped with 100% accuracy), 40% forest (which is mapped with 98% accuracy), 15% agriculture (mapped with 78.66% accuracy), and 5% urban (mapped with 20% accuracy). For the total area this yields a map (at one time) with an accuracy of 92%. But the accuracy for the areas of key interest is actually only 64% - 12.8% of the key 20% is accurately mapped. And if one is looking at differences over time it may get worse. What if one does a map at each of the two times of interest and then compares them? If .92 of the first map is accurate and .92 of the second map is accurate, then combined the two maps are (.92 x .92) = .85 accurate. And the accuracy for the agriculture-urban areas would be (.64 x .64) = .41. Now do the maps look so good for the purpose? No! Much time and effort is spent on measuring accuracy and on determining how much change has actually occurred without comparing two separate maps. As another example, one could have classes of objects that showed greater

variance of some critical distinguishing attribute within the class than between classes. This would obviously not be a good classification or map of the area for the stated purpose.

255. The use of imagery to depict things that are not real is today all around us. One popular image that regularly circulates on the Internet shows Europe and North Africa with a clear demarcation between light (daylight) and dark (night) – the dark areas can be seen in outline along with city lights. The image is a hoax.

256. This International Charter was discussed in Section 3.4.7. www.disasterscharter.org/home.

257. This study was presented by Jim Taylor at, of all places, the Canadian Advisory Committee on Remote Sensing's Agricultural Working Group meeting held in Prince Edward Island in the early-mid 1970s. Ryerson attended the meeting.

258. See the commentary by law firm Ogilvy Renault. www.ogilvyrenault.com/en/resourceCentre_2081.htm.

259. Under UNCLOS four maritime zones are defined: the territorial sea, which stretches from the baseline 12 nautical miles seaward; the contiguous zone, which goes out to 24 miles; the exclusive economic zone, at 200 nautical miles; and the continental shelf, up to 350 miles or more. The nautical mile is the unit of measurement employed in relation to the law of the sea; one mile equals 1852 metres. Maritime zones can be claimed unilaterally as long as the claims do not infringe on other states' maritime zones. For the source of this material and for easily understood details on one application of UNCLOS, see *Asia Surveying and Mapping* magazine. www.asmmag.com/features/mapping-the-indonesia-singapore-border.

260. See www.un.org/apps/news/story.asp?NewsID=28625&Cr=unmis&Cr1.

261. See, for example, the FAO's Country Profiles: www.fao.org/countryprofiles/index.asp?lang=en&iso3=SDN.

262. www.internationalboundarycommission.org/about.html. For further details on the incredible complexity of boundaries see USA/Canada boundary and related maps and activities at www.internationalboundarycommission.org/index-eng.html.

263. Personal communication from Peter Sullivan, Surveyor General and International Boundary Commissioner for Canada, January 4, 2010.

264. By civil engineering infrastructure we refer to roads, sewers, bridges, and water supply – all of which are rapidly deteriorating at an alarming rate in both the developing and developed world.

265. This useful comparison of what a million, billion, and trillion dollars "looks like" is given at www.globalresearch.ca/index.php?context=va&aid=12754.

266. James, P.A., L.M. St. Onge, M.E. van Voorst, and M.P. Walker. Cypress Street Viaducts. www.engineering.com/Library/ArticlesPage/tabid/85/articleType/ArticleView/articleId/73/Cypress-Street-Viaducts.aspx. Accessed June 1, 2010.

267. See www.statcan.gc.ca/ca-ra2006/analysis-analyses/nb-eng.htm, www.statcan.gc.ca/pub/22-008-x/2009003/t005-eng.htm and www.potatoesnb.com/about/corporate.asp.

268. Personal communication from Niall Wallace, CEO of Infonaut Inc. See www.infonaut.ca for details of the application.

269. See, for example, the British Columbia Forest Inventory and Analysis Branch at www.for.gov.bc.ca/hts/ or details on the forest inventory of Finland at www.metla.fi/ohjelma/vmi/nfi.htm.

270. Ryerson, R.A., and R. Arnason 1981.

271. Shao, Y., et al. 2001.

272. Wallen, V.R., and H.R. Jackson 1971.

273. See www.colombo-plan.org/cp/images/cp_ac/cp_ac_ap.asp?cm=3&cp_num=12; McGrath, G., Chapter 18: Canada's International Role, in McGrath, G., and L. Sebert, eds. 1999.

274. The United States Agency for International Development (USAID) uses geospatial to both track and assess their aid programs. USAID has linked their activities to location to better assess where aid is going and the results of that aid. They have developed an Internet-based Management Information System and GIS to track and evaluate its projects, measure results, and tell its story to stakeholders and the public in general. See http://proceedings.esri.com/library/userconf/feduc05/docs/pap194.pdf.

275. For a more thorough discussion of this issue see Ryerson, R.A., and R.J. Batterham 2000.

276. The papers by Shafee, M., R. Ryerson, K. Lim, I. Sadiq, and A. Rasheed 2010; Atwood, D., D. Mayall, and R.A. Ryerson 2009; and Ryerson, R.A., and C. Peanvijarnpong 2008, outline the approach, while papers by Ryerson, R.A., and E. Quiroga 2000, and Ryerson, R.A., and R.J. Batterham 2000, provide useful background.

277. See Shafee, M., R. Ryerson, K. Lim, I. Sadiq, and A. Rasheed 2010.

278. See the excellent examples at the following three web pages: www.usaid.gov/km/seminars/2005/GIS-USAID.pdf GIS as a management and communications tool; www.usaidgiswbg.com/USAID_West_Bank.asp Track work on the West Bank; and www.africastories.usaid.gov/search_details.cfm?storyID=25&countryID=4&sectorID=0&yearID=3 Quantify results in West Africa.

279. The bottom left shows how maps used to be made with rudimentary technology, while the centre depicts Radarsat, a satellite receiving antenna, and a representation of a map of Canada made from this new technology. The idea for this commemoration came from Peter Paul, who worked at the National Atlas, supported by his director, Dr. Bob O'Neil.

280. As this is being written I am looking out a window at a pile of snow that will grow by 25 cm this weekend. Last weekend it was -28 degrees C° – and this is one of the better agricultural areas of the country!

281. Ferguson, J. 2009.

282. ACIL Tasman 2008. The dollar value came from a September 14, 2009 personal communication from Alan Smart, who led the study for ACIL Tasman.

283. Ryerson, R.A., and C. Peanvijarnpong 2008.

284. Atwood, D., D. Mayall, and R.A. Ryerson 2009; Shafee, M., R. Ryerson, K. Lim, I. Sadiq, and A. Rasheed 2010.

285. For example, see Shafee, M., R. Ryerson, K. Lim, I. Sadiq, and A. Rasheed 2010. Our team also won a contract with the Government of Canada to work with a cross-sector team to develop a national strategy.

286. The workshop may be conducted in one day in a small jurisdiction, but may take several days spread over some weeks in more complex environments or where there has yet to surface a common objective for the geospatial program.

287. Shafee, M., R. Ryerson, K. Lim, I. Sadiq, and A. Rasheed 2010.

288. See Stamp (1940) and the more recent explanation of the value of his work in The Maps That Saved Britain from Starvation Win an Online Refuge http://www.port.ac.uk/research/gbhgis/mediaresources/freearticles/filetodownload,23045,en.pdf.

289. See McGrath, G., and L. Sebert, eds. 1999, Chapter 14, for a summary of how education in remote sensing developed in Canada, and other chapters for how other aspects of the field developed.

Additional Illustration Credits

Front cover: University of Wisconsin-Madison Space Science and Engineering Center

Chapter 1 page 1 DigitalGlobe satellite image

Chapter 2 page 7 Beinecke Rare Book and Manuscript Library, Yale University

Chapter 3 page 41 Natural Resources Canada

Chapter 4 page 197 GeoEye satellite image

Chapter 6 page 247 Google, Inc., Imagery © 2010 Cnes/Spot Image, DigitalGlobe, GeoEye

Chapter 5 page 233 Stan Aronoff

Chapter 7 page 343 Natural Resources Canada

INDEX